An understanding of how plants interact with their aerial environment is central to many areas of plant science. This book attempts to provide a soundly based introduction to those features of the atmospheric environment of particular relevance to plants and to describe the physical and physiological principles required for understanding how those features affect plants. The presentation adopted is designed to emphasise the close relationship between the biophysical, physiological and ecological aspects of the adaptation of higher plants to their aerial environment.

This second edition has been fully updated and extended to include information on techniques such as chlorophyll-*a* fluorescence and carbon isotope discrimination which can be applied to plants in the field, as well as coverage of topics of current concern such as global warming and atmospheric pollution.

Plants and microclimate

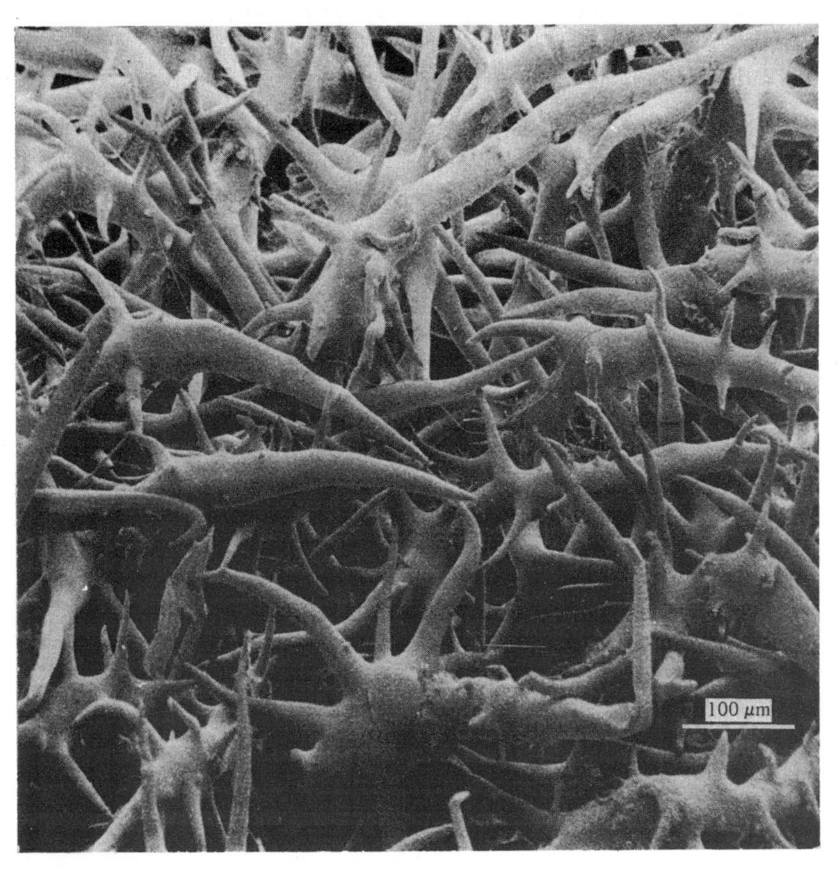

Frontispiece. Scanning electron micrograph of leaf hairs on *Aerva tomentosa* Forsk. (Courtesy of Dr B. M. Joshi, University of Jodhpur.)

Plants and microclimate

A quantitative approach to environmental plant physiology

Second edition

HAMLYN G. JONES

Horticulture Research International, Wellesbourne, Warwick, and Special Professor in Environmental Science, University of Nottingham

Published by the Press Syndicate of the University of Cambridge
The Pitt Building, Trumpington Street, Cambridge CB2 1RP
40 West 20th Street, New York, NY 10011-4211, USA
10 Stamford Road, Oakleigh, Melbourne 3166, Australia

First published 1983
Second edition 1992
Reprinted 1994, 1996

Printed in Malta by Interprint Limited

A catalogue record of this book is available from the British Library

Library of Congress cataloguing in publication data

Jones, Hamlyn G.
Plants and microclimate: a quantitative approach to environmental plant physiology
Hamlyn G. Jones. – 2nd ed.
p. cm.
Includes bibliographical references and index.
1. Vegetation and climate – Mathematical models. 2. Plant–atmosphere relationships – Mathematical models. 3. Plant physiological ecology – Mathematical models. I. Title.
QK754.5J66 1992
581.5′42–dc20 91-24016 CIP

ISBN 0 521 41502 0 hardback
ISBN 0 521 42524 7 paperback

UP

CONTENTS

PREFACE TO THE SECOND EDITION

For this second edition I have retained the successful format of the first edition which aimed to provide an authoritative introduction to environmental plant physiology suitable as a text for upper undergraduate courses and for postgraduates emphasising quantitative aspects of plant response to the aerial environment. The treatment has been extended in a number of areas that are of particular current interest to include, for example, information on plant responses to atmospheric pollutants and the greenhouse effect and with the addition of information on techniques that have only come into their own since the first edition was published (e.g. chlorophyll fluorescence, and carbon isotope discrimination). In other respects coverage is little changed with no attempt being made to cover soil processes or details of the physiology of plant growth and development.

One particular change that has been made is the general adoption of molar units for leaf gas-exchange studies, but because these units are not widely used in micrometeorological literature or in studies of productivity, mass units have been retained for these purposes. Although an ideal treatment might use a consistent treatment throughout there are particular advantages in using units at any level that are consistent with the bulk of the related literature. The coexistence of both types of unit gives rise to some potential confusion, but I have tried to make it clear throughout which units are being used.

In revising the text I am indebted to many colleagues around the world for their helpful and constructive comments on the first edition, and I am particularly grateful to those who have read and commented on sections of this revised text: these include Dr J. E. Corlett, Dr W. J. Davies, Dr M. Malone, Dr A. Massacci, Professor T. A. Mansfield, Dr J. A. C. Smith, Dr B. Thomas, and Professor M. H. Unsworth. I am also especially grateful to my wife Amanda, both for drawing many of the figures and also for her support throughout the project.

Warwick, February 1991 H.G.J.

PREFACE TO THE FIRST EDITION

Plant growth, productivity and survival are intimately coupled to the aerial environment, with processes such as energy exchange, the uptake of carbon dioxide in photosynthesis and the loss of water vapour in transpiration being central to all plant sciences. In this book I have attempted to provide an introduction to the basic physical and physiological principles necessary for understanding the interactions between vegetation and the aerial environment. This information is applied to an analysis of plant responses and adaptations to specific conditions and to the consideration of some agricultural and ecological problems. I have adopted a quantitative approach throughout because I believe that this provides the key to future advances in environmental physiology, as well as providing a framework for summarising current knowledge. However, recognising the background and inclination of many biology students, I have limited mathematics to simple algebra and elementary calculus.

The contents are based on courses in Environmental Physiology and the Physical Environment of Plants given to advanced undergraduates at Glasgow University and during a sabbatical period at the University of Toronto in 1981, though the material has been expanded so as to be of more general value as a text and reference for students and researchers in related fields of plant science including ecology, physiology and agricultural or horticultural science.

In a book of this length it has not been possible to cover all aspects of environmental physiology so examples have been limited to higher plants and no attempt has been made to discuss, for example, pathological aspects or air pollution. Similarly, although I have emphasised the importance of plant developmental responses, physiological aspects of the control of development have been omitted. I hope that what has been included will provide a stimulus for a rigorous and quantitative approach to further plant–environment studies, while giving enough information for those students who might not wish to go deeper into the subject.

The symbol set that I have used throughout the book was chosen to conform with accepted practice as much as possible, it was necessary, however, to introduce some non-standard forms in order to minimise the confusion that arises where one symbol can have multiple meanings.

I am very grateful to many colleagues and friends throughout the world who have contributed either directly or indirectly to this book: particularly to Dr Clifford Evans, Professor Ralph Slatyer and Professor Barry Osmond who stimulated my interest in environmental physiology; to all those who have commented on drafts of one or more chapters including Dr D. J. Avery, Professor D. A. Baker, I. G. Cummings, Dr W. Day, Dr G. C. Evans, Dr M. F. Hipkins, Dr Ch. Körner, K. J. Martin, Professor J. L. Monteith, Dr J. W. Palmer and Dr S. A. Quarrie; and especially to my wife Amanda who helped in all phases of the project including preparation of illustrations and typing of the manuscript.

July 1982 H.G.J.

ACKNOWLEDGEMENTS

I am very grateful to Dr M. L. Parker and the Plant Breeding Institute, Cambridge, for permission to reproduce photographs in Figs. 6.2 and 7.1 and to Dr B. M. Joshi for permission to use photographs in the Frontispiece and Fig. 6.1. I am also grateful to the appropriate authors and to the following for permission to use published material: Carnegie Institution of Washington (Figs. 7.15, 9.13); The Royal Society (Fig. 7.22); British Ecological Society (Fig. 7.18); The Director, National Institute for Agricultural Botany, Cambridge (Fig. 12.2). I am also grateful to Dr P. Valko for providing the data for Fig. 11.7.

MAIN SYMBOLS & ABBREVIATIONS

The individual superscripts and subscripts given for each main symbol may be combined to make compound symbols.

a	activity (dimensionless) *subscript*: a_W water
A	area (m^2) *subscript*: A_{mes} mesophyll surface; A_0 initial leaf
A	absorbance (= ln $[I_o/I]$)
A	amp (coulomb s^{-1})
ABA	abscisic acid
ATP	adenosine triphosphate
c	concentration (kg m^{-3} or mol m^{-3}) *subscripts*: c_C carbon dioxide; c_H heat (J m^{-3} = $\rho c_p T$); c_M momentum (kg m^{-2} s^{-1} = ρu); c_O oxygen; c_W water; c_a in the outside air; c_e in inlet air; c_i in the intercellular spaces at the surface of the cell wall; c_o in outlet air; c_ℓ leaf; c_s solute; c_x at carboxylase *superscripts*: c^m molar; c' carbon dioxide (= c_C)
c_p	specific heat capacity (J kg^{-1} K^{-1} = 1012 for air) *superscript*: c_p^* 'leaf' specific heat
c_D	drag coefficient (dimensionless)
c	speed of light (2.998×10^8 m s^{-1})
C	capacitance (m MPa^{-1} or m^3 MPa^{-1})
C_r	relative capacitance (MPa^{-1})
C	control coefficient
C	sensible heat flux (W m^{-2}) *subscripts*: $\mathbf{C}_{(d)}$ dry; $\mathbf{C}_{(w)}$ wet
CFCs	chlorofluorocarbons
C_3	three-carbon photosynthetic pathway
C_4	four-carbon photosynthetic pathway
CAM	crassulacean acid metabolism
CGR	crop growth rate (kg m^{-2} day^{-1})

d zero plane displacement (m)
d diameter (m)
d characteristic dimension or distance to leading edge (m)
D growing degree-days or temperature sum (°C day)
subscript: D_{eff} effective day degrees
$\boldsymbol{D}$ diffusion coefficient ($m^2\ s^{-1}$)
subscripts: $\boldsymbol{D}_A$ air; $\boldsymbol{D}_C$ carbon dioxide; $\boldsymbol{D}_H$ heat; $\boldsymbol{D}_i$ the *i*th species; $\boldsymbol{D}_M$ momentum; $\boldsymbol{D}_O$ oxygen; $\boldsymbol{D}_W$ water; $\boldsymbol{D}_X$ pollutant
superscript: $\boldsymbol{D}^\circ$ reference value
$\mathscr{D}$ dielectric constant (dimensionless)

e water vapour pressure (Pa)
subscripts: e_a in the bulk air; e_e in inlet; e_ℓ saturation at leaf temperature; e_o in outlet; e_s surface; e_s saturation; $e_{s(T\ell)}$ saturation at leaf temperature ($= e_\ell$)
e equation of time
e base for natural logarithm (2.71828)
e^- electron
E energy (J)
subscript: E_a activation energy
E evaporation rate ($kg\ m^{-2}\ s^{-1}$ or $mm\ h^{-1}$)
subscript: $\mathbf{E}_\ell$ transpiration; $\mathbf{E}_o$ free water evaporation; $\mathbf{E}_p$ potential evaporation
superscript: $\mathbf{E}^m$ molar

f allocation to leaves
f enhancement factor
f ratio of Pfr/Pr
f fraction of water in tissue that is liquid (dimensionless)
F fluorescence ($mol\ m^{-2}\ s^{-1}$)
subscripts: F_m maximum; F_o basal with open reaction centres; F_v variable
F force (N)
$FADH_2$ reduced flavin adenine dinucleotide

g conductance ($mm\ s^{-1}$)
subscripts: g_A canopy boundary layer conductance; g_L canopy physiological conductance; g_C carbon dioxide; g_H heat; g_M momentum; g_O oxygen; g_R radiation ($= 4\varepsilon\sigma T_a^3/\rho c_p$); g_{HR} parallel heat and radiative transfer; g_W water; g_a boundary layer; g_c cuticle; g_ℓ leaf; g_i intercellular space; g_m mesophyll; g_o reference value; g_s stomatal
superscripts: g^m molar conductance ($= \mathbf{g}$); g' carbon dioxide ($= g_C$)

g	molar conductance ($= g^m$) (mol m^{-2} s^{-1})
	subscripts: as for g
	superscripts: g' carbon dioxide ($= g_C$)
$\mathbf{g}$	acceleration due to gravity (m s^{-2} = 9.8 at sea level)
G	Gibbs free energy (J)
$\mathbf{G}$	soil heat storage or heat loss by conduction (W m^{-2})
h	hour angle of the sun (degree or radian)
h	Planck's constant (6.626×10^{-34} J s)
h	relative humidity (dimensionless)
h	height, thickness (m)
h	hour
ha	hectare ($= 10^4$ m^2)
I	moment of inertia (kg m^{-2})
I	electric current (A)
$\mathbf{I}$	irradiance (W m^{-2} or mol m^{-2} s^{-1})
	subscripts: $\mathbf{I}_A$ extraterrestrial; $\mathbf{I}_L$ longwave; $\mathbf{I}_{PAR}$ in photosynthetically active region; $\mathbf{I}_S$ shortwave; $\mathbf{I}_{S(diff)}$ diffuse; $\mathbf{I}_{S(dir)}$ direct; $\mathbf{I}_c$ light compensation point; $\mathbf{I}_e$ energy (usually assumed); $\mathbf{I}_p$ photon irradiance or incident photon flux density (mol m^{-2} s^{-1}); $\mathbf{I}_o$ unattenuated reference value
IAA	indole-3-acetic acid
IR	infra-red
IRGA	infra-red gas analyser
J	joule (1 kg m^2 s^{-2})
$\mathbf{J}$	mass transfer rate per unit area or flux density (kg m^{-2} s^{-1})
	subscripts: $\mathbf{J}_v$ volume flux density (m s^{-1}); for other subscripts see $\boldsymbol{D}$
k	thermal conductivity (W m^{-1} K^{-1})
k	rate constant, or other constant
	subscripts: k_D thermal dissipation; k_F fluorescence; k_P photochemistry; k_T energy transfer to PSI; k_d rate of development ($= 1/t$)
$\mathscr{k}$	von Karman's constant (= 0.41, dimensionless)
$\mathscr{k}$	extinction coefficient (dimensionless)
kg	kilogram
K_m	Michaelis constant (dimensions as for concentration or irradiance)
	superscripts: K_m^C for carbon dioxide; K_m^I for light
K	Kelvin
$\boldsymbol{K}$	transfer coefficient (m^2 s^{-1})
	subscripts: as for $\boldsymbol{D}$
ℓ	length, thickness (m)
	superscript: ℓ^* 'leaf' thickness

ℓ photosynthetic limitation ($s\ m^{-1}$)
subscripts: as for resistance (r)

ℓ' relative photosynthetic limitation (dimensionless)
subscript: ℓ'_g gas-phase

ln natural logarithm

log logarithm to base 10

L hydraulic conductivity ($m^2\ s^{-1}\ Pa^{-1}$)

L_p hydraulic conductance ($m\ s^{-1}\ Pa^{-1}$)

L_v volumetric hydraulic conductance ($s^{-1}\ Pa^{-1}$)

L leaf area index (dimensionless)

L' leaf area index expressed per unit area of ground shaded by non-transmitting plants (dimensionless)

LAR leaf area ratio ($m^2\ kg^{-1}$)

LD long day

LHC light harvesting complex

m mass fraction (dimensionless)
subscripts: as for ***D***
superscript: m' carbon dioxide

m mass (kg)

m air mass (dimensionless)

m metre

mol mole (amount of substance containing Avogadro's number of particles)

M molecular weight (kg)
subscripts: as for ***D***

M metabolic heat storage ($W\ m^{-2}$)

n hours bright sunshine (h)

n number of moles (mol)
subscripts: n_s solute; n_p photons; other subscripts as for ***D***

$n(E)$ number of moles with energy equal to or exceeding E (mol)

n number (e.g. of stomata)

n exponent

N daylength (h)

N newton ($1\ kg\ m\ s^{-2}$)

NADPH reduced nicotinamide adenine dinucleotide phosphate

$NADP^+$ oxidised nicotinamide adenine dinucleotide phosphate

NAR net assimilation rate or unit leaf rate ($kg\ m^{-2}\ day^{-1}$)

NDVI normalised difference vegetation index (dimensionless)

OAA oxaloacetate

p partial pressure (Pa)
subscripts: as for $\mathbf{D}$ and c
superscript: p' carbon dioxide

pH negative logarithm of hydrogen ion activity

P pressure (Pa)
superscripts: P^{o} reference; P^{*} balance pressure

P period of oscillation

$\mathbf{P}$ photosynthetic rate (mg m^{-2} s^{-1} or μmol m^{-2} s^{-1})
subscripts: $\mathbf{P}_n$ net photosynthesis; $\mathbf{P}_g$ gross photosynthesis: $\mathbf{P}_{max}$ maximum photosynthesis with either light or CO_2 saturating
superscripts: $\mathbf{P}^m$ molar; $\mathbf{P}^{max}$ maximum photosynthesis with saturating light and CO_2; $\mathbf{P}^{o}$ reference value at infinite conductance

Pr red light absorbing form of phytochrome

Pfr far-red absorbing form of phytochrome

Pa pascal (N m^{-2} = kg m^{-1} s^{-2})

PAR photosynthetically active radiation (usually 400 to 700 nm)

PEP phosphoenolpyruvate

PCO photorespiratory carbon oxidation cycle

PCR photosynthetic carbon reduction cycle (Calvin cycle)

PGA phosphoglyceric acid

PS I photosystem one

PS II photosystem two

q fluorescence quenching
subscripts: q_I photoinhibition; q_N non-photochemical; q_o F_o quenching; q_T state transition

qr quantum requirement

Q_{10} temperature coefficient for 10 K increase in temperature (dimensionless)

$\mathbf{Q}_e$ radiant flux (W = J s^{-1})

$\mathbf{Q}_p$ photon flux (mol s^{-1})

r radius (m)

r resistance (s mm^{-1})
subscripts: r_A canopy boundary layer resistance; r_L canopy physiological resistance; r_X canopy resistance to pollutant transfer; r_C carbon dioxide; r_H heat; r_M momentum; r_O oxygen; r_R radiation $(= \rho c_p/4\varepsilon\sigma T_a^3)$; r_{HR} parallel heat and radiative transfer; r_W water; r_a boundary layer; r_c cuticle; r_g gas phase; r_ℓ leaf; r_i intercellular space; r_m mesophyll; r_o reference value; r_r residual; r_s stomatal; r_w wall; r_* $dc'_w/d\mathbf{P}_n$ at normal ambient CO_2
superscripts: r^m molar resistance; $(= r)$; r' carbon dioxide $(= r_C)$

r molar resistance ($= r^m$) (m^2 s mol^{-1})
subscripts: r_C carbon dioxide; r_O oxygen; r_W water; r_a boundary layer; r_c cuticle; r_ℓ leaf; r_i intercellular space; r_m mesophyll; r_o reference value
superscripts: r' carbon dioxide ($= r_C$)

R liquid phase hydraulic resistance (MPa s m^{-1} or MPa s m^{-3})
subscripts: R_ℓ leaf; R_p plant; R_r root; R_s soil; R_{st} stem

R electrical resistance (ohm = W A^{-2})

$\mathbf{R}$ respiration rate (mg m^{-2} s^{-1} or μmol m^{-2} s^{-1})
subscripts: $\mathbf{R}_d$ dark; $\mathbf{R}_\ell$ light; $\mathbf{R}_g$ growth; $\mathbf{R}_m$ maintenance; $\mathbf{R}_{non\text{-}ps}$ respiratory loss from non-photosynthesising tissue

$\mathscr{R}$ gas constant (8.3144 J mol^{-1} K^{-1} or Pa m^3 mol^{-1} K^{-1})

RGR relative growth rate (day^{-1})

RuBP ribulose bis-phosphate

Rubisco ribulose bis-phosphate carboxylase

s slope of the curve relating saturation vapour pressure to temperature (Pa K^{-1})

s second

sr steradian

S stress (Pa)

$S_i(z)$ source density profile of entity i with height

$S(t)$ state of development at time t

S rate of physical heat storage (W m^{-2})

SD short day

t time (s or h)

T temperature (°C or K)
subscripts: $T_{(d)}$ dry; $T_{(w)}$ wet; T_a air; T_{base} non-water-stressed baseline temperature (for calculation of stress index – Chapter 10); T_d dewpoint; T_e equilibrium; T_ℓ leaf; T_{max} maximum temperature (for calculation of stress index – Chapter 10); T_m mean; T_o optimum; T_s surface; T_{sky} apparent radiative temperature of the sky; T_t threshold; T_w wet bulb
superscript: T^o reference temperature

$\mathbf{T}$ torque (N m or J)

$\mathscr{T}$ transmission by canopy (dimensionless)

$\mathscr{T}_f$ fraction of radiation that would reach the ground if plants are non-transmitting (dimensionless)

u molar flow (mol s^{-1})
subscripts: u_e entering chamber; u_o leaving chamber

u windspeed (m s^{-1})
subscripts: u_z windspeed at height z; u_* friction velocity

UV ultraviolet

v flow rate ($m^3 s^{-1}$)
V volume (m^3)
subscripts: V_e expressed sap; V_o turgid volume of cell
$\bar{V}_w$ partial molal volume of water (18.048×10^{-6} $m^3 mol^{-1}$ at 20 °C)
V potential difference (volt = W A^{-1})
vpm volume parts per million

w mixing ratio (dimensionless)
subscripts: as for ***D***
W water content (kg m^{-2})
subscripts: W_ℓ leaf; W_{max} maximum
W leaf weight in CO_2 equivalents (g m^{-2})
W watt (J s^{-1})

x mole fraction (dimensionless)
subscripts: x_e entering; x_o outgoing; x_s saturating; other subscripts as for ***D*** and c
superscript: x' carbon dioxide
x distance (m)

Y yield threshold (Pa)
Y economic yield (tonne ha^{-1} or kg m^{-2})
subscript: Y_d dry matter yield

z distance, height, elevation (m)
subscript: z_o roughness length

α contact angle (degree or radian)
α absorptivity, absorption coefficient or absorptance (dimensionless)
subscripts: define waveband i.e. α_{660} absorptivity at 660 nm; α_S absorption coefficient for solar radiation

β solar elevation (degree or radian)
$\boldsymbol{\beta}$ Bowen ratio (dimensionless = $\mathbf{C}/\lambda\mathbf{E}$)

γ psychrometer constant (Pa K^{-1} = $P c_p/0.622 \lambda$)
γ_w activity coefficient for water (dimensionless)
Γ CO_2 compensation concentration (mg m^{-3} or mmol m^{-3})

δ solar declination (degree or radian)
δ deviation of isotope abundance from ratio in a standard sample
subscripts: δ_a air; δ_p plant
δ average thickness of a laminar boundary layer (m)
subscript: δ_{eq} thickness of equivalent boundary layer (m)

δc_{W} absolute humidity deficit of the air (kg m^{-3})
δe water vapour pressure deficit (Pa)
subscript: δe_{ℓ} leaf to air vapour pressure difference
∂ partial differential
Δ carbon isotope discrimination (dimensionless)
Δ finite difference
ΔF difference between steady state and maximal fluorescence
ΔT_{f} freezing point depression (K)

ε increase of latent heat content per increase of sensible heat content of saturated air at ambient temperature ($= s/\gamma$, dimensionless)
ε emissivity (dimensionless)
ε_{p} efficiency of utilisation of incident light in photosynthesis (dimensionless)
ε_{q} quantum efficiency (dimensionless) (see also ϕF)
ε_{B} bulk modulus of elasticity (Pa)
ε_{Y} Young's modulus (Pa)

ζ ratio of photon flux densities in red (655–665 nm) and far-red (725–735 nm) (dimensionless)

η dynamic viscosity (N s m^{-2})

θ angle from beam to normal (degree or radian)
θ relative water content (dimensionless)

λ latitude (degree or radian)
λ wavelength of light (m)
subscript: λ_{m} peak of Planck distribution
λ latent heat of vaporisation (J kg^{-1})
$\boldsymbol{\lambda}$ constant value of $\partial \mathbf{E}_{\ell}/\partial \mathbf{P}_{\mathrm{n}}$

μ chemical potential (J mol^{-1})
subscript: μ_{W} water
superscript: μ^{o} reference value

ν frequency of oscillation (hertz)
ν frequency of stomata (mm^{-2})
ν_{s} sedimentation velocity (m s^{-1})
$\boldsymbol{\nu}$ kinematic viscosity (m^{2} s^{-1} $= \boldsymbol{D}_{\mathrm{M}}$)

π ratio of circumference of a circle to its diameter (3.14159)
Π osmotic pressure (MPa $= -\Psi_{\pi}$)

ρ density (kg m^{-3})
subscripts: ρ_{a} dry air; ρ_{i} i-th component of a mixture; ρ_{p} particle
superscript: ρ^{*} density of leaf or leaf replica

ρ reflectivity, reflection coefficient or reflectance (dimensionless)
subscripts: as for α

σ Stefan–Boltzmann constant (5.6703×10^{-8} W m^{-2} K^{-4})
σ reflection coefficient for a solute (dimensionless)
$\boldsymbol{\sigma}$ surface tension (N m^{-1} = 7.28×10^{-3} for water at 20 °C)

τ time constant (s)
τ transmissivity or transmittance (dimensionless)
subscripts: as for α
$\boldsymbol{\tau}$ shearing stress (Pa = N m^{-2})
subscript: $\boldsymbol{\tau}_f$ form drag

ϕ phase lag (s)
ϕ photoequilibrium or ratio of Pfr to total phytochrome (dimensionless)
ϕ extensibility (s^{-1} Pa^{-1})
ϕF quantum yield for fluorescence (dimensionless)
ϕ_{PSII} quantum yield of PSII
$\boldsymbol{\phi}_C$ CO_2 losses at night and from the roots expressed as a fraction of assimilation through stomata in the day (dimensionless)
$\boldsymbol{\phi}_W$ water losses from the soil or cuticle as a fraction of stomatal transpiration (dimensionless)
$\boldsymbol{\Phi}$ radiant ($\boldsymbol{\Phi}_e$) or photon ($\boldsymbol{\Phi}_p$) flux density (W m^{-2} or mol m^{-2} s^{-1})
subscripts: $\boldsymbol{\Phi}_{(d)}$ dry; $\boldsymbol{\Phi}_{(w)}$ wet; $\boldsymbol{\Phi}_d$ downward; $\boldsymbol{\Phi}_u$ upward; $\boldsymbol{\Phi}_n$ net radiation; $\boldsymbol{\Phi}_{ni}$ isothermal net radiation; $\boldsymbol{\Phi}_p$ perpendicular to beam; $\boldsymbol{\Phi}_{pA}$ solar constant ($\simeq$ 1370 W m^{-2}); $\boldsymbol{\Phi}_\lambda$ at wavelength λ; other subscripts as for **I**

Ψ water potential (MPa)
subscripts: Ψ_p pressure potential; Ψ_π osmotic potential ($= -\Pi$); Ψ_g gravitational potential; Ψ_m matric potential; Ψ_ℓ leaf; Ψ_s soil; Ψ_o initial value

Ω solid angle (st)
Ω decoupling coefficient (dimensionless)

$|x|$ absolute value of x

1

A quantitative approach to plant–environment interactions

Progress in environmental plant physiology, as in other scientific disciplines, involves repeated cycles of observation or experimentation followed by data analysis and the construction and refining of hypotheses concerning the behaviour of the plant-environment system. This process is illustrated in very simplified form in Fig. 1.1. At any stage the information and hypotheses may be qualitative or quantitative, and there may be more or less emphasis on the use of controlled experiments for providing the necessary data.

The initial stages of an investigation tend to provide a more qualitative description of system behaviour: much early ecological research, for example, was concerned with the description and classification of vegetation types, with a relatively small proportion of effort being devoted to understanding the underlying processes determining plant distribution. Further improvements in the understanding of any system, however, require a more quantitative approach based on a knowledge of the underlying mechanisms.

It is at this second level that this book is aimed: I have attempted to provide an introduction to environmental biophysics and to the physiology of plant responses that can be used to provide a quantitative basis for the study of ecological and agricultural problems. Further information on specific topics may be found in specialised texts referred to throughout the book.

Modelling

Mathematical modelling provides a particularly powerful tool for the formulation of hypotheses and the quantitative description of plant function. As modelling techniques are being used increasingly in all areas of plant science, and because they are used throughout this book, it is necessary to start with a simple introduction to mathematical modelling. Further details of modelling techniques may be found in appropriate texts

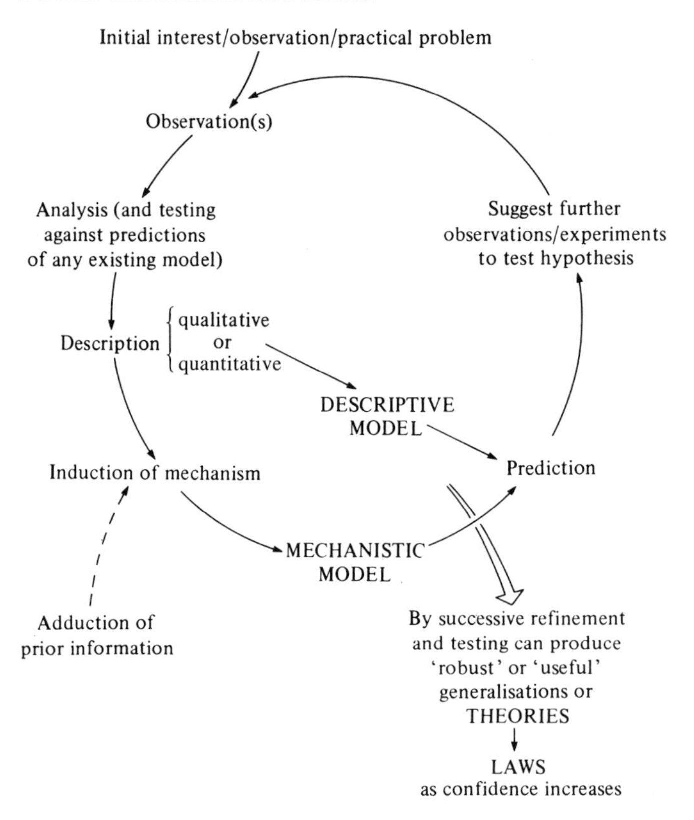

Fig. 1.1. The role of models in scientific method.

(e.g. Thornley 1976; Gold 1977; Rose & Charles-Edwards 1981), as may descriptions of their application to ecological (Pielou 1969; Patten 1971; de Wit & Goudriaan 1978) and physiological (Thornley 1976; Hesketh & Jones, 1980; Rose & Charles-Edwards 1981; France & Thornley 1984; Thornley & Johnson 1990) problems.

In the present context, a model is any representation of a real system, such as a plant, that can be used to simulate certain features of the more complex real system. For example light penetration in plant canopies is extremely complex (Chapter 2), but useful advances have been made by setting up a simple model where the actual canopy with its individual leaves is replaced by a layer of homogeneous absorber. This model can be either an actual object (e.g. a solution of chlorophyll) whose properties can be studied empirically, or a mathematical abstraction that simulates those properties. A mathematical model, therefore, can constitute a concise formulation of an hypothesis (in this case that light penetration through a canopy is the same as through an homogeneous absorber). As such it can be readily used to generate testable predictions (e.g. of the effect of altering the angle of incident radiation). The results of these tests can then be used to refine, confirm or refute the initial hypothesis (Fig. 1.1). In the present example, the accuracy with which the equations for an homogeneous absorber predict the penetration of light in a real canopy would be used to determine the adequacy of the model.

Because biological systems are so complex, one can rarely achieve complete mathematical descriptions of their behaviour. It is necessary, therefore, to make simplifying assumptions about the system behaviour and concerning the relevant components for inclusion in any study. This selection of variables is perhaps the most difficult task in the development of any mathematical model. An equally important step, however, in the development of useful models is their validation and testing. Some of the main advantages of mathematical models and the ways in which they can be used are summarised below.

(1) They constitute precise statements of our hypotheses.
(2) They are inherently testable.
(3) They can 'explain' or describe a large number of separate observations in concise form.
(4) They help to identify those areas where knowledge is lacking and further experiments or observations are required.
(5) They can be used to predict system behaviour in untried combinations of conditions.
(6) They can be used as management tools, for example for scheduling crops and management operations so as to maximise profit.
(7) They can be used in diagnosis, for example in identifying crop diseases.

This last application has only been developed relatively recently with the advent of 'expert systems'. These attempt to encapsulate the knowledge of human experts into a set of rules that can be applied, among other things, to the diagnosis of disorders. A feature of this approach is that it can take account of uncertainty in any of the answers and weight them accordingly in coming to a conclusion. Although mathematical modelling has been widely used in the more physically based sciences, such as meteorology, it has, at least until recently, been underutilised in physiological and ecological studies.

Types of model

Various types of mathematical model will be encountered throughout this book. These vary from relatively untested hypotheses (such as the models used in studies of 'optimum' stomatal behaviour – see Chapter 10), through partially tested models (i.e. theories), to well-tested models (i.e. laws – such as those dealing with well-known physical processes such as diffusion) where, given certain conditions, one can say with some certainty that a particular consequence will always ensue.

Mathematical models may be empirical, where no attempt is made to describe the mechanisms involved and minimal information is used *a priori* in their development. Alternatively, an attempt may be made to make the model mechanistic, using knowledge from previous work. A mechanistic model usually attempts to explain a phenomenon at a more detailed level of organisation. The choice of modelling approach depends on the particular research objective. Although both types of model may be used for predictions, the mechanistic approach probably has greater scope for generalised application and can lead to important advances in understanding. In the long run it is also likely to provide the more accurate predictions of system behaviour under a wide range of conditions.

An example of the empirical approach is the use of relatively objective statistical regression techniques to describe and predict variation in crop yield in terms of weather variation from year to year (an example is given in Chapter 9 for hay yields in Iceland). This type of model can provide a useful description of the system by using routine techniques without the need for any physiological knowledge. The approach can, however, be made significantly more efficient with input of physiological knowledge to select the weather variates studied and to suggest appropriate forms for the relationships. It follows, therefore, that this approach is not completely distinct from the mechanistic approach, and indeed many empirical models tend to develop into mechanistic ones as they are refined.

In addition to being empirical or mechanistic, models may be either deterministic or stochastic, and dynamic or static. In deterministic models, the output is defined once the inputs are known, while stochastic models incorporate an element of randomness as part of the model. Most models in physiological ecology are deterministic, mainly because of their greater simplicity and convenience, but some stochastic models have been used, for example to simulate random weather sequences, light penetration in canopies, spread of pathogens or ovule fertilisation (see Jones 1981*c*).

Dynamic models include treatment of the time dependence of a process and are therefore particularly appropriate for simulating processes such as plant growth and yield production that integrate developmental and environmental changes over long periods. Many large-scale dynamic ecosystem and crop simulation models have been developed. These complex computer simulations, however, can rarely be tested in the sense that physicists use the word, because of the large numbers of variables and assumptions used in their construction. They can, nevertheless, provide useful information on the sensitivity of crops or other systems to environmental variables.

Static models, in contrast to dynamic models, are used for steady-state systems or for simple descriptions of a final result. For example, many of the transport models described in this book consider only the steady state, so can be regarded as static models, as can those yield models where final yield is predicted by means of a simple regression equation between yield and certain weather variates during the season.

In addition to mathematical models, there are several examples where physical models can be used. For example, electrical circuits can be used to model diffusion and other transport processes, and with complex systems they may be easier to use than the corresponding mathematical abstractions.

Another class of models, which, although not necessarily quantitative, can contribute greatly to the development of understanding are what might be termed conceptual models. These include general concepts such as the classification of plants into 'pessimistic' and 'optimistic' on the basis of their response to drought (Chapter 10), or more generally the development of what has been termed 'plant strategy theory' (see Grime 1979, 1989). This approach provides a valuable method for rationalising the vast array of evolutionary and ecological specialisations in plants, and involves the assumption that there are a limited number of what have been called 'primary strategies' available to plants; furthermore one type of specialisation for one type of existence and habitat condition tends to preclude success in other environments. The 'Competitor–Stress tolerator–Ruderal' or C–S–R model is a particularly powerful example of the application of this approach and can explain and predict stress responses very successfully (see Grime 1989). Although primarily a conceptual model it is amenable to quantitative analysis, since the equilibria between competition, stress and disturbance in vegetation may be readily quantified and represented graphically.

Fitting models

Any observations that one makes need reducing to a simple framework, if they are to be of value in the development of a hypothesis or for predicting future behaviour of the system. In general some form of curve-fitting procedure is adopted in order to derive a concise mathematical summary of the data. The summarising equation can be used to predict further values, as well as providing information to confirm or refute a theoretical model.

If, for example, a series of observations of photosynthetic rate at different irradiances has been made, a first step in the analysis might be to plot a graph of photosynthesis (on the ordinate, since it is likely to be the dependent variable) against irradiance (on the abscissa). One could then attempt to fit a line through the observations assuming that the points are particular examples of a general relationship. However, it is unlikely that all points will fall on the line because some other factor (such as temperature) is also varying. The equation to the best-fit line (together with some description of the error) provides a useful mathematical summary of the observations.

For any particular set of points there may be an infinite number of equations that fit them and, although many may be far too complex for serious consideration, there may be several simple types that fit the observations satisfactorily. However, one must bear in mind Occam's Razor (the principle that 'hypotheses must not be multiplied beyond necessity'). That is, when faced with the choice between two equally adequate models or hypotheses, one should take the simpler.

Useful introductions to the techniques for fitting curves may be found in appropriate statistical textbooks such as those by Box *et al.* (1978) and Gilchrist (1984). A particularly powerful and appropriate computer package for performing the necessary analyses is GENSTAT 5; its facilities are well described in the reference manual (Payne *et al.* 1987).

Use of experiments

The observational and experimental phases of research are equally important as the modelling phase. Purely observational studies, of the type that has characterised much ecological research in the past, where one relies on natural variation in the environmental factors of interest, can be restrictive and difficult to interpret. This is because of the inherent complexity of the natural system and the tendency for correlations to occur between factors such as temperature and sunshine. For this reason it is usually necessary to be able to manipulate the various environmental factors independently in controlled experiments.

It is possible to perform experiments with either more or less interference with the natural environment (Table 1.1) and either more or less precise control of certain variates. In general there is a trade-off between good control of environment and minimal interference with the natural environment, with combinations nearer the top left in this table providing more precise, but not necessarily more accurate, information on plant response to individual factors. Field experiments may suffer from poor environmental control but, because the conditions are likely to be closer to natural than those in glasshouses or controlled environment chambers, any results obtained in the field are generally more reliable.

In addition to varying degrees of modification of the physical

Table 1.1. *Differing degrees of experimental modification of root and aerial environment* (*modified after Evans* 1972)

		Wholly artifical ⟵	Aerial environment		⟶	Wholly natural
		Con-trolled environ-ments	Daylit cabinets	Glass-house, enclosures	Shelter, neutral screens	Field
Wholly artificial	Nutrient solution	✓	✓	✓	✓	×
↑	Inert base + nutrient solution	✓	✓	✓	✓	×
Root environ-ment	Soil in pots	✓	✓	✓	✓	✓
	Field with fertilisation or irrigation	×	×	✓	✓	✓
↓	Transplant experiments	×	×	✓	✓	✓
Wholly natural	Natural	×	×	✓	✓	Observation only

× = impractical combination

environment, the results obtained depend to some degree on the biotic environment (competition, pathogens, etc.). Most of the studies described in this book can be classified as autecological, that is they consider the behaviour of one species in isolation. Although much valuable ecological information can be obtained from such studies, they can go part-way towards an 'explanation' of any ecological phenomenon. At least in many important agricultural ecosystems, the most important type of biological competition is that from plants of the same species, while other biotic factors such as pests and diseases may be effectively controlled.

In practice, the choice of experimental system depends on the specific objectives. The more detailed a mechanistic explanation or model that is required for any phenomenon, the greater will be the need for controlled experiments. However, it then becomes important to minimise the interference with normal plant growth, or to become skilled in what Evans

(1972) has called 'plant stalking'. It is usually necessary to carry out a range of types of experiment, from those in tightly controlled conditions to some in the field. The latter are necessary to confirm any model derived in controlled environments.

Several examples of the dangers of relying too much on controlled environments will be encountered in what follows. There is now extensive evidence that field-grown material behaves very differently from that grown in controlled environments and only in a few cases is the reason for this difference fully understood. One example is provided by the very different stomatal response to plant water potential shown by field- and controlled environment-grown plants (see Chapter 6). The problems caused by the different 'coupling' of plants to their environment in controlled environments and in the field, and the consequences for studies of the control of evaporation by stomata, are discussed in Chapter 5. Plant morphology is also markedly different between these environments as a result of different irradiances and spectral distribution (see Chapter 8). Another example, as yet unresolved, is my unpublished observation that certain genotypes of wheat showed marked leaf rolling in a dry season in the field. Attempts to investigate this phenomenon in controlled environments have not been successful, apparently because of differences in leaf morphology in the two environments.

Some features of the environment are more easy to control than others. Field studies on plant nutrition and water status, for instance, have been conducted for over 100 years, but it is only in the last 20 or so that any useful attempts have been made to control temperature in the field. But even now, temperature studies involve enclosing the plant canopy and altering a wide range of other factors at the same time. The use of reciprocal transplant experiments, such as those conducted at the Carnegie Institution's research gardens (Björkman *et al.* 1973), provides a useful technique for studying the effect of the aerial environment without using controlled environments. Growing material at all sites in soil from the same source maximises the potential for studying the aerial environment in this type of experiment.

Whether one should attempt to use controlled environments to mimic all the features of the natural environment is, however, still controversial. Many elaborate systems have been set up to simulate the detailed daily trends of temperature and radiation (e.g. Rorison 1981), but their advantages have not been demonstrated convincingly. The increased environmental complexity tends to negate the main advantage of a controlled environment.

2

Radiation

There are four main ways in which radiation is important for plant life:

1. Thermal effects. Radiation is the major mode of energy exchange between plants and the aerial environment: solar radiation provides the main energy input to plants, with much of this energy being converted to heat and driving other radiation exchanges and processes such as transpiration, as well as being involved in determining tissue temperatures with consequences for rates of metabolic processes and the balance between them (see particularly Chapters 5 and 9).

2. *Photosynthesis*. Some of the solar radiation absorbed by plants is used for the synthesis of energy-rich chemical bonds and reduced carbon compounds. This process (photosynthesis) is characteristic of plants and provides the main input of free energy into the biosphere (see Chapter 7).

3. *Photomorphogenesis*. The amount and spectral distribution of shortwave radiation also plays an important role in the regulation of growth and development (see Chapter 8).

4. *Mutagenesis*. Very shortwave, highly energetic radiation, including the ultra-violet, as well as X- and γ-radiation, can have damaging effects on living cells, particularly affecting the structure of the genetic material and causing mutations.

This chapter introduces the basic principles of radiation physics that are needed for an understanding of environmental physiology, and describes various aspects of the radiation climate within plant stands. The extreme complexity of the radiation climate means that inevitably much of the treatment is concerned with the derivation of useful simplifications or models that can be used by ecologists or crop scientists. More detailed discussion of radiation physics and the radiation climate may be found in texts by Coulson (1975), Gates (1980) and Monteith & Unsworth (1990).

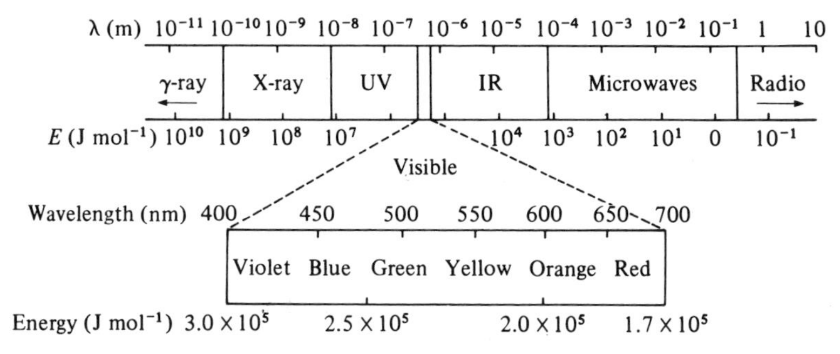

Fig. 2.1. The electromagnetic spectrum.

Radiation laws

Nature of radiation

The wavelengths of radiation that are of primary concern in environmental plant physiology lie between about 300 nm and 100 μm and include some of the ultraviolet (UV), the photosynthetically active radiation (= PAR, which is broadly similar to the visible) and the infra-red (IR) (see Fig. 2.1). Radiation has properties of both waves (e.g. it has a wavelength) and of particles (energy is transferred as discrete units termed quanta or photons). The energy (E) of a photon is related to its wavelength (λ) or its frequency of oscillation (v) by

$$E = hc/\lambda = hv \tag{2.1}$$

where h is Planck's constant (= 6.63×10^{-34} J s) and c is the speed of light (= 3×10^{8} m s^{-1}). An alternative measure of frequency that is commonly used is the wave number (= λ^{-1}, in cm^{-1}).

Using equation 2.1 it can be calculated that a photon of red light (for example with $\lambda = 650$ nm) would have an energy, $E = 3.06 \times 10^{-19}$ J (i.e. $6.63 \times 10^{-34} \times 3 \times 10^{8}/(6.5 \times 10^{-7})$), while for a photon of blue light ($\lambda =$ 450 nm), $E = 4.42 \times 10^{-19}$ J; that is 44 % greater than the longer wavelength. It is often more convenient to refer to the energy in a mole (i.e. Avogadro's number, = 6.023×10^{23}) of photons. Therefore for radiation of 650 nm, the energy per mole would be 1.84×10^{5} J mol^{-1} (i.e. $3.06 \times 10^{-19} \times 6.023 \times 10^{23}$). The variation with wavelength of energy per mole is given in Fig. 2.1. Although the Einstein is sometimes used to describe a mole of photons, it is an ambiguous term that is not part of the SI system, so should be avoided.

Some of the more useful measures of radiation for plant physiology are summarised in Table 2.1. For further discussion see Bell & Rose (1981). The amount of radiant energy (in J) emitted, transmitted or received by a surface per unit time is called the **radiant flux** (**$\mathbf{Q}_e$**, which has units of power,

Table 2.1. *Terminology for radiation measurement*

The various terms may be further qualified to include only radiation in certain wavebands (e.g. PAR, 'shortwave', etc.). Corresponding photometric terms based on luminous flux (lumen) are available; these are weighted for the spectral sensitivity of the human eye.

Term	Unit	Symbol and definition	
Radiant energy	J	E	Energy in the form of electromagnetic radiation
Number of photons	mol	n_p	
Radiant exposure	$J\ m^{-2}$	$\int \mathbf{I}_e dt$	Incident radiant energy per unit area
Photon exposure	$mol\ m^{-2}$	$\int \mathbf{I}_p dt$	Number of incident photons per unit area
Radiant flux	$J\ s^{-1} = W$	$\mathbf{Q}_e = E/dt$	Radiant energy emitted or absorbed by a surface per unit time
Photon flux	$mol\ s^{-1}$	$\mathbf{Q}_p = n_p/dt$	Number of photons emitted or absorbed by a surface per unit time
Radiant flux (area)[a] density	$W\ m^{-2}$	$\mathbf{\Phi}_e = \mathbf{Q}_e/dA$	Net radiant flux through a plane surface per unit area
Photon flux (area) density	$mol\ m^{-2}\ s^{-1}$	$\mathbf{\Phi}_p = \mathbf{Q}_p/dA$	Net photon flux through a plane surface per unit area
Irradiance[b]	$W\ m^{-2}$	$\mathbf{I}_e = \mathbf{Q}_e/dA$	Radiant flux incident on unit area of a plane surface
{ Incident photon flux (area) density[c] / Photon irradiance	$mol\ m^{-2}\ s^{-1}$	$\mathbf{I}_p = \mathbf{Q}_p/dA$	Photon flux incident on unit area of a plane surface
Emittance	$W\ m^{-2}$	$\mathbf{Q}_e/dA$	Radiant flux emitted from unit area of a plane surface
Radiant intensity	$W\ sr^{-1}$	$\mathbf{Q}_e/d\Omega$	Radiant flux from a source into unit solid angle (Ω)
Fluence	$mol\ m^{-2}$		Number of photons across unit area (incident on a spherical surface)

[a] The qualifying term area is usually omitted.

[b,c] Also called energy (*b*) or photon (*c*) fluence rate, though more correctly fluence rate is the flux per unit cross-sectional area incident on a spherical volume element (therefore it requires a spherical detector). See Bell & Rose (1981) for criticism of these terms.

i.e. J s^{-1} or W). The **net radiant flux** through a unit area of surface is the **radiant flux density** ($\mathbf{\Phi}_e$, W m^{-2}). That component of the flux incident on a surface is termed **irradiance** ($\mathbf{I}_e$, W m^{-2}) while that emitted by a surface is termed the **emittance** (or radiant excitance) (W m^{-2}). The subscript 'e' will be used where it is necessary to distinguish an energy flux from a flux of photons, identified by subscript 'p' (see Table 2.1), for example a photon flux density ($\mathbf{\Phi}_p$) has units mol m^{-2} s^{-1}.

Two other terms that occur in the plant physiological literature are also worth mentioning. The term radiation **intensity** is often used rather loosely as a synonym for flux density, but it is correctly defined as a flux per unit solid angle emitted from a point source (and so has units of watts per steradian) and its use should be restricted to that sense. The term **fluence rate** is also sometimes used as a synonym for flux density, though the two terms are not identical as fluence rate measures the flux per unit cross-sectional area incident from all directions on a spherical volume element, and therefore requires the use of a spherical detector. For some purposes, for example where one is concerned with light incident on a chloroplast, fluence rate may be the most appropriate measure of incident light, though the necessary spherical sensors are rare.

Black body radiation

The process of radiation emission or absorption requires a corresponding change in the potential energy of the material. The wavelength of the radiation depends on the magnitude of the energy change (equation 2.1), and thus on the possible transitions between available energy states. In atoms, the transitions occur between the limited number of allowed states for the orbital electrons thus giving rise to their characteristic spectra that consist of sharp lines at particular wavelengths that correspond to the particular electronic transitions. In molecules, transitions between different vibrational or rotational states are also allowed so that the spectra are much more complex and the vast numbers of possible transitions can give rise to broad bands of absorption or emission. A chemically complex body thus may have an infinite number of possible energy transitions covering all wavelengths, so that it should have a more or less continuous absorption or emission spectrum. An ideal material that is a perfect absorber or emitter of radiation at all wavelengths is termed a black body.

Because the energy transitions involved in emission and absorption of radiation are the same (but in opposite directions), it follows that absorption spectra correspond to emission spectra and that a good absorber at a particular wavelength will also be a good emitter at that wavelength. The **absorptivity** (or **absorptance**) (α) of a material is defined as the fraction of the incident radiation at a specified wavelength, or over a specified waveband, that is absorbed by a material. The appropriate wavelength, or wavelength interval, is usually indicated by a subscript. Similarly the **emissivity** (ε) at a particular wavelength is defined as the radiation emitted as a fraction of the maximum possible radiation at that wavelength that can be emitted by a body at that temperature. The maximum possible emittance is called the **black body radiation**.

The energy distribution for emission from a true black body ($\varepsilon = 1$ at all wavelengths) is given by *Planck's Distribution Law*. Examples of the spectral distributions given by this law for black bodies at 6000 K (approximately equivalent to the sun) and 300 K (approximately equivalent to the earth) are given in Fig. 2.2.

The peak wavelength (λ_m, μm) of the Planck distribution is a function of temperature and given by *Wien's Law*:

$$\lambda_m = 2897/T \tag{2.2}$$

As shown in Fig. 2.2 λ_m for a black body at 6000 K (close to the apparent surface temperature of the sun) is within the visible region of the spectrum at 483 nm, while λ_m for a radiator at a typical terrestrial temperature of 300 K is 9.65 μm (well into the IR). There is negligible overlap between the solar spectrum and the thermal radiation emitted by objects at normal terrestrial temperatures (Fig. 2.2). It is convenient, therefore, to distinguish between shortwave radiation that falls between 0.3 and 3 μm and mainly comprises radiation originating from the sun, and longwave radiation (sometimes called terrestrial or thermal radiation) between 3 and 100 μm that is emitted by bodies at normal terrestrial temperatures. The symbols S and L will be used throughout as subscripts to **Q**, **Φ** or **I** to distinguish shortwave and longwave radiation fluxes.

As well as being emitted or absorbed by a body, radiation can be reflected or transmitted. The **reflectivity** (or **reflectance**) (ρ) can be defined as the fraction of incident radiation of a specified wavelength that is reflected. Similarly the **transmissivity** (or **transmittance**) (τ) is the fraction of incident radiation at a given wavelength that is transmitted by an object. The sum of $\alpha+\rho+\tau$ at any wavelength equals 1.

When describing the absorption, reflection or transmission over a broad waveband, such as for solar radiation, the terms **absorption coefficient** (α_S), **reflection coefficient** (ρ_S) and **transmission coefficient** (τ_S) are used. Each of these represents an average absorptivity (or reflectivity or transmissivity) over the relevant wavelengths (0.3 to 3.0 μm for solar radiation) weighted by the distribution of radiation in the solar spectrum. The reflection

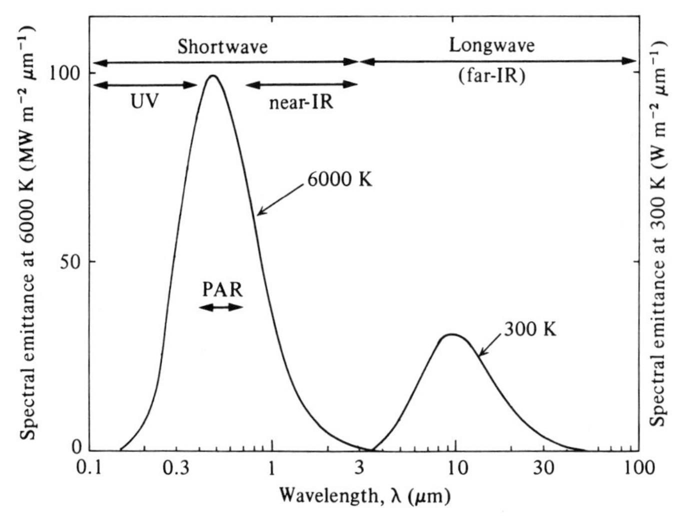

Fig. 2.2. Spectral distribution of radiation emitted from black bodies at temperatures approximately equivalent to the sun (6000 K) and the earth (300 K).

coefficient for solar radiation of natural surfaces is sometimes called the **albedo**.

It is necessary to distinguish between radiation emitted by a body and that reflected. Snow, for example, is white because it reflects well in the visible wavelengths but between 3 and 100 μm (the longwave region) it behaves almost like a black body being a good absorber and radiator. In fact most natural objects (plants, soils, water) have emissivities close to one in the longwave part of the spectrum. Colour, however, depends on the wavelengths that are reflected in the visible, leaves being green because they reflect predominantly green light.

Stefan–Boltzmann Law: The total amount of radiant energy emitted per unit area per unit time ($\mathbf{\Phi}_e$) by a material is strongly dependent on temperature and is given by

$$\mathbf{\Phi}_e = \varepsilon\sigma T^4 \qquad (2.3)$$

where σ is the Stefan–Boltzmann constant ($= 5.67 \times 10^{-8}$ W m^{-2} K^{-4}), and T is the Kelvin temperature. This equation gives the total area under the curves in Fig. 2.2. For a black body $\varepsilon = 1$, but where $\varepsilon \neq 1$ the value of the exponent may not be exactly 4.

Attenuation of radiation

Parallel monochromatic radiation passing through a homogeneous medium is attenuated according to *Beer's Law*:

$$\mathbf{\Phi}_\lambda = \mathbf{\Phi}_{\lambda o} e^{-\mathscr{k}x} \tag{2.4}$$

where $\mathbf{\Phi}_{\lambda o}$ is the flux density at the surface, x is the distance travelled in the medium and $\mathscr{k}$ is an extinction coefficient. This form of equation can be used to describe the attenuation of radiation in the atmosphere and in water, and approximates the light profile within leaves and in plant canopies. Although it is only precise for monochromatic radiation, it can be used with reasonable accuracy over any wavelength band where $\mathscr{k}$ is approximately constant.

It should be noted that similar terms are used with rather different meanings in chemistry texts concerned with absorption of radiation by solutions. In particular, it is common to take logarithms of equation 2.4 to give

$$\ln(\mathbf{\Phi}_\lambda/\mathbf{\Phi}_{\lambda o}) = -\mathscr{k}x \tag{2.5}$$

Which, for convenience, is usually inverted to eliminate the negative sign, giving

$$\ln(\mathbf{\Phi}_{\lambda o}/\mathbf{\Phi}_\lambda) = A = \mathscr{k}x \tag{2.6}$$

where A is known as the **absorbance**. The extinction coefficient in this equation is the product of the molar concentration and what is called a molar absorptivity, or a molar extinction coefficient.

Lambert's Cosine Law

The irradiance at a surface depends on its orientation relative to the radiant beam according to

$$\mathbf{I} = \mathbf{I}_o \cos\theta = \mathbf{I}_o \sin\beta \tag{2.7}$$

where $\mathbf{I}$ is the flux density at the surface, $\mathbf{I}_o$ is the flux density normal to the beam, θ is the angle between the beam and the normal to the surface and β is the complement of θ (see Fig. 2.3). The more the beam is at an angle to the surface, the larger the area it is spread over, so the irradiance decreases. The same relationship holds for the spatial distribution of the radiation emitted from a black body.

Spectral distribution and radiation units

There are several possible ways of expressing radiation that depend on the spectral response of the detector, each being of particular value for different purposes. For example, when the total energy exchange is of concern as in energy balance studies (see Chapters 5 and 9), measurements of total energy

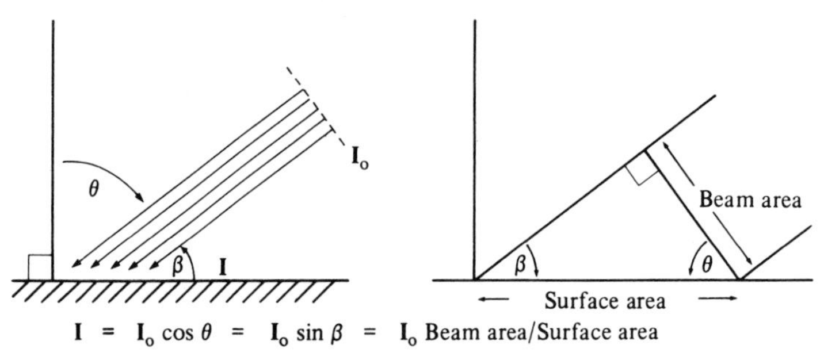

Fig. 2.3. Lambert's Cosine Law.

(made with detectors equally sensitive to all wavelengths), are most relevant with $\mathbf{Q}_e$, $\mathbf{\Phi}_e$ or $\mathbf{I}_e$ being summed over all wavelengths. In other cases, such as photosynthesis (Chapter 7) or morphogenesis (Chapter 8) only a limited range of wavelengths is effective, so it is usual to restrict measurements to the appropriate waveband. For photosynthesis, the photosynthetically active region (PAR) is usually defined as the waveband between 400 and 700 nm (though slightly different wavebands are used by some authors).

Most processes, however, including photosynthesis, are not equally responsive over the whole of the relevant waveband. The absorption spectra of some plant pigments important in photosynthesis and morphogenesis, together with an action spectrum describing the relative efficacy of different wavelengths for photosynthesis and the spectral sensitivity of the human eye are plotted in Fig. 2.4. An ideal detector would have the same spectral sensitivity as the process under consideration. An example where such detectors are used is in illumination measurement using photometric units (e.g. candela, lumen or lux), which are based on the spectral response of the human eye (Fig. 2.4). Using such a detector, different sources with widely differing spectral distributions (e.g. sunlight or a fluorescent tube), would appear equally bright to a human observer when the luminous flux densities (in lux) are equal. In contrast, the radiant flux densities for the same sources in the same conditions would differ widely. Because the spectral sensitivity of the eye does not correspond to that of any plant process, it is best to avoid photometric units in plant studies.

Often the effect of radiation is more dependent on the total number of photons absorbed, than on their energy. In these cases it is more appropriate to express radiation as a photon flux density (mol m^{-2} s^{-1}). The photon flux density within the PAR is commonly used in photosynthetic studies.

Table 2.2 shows how critically the relative values for different radiation measures depend on the light source and its spectral distribution. It can be

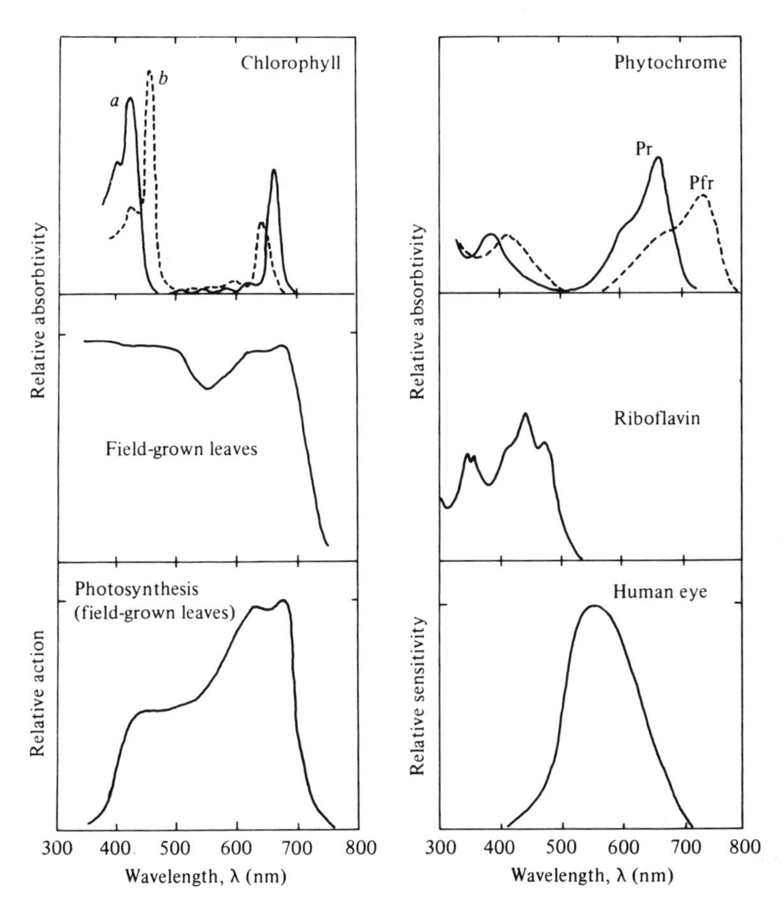

Fig. 2.4. Absorption spectra of various plant pigments (chlorophyll a and b, the far-red light absorbing (Pfr) and red light absorbing (Pr) forms of phytochrome, and riboflavin – after Smith 1981*b*) and of field-grown leaves (data from McCree 1972*a*). Also shown is the action spectrum (in terms of absorbed energy) for leaf photosynthesis and, for comparison, the sensitivity of the human eye.

seen that the photon irradiance ($\mathbf{I}_{p(PAR)}$) gives the best measure of photosynthetic effectiveness of the different sources, being no more than 5% in error for any source, while irradiance in the PAR ($\mathbf{I}_{e(PAR)}$) may be as much as 15% in error, incident luminous flux 53% in error and total shortwave irradiance ($\mathbf{I}_{eS}$) even worse (particularly for quartz-iodine or tungsten lamps which have a very high output in the near-IR).

Table 2.2. *A comparative table showing the values of different radiation measures for a range of different light sources, each providing an irradiance* ($\mathbf{I}_{e(PAR)}$) *in the PAR* (400–700 *nm*) *of* 100 $W\,m^{-2}$

In addition to giving the absolute values of the different measures for each light source (total energy in the shortwave, photon irradiance in the PAR and luminous flux), the table also gives the values for each measure relative to the value for sun + sky (as a percentage). Also shown are the relative photosynthetic efficiencies of the different sources, weighted for the average spectral response of photosynthesis. (Data from McCree 1972*b*, except for $\mathbf{I}_{eS}$ which was estimated for similar sources using a Kipp solarimeter with or without a Schott RG715 filter.)

Light source	Relative Photosynthetic Effectiveness	Corresponding $\mathbf{I}_{eS}$		Corresponding $\mathbf{I}_{p(PAR)}$		Corresponding incident luminous flux	
	(%)	($W\,m^{-2}$)	(%)	($\mu mol\,m^{-2}\,s^{-1}$)	(%)	(klux)	(%)
Sun + sky	100	200	(100)	457	(100)	25.2	(100)
Blue sky	93	152	(76)	425	(93)	20.4	(81)
Metal arc	114	210	(105)	498	(109)	36.0	(143)
Cool white	101	146	(73)	466	(102)	38.8	(154)
Warm white	97	n.a.	n.a.	457	(100)	36.5	(145)
Mercury	102	208	(104)	471	(103)	27.7	(110)
Quartz–iodine	115	550	(275)	503	(110)	25.2	(100)

n.a. data not available.

Radiation measurement

Details of the instruments available for radiation measurement are well described elsewhere (Bainbridge *et al.* 1966; Fritschen & Gay 1979; Unwin 1980; Rosenberg *et al.* 1983; Marshall & Woodward 1985; Pearcy *et al.* 1989). The purpose of this section is to emphasise some important considerations that influence the choice of method of radiation measurement. The commonest techniques involve either (*a*) photoelectric detectors (e.g. silicon cells, cadmium sulphide photoresistive cells, or selenium cells) or (*b*) thermal detectors that measure the temperature difference between surfaces that differentially absorb the incident radiation (one at least is usually matt black and therefore a good absorber at all wavelengths). Photoelectric devices generally have a faster response than thermal detectors, but the latter tend to be preferable for IR because of their wide wavelength response.

Spectral sensitivity. Because the spectral responses of various physiological processes differ, any measurement instrument must have an appropriate spectral sensitivity. This is particularly important when the spectral distribution of the radiation changes as it does, for example, when being filtered through vegetation (see below). The possible magnitude of the error that can occur when using inappropriate detectors is well illustrated by the data in Table 2.2.

It is possible to select combinations of photocells and filters (McPherson 1969) that approximate the responses needed to measure either irradiance in the PAR or photon irradiance in the PAR. These are often called photosynthetic energy and quantum sensors, respectively. For measurements of the total energy in the shortwave, near-IR or longwave, however, thermal detectors with appropriate filters are preferable. Instruments that measure radiant flux density are called radiometers; particular types include net radiometers that measure the difference between energy fluxes in opposite directions across a plane, and solarimeters (pyranometers) that measure total shortwave radiation incident upon a surface.

Directional sensitivity. Conventionally radiation is usually measured on a horizontal surface, so that irradiance, for example, usually refers to the radiant flux incident per unit area of horizontal surface. In certain cases, however, other sensor orientations are needed. For example, in leaf photosynthesis studies where leaves may have different orientations but the same angle to the vertical, the irradiance at the leaf surface may be more relevant than that on a horizontal surface.

For most purposes, it is important that the detector has a good cosine response. That is the irradiance measured for a beam at an angle θ to the normal should be proportional to $\cos\theta$ as expected from Lambert's Law (equation 2.7). Many sensors show a more severe reduction in sensitivity with grazing angles of incidence. The resulting errors can be particularly significant when the sun is low in the sky.

Averaging. The radiation climate within plant communities is extremely variable, both spatially and temporally. This heterogeneity can lead to large sampling errors. Various techniques for minimising the problem have been suggested. One approach is to use large radiation sensors, such as the long tube solarimeters described by Szeicz *et al.* (1964) that average over a large area. Another is to use a large numbers of sensors, as in the transmission meter developed by Williams & Austin (1977) or the 'Sunfleck Ceptometer' (Decagon Instruments). These instruments consist of a large number of small sensors on a linear probe that can be inserted in the canopy, and in the case of the latter instrument, can be used to obtain a readout either of the average irradiance or of the fraction illuminated by high irradiance sunflecks. Alternatively one can move sensors through the crop taking readings at many positions (e.g. Norman & Jarvis 1974). It is necessary to remember, however, that although the mean irradiance may provide a general indication of light penetration into a canopy, more detailed information on the spatial and temporal variation is required for some purposes. For example photosynthesis does not respond linearly to light, with the light within sunflecks often not being used efficiently, so that the average irradiance could not be expected to predict CO_2 uptake well. The high irradiance within sunflecks may even cause photoinhibitory damage to the choroplasts (*see* Chapter 7). It follows from this that for many purposes, especially in studies of photosynthesis, the average irradiance on a horizontal surface at any depth in the canopy is not the value of interest. Of more interest is the distribution of leaf area in different irradiance classes. It is necessary to integrate this over the whole diurnal cycle if one is, for example, to predict photosynthesis.

When irradiance measurements are being made in conjunction with photosynthesis measurements it may be appropriate to set the time constant of the light sensors similar to that of the photosynthetic system (seconds to minutes), but when relating irradiance to growth or phenology daily totals are likely to be adequate.

Estimation. Unfortunately radiation is only measured at relatively few meterological sites so that it is frequently necessary to estimate solar or net radiation from other measurements, such as the duration of bright sunshine

(n). The conversion from sunshine duration to total solar or net radiation depends on site, type of cloud and time of year.

To a reasonable approximation, the average of $\mathbf{I}_S$ over periods of weeks or longer may be obtained from the Ångstrom equation

$$\mathbf{I}_S = \mathbf{I}_A[a + (bn/N)] \quad (2.8)$$

where n/N is the fraction of the daylength with bright sunshine, $\mathbf{I}_A$ is the extraterrestrial irradiance on a horizontal surface appropriate for the time of year and latitude (see Appendix 7 for calculation of N and $\mathbf{I}_A$), and a and b are constants that depend, for example, on site, atmospheric pollutants and time of year. Published values of these constants vary over a wide range (Martínez-Lozano *et al.* 1984). In England, for example, a is approximately 0.24 and b varies between 0.50 in winter and 0.55 in summer, with a small amount of additional variation around the country (Thompson *et al.* 1982). Net radiation (see below) may also be estimated from sunshine duration (see Linacre 1969).

The conversions between different radiation measures for different light sources can be obtained from Table 2.2. For example, this table shows that for average sun + sky light, $\mathbf{I}_{PAR} \simeq 0.5\,\mathbf{I}_S$.

Radiation in natural environments

Shortwave radiation

The radiant flux density normal to the solar beam at the top of the atmosphere at the mean distance of the earth from the sun is called the solar constant ($\mathbf{\Phi}_{pA}$) and is approximately 1370 W m^{-2}. The actual value of the flux density at the top of the atmosphere varies by about ±3.5% between July and January (when the sun is closest to the earth). The radiation that actually reaches the earth's surface is much modified in terms of quantity, spectral properties and angular distribution as a result of absorption or scattering by molecules in the atmosphere and by scattering or reflection from clouds and particulates. Reflection from and transmission through terrestrial objects such as leaves also modify the radiation climate. A useful simplification when discussing solar radiation is to distinguish between the relatively unmodified parallel radiation in the direct beam – direct solar radiation ($\mathbf{\Phi}_{S(dir)}$), and the diffuse shortwave radiation ($\mathbf{\Phi}_{S(diff)}$) that includes reflected and scattered radiation from all portions of the sky. The sum of the direct and diffuse radiation incident on a horizontal surface is often called global radiation.

Scattering. Some of the direct solar beam is scattered by molecules or particles in the atmosphere. Rayleigh scattering is by molecules smaller

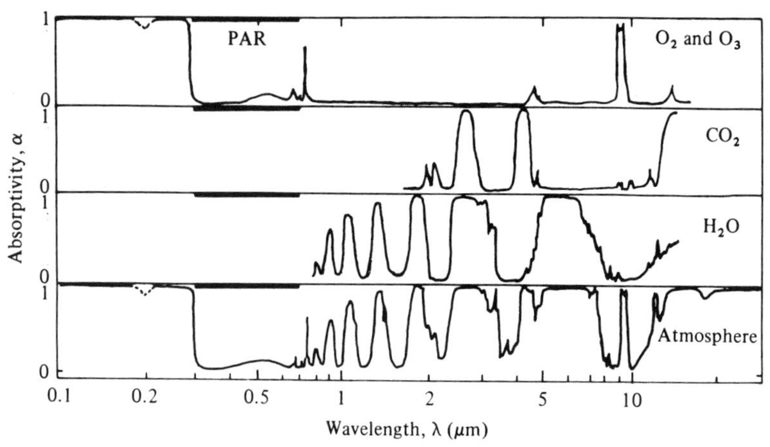

Fig. 2.5. Absorption spectra for the atmosphere and some pure atmospheric components. (After Fleagle & Businger 1963.)

than the wavelength of light, being most effective for shorter wavelengths (blue), while Mie scattering by larger particles such as dust and water droplets is relatively wavelength independent.

Absorption in the atmosphere. Radiation absorption in the atmosphere is a function of the pathlength through the atmosphere and the content of absorbers, particularly water vapour. Absorption spectra for some of the more important absorbers, as well as the whole atmosphere, are given in Fig. 2.5. This figure shows that there is a 'window' in the visible/PAR where the atmosphere is relatively transparent. Biologically important absorption bands include those in the UV (primarily ozone) that reduce the quantity of mutagenic UV radiation reaching the earth's surface, and those in the IR (particularly water vapour and CO_2 – see Chapter 11).

Transmission through the atmosphere. Atmospheric transmission is a function of the optical air mass (m), which is the ratio of the mass of atmosphere traversed per unit cross-sectional area of the actual solar beam to that traversed for a site at sea level if the sun were overhead. The value of m therefore decreases with altitude (in proportion to the atmospheric pressure, P) and with increasing solar elevation (β), approximately according to (see Fig. 2.6)

$$m \simeq (P/P_o)/\sin\beta = (P/P_o)\operatorname{cosec}\beta \tag{2.9}$$

where P_o is the atmospheric pressure at sea level. If the transmissivity (τ) of the atmosphere when free of dust and clouds is defined as the fraction of

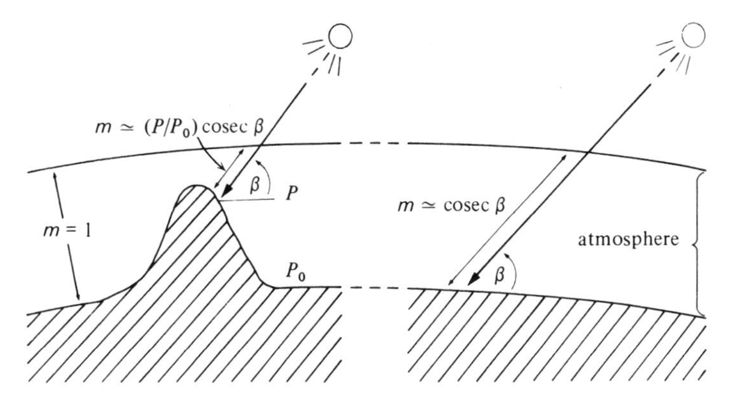

Fig. 2.6. The calculation of optical air mass (m) in relation to solar elevation (β) and atmospheric pressure (P).

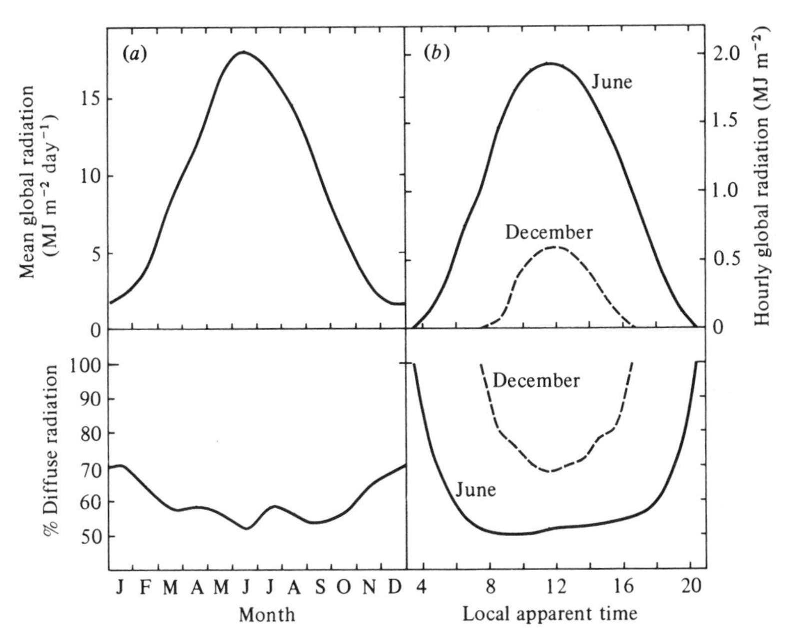

Fig. 2.7. Mean values (1959–1975) of shortwave irradiance (MJ m^{-2}) on a horizontal surface at Kew (51.5° N), and the proportion that is diffuse: (*a*) annual trend in mean daily global irradiance and (*b*) hourly values for June and December. (Data from Anon. 1980.)

incident solar radiation that is transmitted when $m = 1$, a combination of Beer's and Lambert's Laws gives the direct irradiance on a horizontal surface ($\mathbf{I}_{S(dir)}$) as

$$\mathbf{I}_{S(dir)} = \mathbf{\Phi}_{pA}\, \tau^m \sin\beta \tag{2.10}$$

where the first term represents attenuation by absorbers in the atmosphere and the second term represents the cosine correction (equation 2.7) expressed in terms of the complement of the zenith angle. For 'clear sky' τ is commonly between about 0.55 and 0.7, though higher values can be obtained for very clear dry sky particularly at higher elevations.

Diffuse radiation. Although equations such as 2.9 and 2.10 are useful for modelling the direct radiation environment, even on a clear day diffuse radiation contributes between about 10 and 30 % of total solar irradiance. In fact in England, for example, cloud cover is such that there is bright sunshine only about 34 % of the time that the sun is above the horizon, with diffuse radiation contributing on average between 50 and 100 % of the shortwave radiation depending on season and time of day (Fig. 2.7). In drier climates, the proportion of diffuse radiation is much lower. For example in Yuma, Arizona, USA, the sunshine duration reaches 91 % of the possible amount, so that the proportion of diffuse is probably correspondingly low. The proportion of diffuse radiation can be estimated with adequate precision for many purposes from the ratio of global radiation at the surface to the maximum potential (as measured by the irradiance at the top of the atmosphere, $\mathbf{I}_A$). This relationship is illustrated in Fig. 2.8.

On a clear day, most diffuse radiation comes from the region near the sun (as a result of forward scattering). With an overcast sky, although it is significantly brighter near the zenith, it can be assumed, to a first approximation, that the sky is equally bright in all directions (called a uniform overcast sky). In practice, for an overcast sky, the radiance at the zenith is usually about 2.1 to 2.4 times the radiance of the sky at the horizon (see Monteith & Unsworth 1990).

Although about 45 % of the energy in the direct solar beam at the earth's surface is in the photosynthetically active wavelengths, the average proportion of direct plus diffuse radiation in the PAR is approximately 50 %. This is because diffuse radiation tends to be enriched in the visible wavelengths, particularly when the sun is low in the sky. This compensates for any depletion of the visible content of the direct beam at low solar elevations (as a result of Rayleigh scattering).

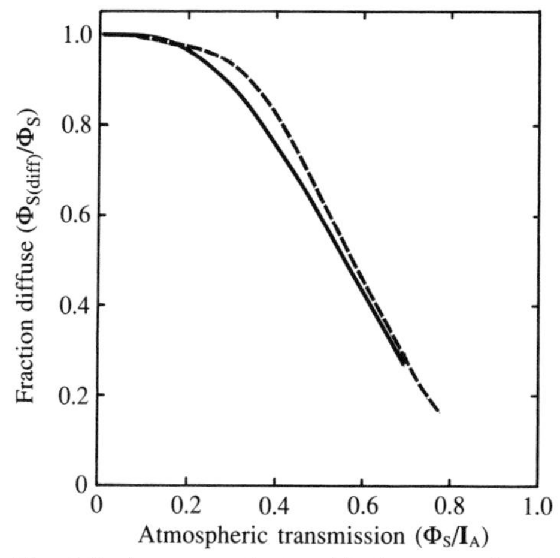

Fig. 2.8. Average relationship between the fraction of total global radiation that is diffuse ($\mathbf{\Phi}_{S(diff)}/\mathbf{\Phi}_S$) and the atmospheric transmission ($\mathbf{\Phi}_S/\mathbf{I}_A$), where $\mathbf{I}_A$ is the amount of radiation received on a horizontal surface at the top of the atmosphere (see Appendix 7), over a range of environments for hourly (-----) or daily data (———) (summarised from data collated by Spitters *et al.* 1986).

Radiation at different sites. The amount of radiation received on a horizontal surface at the top of the atmosphere ($\mathbf{I}_A$) is a simple function of latitude, time of day and time of year and can be calculated using readily available equations (see Appendix 7). Amounts received range from zero in the polar regions in winter to more than 40 MJ m^{-2} day^{-1} in mid-summer for latitudes north of 40° N (June) or south of 15° S (December). Similar geometrical considerations enable one to calculate the radiation receipts on slopes of different aspect or angle (Fig. 2.9).

Annual and daily trends in actual total solar irradiance (on a horizontal surface) at a site in southeast England are shown in Fig. 2.7. These values range from a mean of 1.8 MJ m^{-2} day^{-1} in December (approximately 25% of the extraterrestrial radiation), to 18 MJ m^{-2} day^{-1} in June (about 40% of the extraterrestrial radiation). In contrast, the mean daily radiation in June at China Lake, California, USA, is as high as 34.2 MJ m^{-2} day^{-1} and mean values greater than 25 MJ m^{-2} day^{-1} occur over most of the western and southern United States (Anon. 1964).

The maximum irradiance (at midday) commonly lies between about 800 and 1000 W m^{-2} during the growing season over much of the earth's surface. For many purposes the diurnal trend in irradiance, at least on clear days, can be approximated by a sine curve:

$$\mathbf{I}_{St} = \mathbf{I}_{S(max)} \sin(\pi t/N) \qquad (2.11)$$

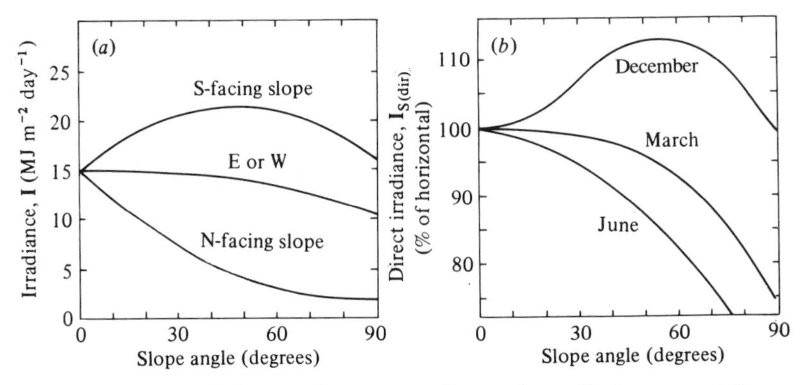

Fig. 2.9 Influence of slope and aspect on direct solar radiation potentially received at a latitude of 53° 15′ N with $\tau = 0.7$ (after Pope & Lloyd 1975). (*a*) Annual mean of potential daily direct irradiance (actual receipts will be considerably less than shown because of cloud). (*b*) Seasonal variation in potential irradiance relative to that on a horizontal surface, for east- or west-facing slopes.

where $\mathbf{I}_{St}$ is the irradiance t hours after sunrise and N is the daylength in hours (see also Appendix 7). Integration of equation 2.11 gives an estimate of the daily, integral of irradiance as $(2n/\pi)\mathbf{I}_{S(max)}$. For example, with a maximum irradiance typical of a cloudless day in southern England in June of 900 W m^{-2} and a daylength of 16 h (5.8×10^4 s) this formula gives a daily insolation of 33 MJ m^{-2} compared with a measured maximum of about 30 MJ m^{-2}.

Variation of annual mean daily solar radiation over the earth's land surface is shown in Fig. 2.10. This illustrates that the greatest annual totals occur in the mid-latitudes.

Longwave and net radiation

In addition to shortwave radiation from the sun and sky, an important contribution to the radiation balance of plants is made by longwave radiation. For example, the sky is an important source of longwave radiation: this is emitted by the gases (especially water vapour and CO_2) present in the lower atmosphere. However, as these atmospheric gases are not perfect emitters in the longwave (i.e. the emissivity of the atmosphere is < 1; compare Fig. 2.5), the apparent radiative temperature of the atmosphere (T_{sky}) is less than its actual temperature. Empirical results suggest that the sky behaves approximately as a black body with its temperature about 20 K below the temperature measured in a meterological

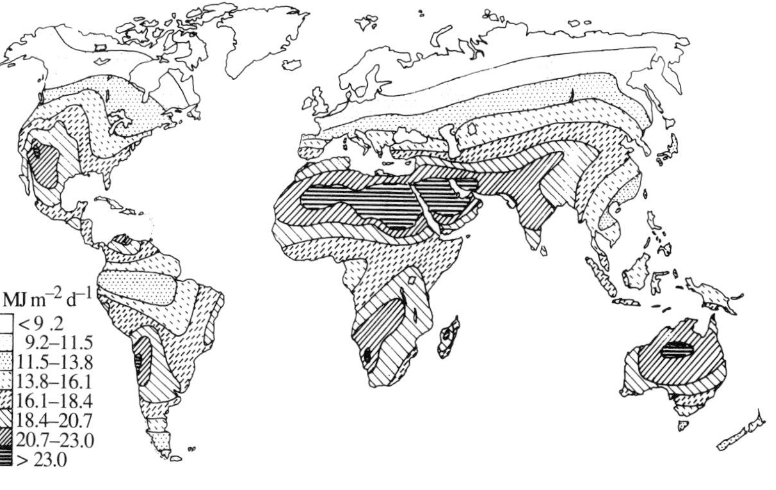

Fig. 2.10. Annual mean daily total shortwave irradiance at the earth's surface. (Data from Landsberg 1961 and Anon. 1964.)

screen, so that the downward flux of longwave radiation ($\mathbf{\Phi}_{Ld}$) under clear skies is given approximately by

$$\mathbf{\Phi}_{Ld} \simeq \sigma (T_a - 20)^4 \tag{2.12}$$

where T_a is the air temperature (K). The difference between the downward longwave radiation and that emitted upwards by the surface ($\mathbf{\Phi}_{Lu} \simeq \sigma T_a^4$) is the net longwave radiation ($\mathbf{\Phi}_{Ln}$). With clear skies this net radiation ($\mathbf{\Phi}_{Ld} - \mathbf{\Phi}_{Lu}$) is approximately constant throughout the day and throughout the year at a net loss of 100 W m^{-2}.

The presence of clouds increases the downward flux of longwave radiation since clouds are more effective emitters and have a mean radiative temperature that averages only 2 K below the mean screen temperature. This corresponds to a net upward longwave flux of only 9 W m^{-2} in cloudy conditions. More precise methods for estimating longwave radiation fluxes are outlined by Sellers (1965) and Gates (1980).

The net flux of all radiation (shortwave and longwave) across unit area of a plane is called the **net radiation ($\mathbf{\Phi}_n$)**. Alternatively the net radiation absorbed by an object is the sum of all incoming radiation fluxes minus all outgoing radiation (see Fig. 2.11). Incoming radiation includes all incident direct and diffuse solar radiation and that reflected from the surroundings, as well as incident longwave emitted by the sky and the surroundings. Radiation losses include thermal radiation emitted, as well as any incident radiation that is reflected or transmitted by the object.

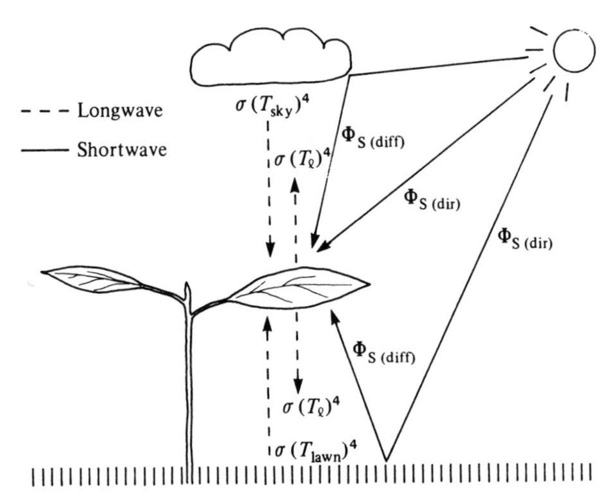

Fig. 2.11. Schematic diagram of the longwave and shortwave radiation exchanges between a leaf and its environment.

Therefore, for a lawn (or a crop) which may be treated as a horizontal plane surface that does not transmit radiation, one can write

$$\begin{aligned}\mathbf{\Phi}_n &= \mathbf{I}_S + \mathbf{I}_{Ld} - \mathbf{I}_S\,\rho_{S(lawn)} - \varepsilon\sigma(T_{lawn})^4 \\ &\simeq \mathbf{I}_S(\alpha_{S(lawn)}) + \mathbf{I}_{Ld} - \sigma(T_{lawn})^4\end{aligned} \tag{2.13}$$

where α_S and ρ_S, respectively, are the absorption coefficient and the reflection coefficient (of the lawn). All fluxes are referred to a horizontal surface. ε is usually assumed to equal 1.

For an object with two sides, such as an isolated horizontal leaf exposed above the lawn

$$\mathbf{\Phi}_n = (\mathbf{I}_S + \mathbf{I}_S\,\rho_{S(lawn)})\,\alpha_{S(leaf)} + \mathbf{I}_{Ld} + \sigma(T_{lawn})^4 - 2\sigma(T_{leaf})^4 \tag{2.14}$$

Note the additional terms for solar radiation reflected by the lawn and for longwave radiation from the lawn, and the factor of 2 in the term for longwave losses. Note also that all radiation fluxes are expressed per unit projected area. For a lawn or a crop, that equals the ground area, but for a leaf it is the area of *one* side. This convention will be used throughout the book for treatment of heat, mass and radiation transfer.

The use of equations 2.13 and 2.14 for one- and two-sided surfaces is best illustrated by the examples in Table 2.3, where the net radiation balance of a lawn and of an isolated leaf are calculated for different weather conditions. As might be expected $\mathbf{\Phi}_n$ is lower on cloudy than on sunny days for both surfaces. At night, when there are no shortwave exchanges, $\mathbf{\Phi}_n$ is negative. On a cloudy night, however, net radiation would be close to zero.

Table 2.3. *Comparison of net radiation balance (where* $\mathbf{\Phi}_n = \Sigma\mathbf{\Phi}_{absorbed} - \mathbf{\Phi}_{emitted}$) *for a lawn and an isolated horizontal leaf exposed above the lawn for different conditions, assuming that in the shortwave* $\alpha_{lawn} = 0.77$, $\alpha_{leaf} = 0.5$, $\rho_{lawn} = 0.23$, *and longwave emissivity of lawn and leaf are unity. All fluxes expressed per unit projected area* ($W\ m^{-2}$)

Assumed conditions	Sunny day	Cloudy day	Clear night
Shortwave irradiance on horizontal surface ($\mathbf{I}_S$, W m^{-2})	900	250	0
T_a(K)	293	291	283
T_{sky}(K)	273	289	263
T_{lawn}(K)	297	288	279
T_{leaf}(K)	297	288	277
Lawn (equation 2.13)			
$\mathbf{\Phi}_{S(absorbed)} = \mathbf{I}_S(\alpha_{S(lawn)})$	693	193	0
$\mathbf{\Phi}_{L(absorbed)} = \sigma(T_{sky})^4$	309	389	266
$\mathbf{\Phi}_{L(emitted)} = \sigma(T_{lawn})^4$	433	383	337
$\mathbf{\Phi}_n$	569	199	−71
Leaf (equation 2.14)			
$\mathbf{\Phi}_{S(absorbed)} = \mathbf{I}_S(\alpha_{S(leaf)})(1+\rho_{S(lawn)})$	554	154	0
$\mathbf{\Phi}_{L(absorbed)} = \sigma(T_{sky})^4 + \sigma(T_{lawn})^4$	742	772	604
$\mathbf{\Phi}_{L(emitted)} = 2\sigma(T_{leaf})^4$	866	766	656
$\mathbf{\Phi}_n$	430	160	−52

The net radiation for leaf is always smaller than that for the lawn. Note that the precise values for the different fluxes depend on the actual temperatures which depend on the complete heat balance including convection and evaporation (see Chapter 5).

For a non-horizontal leaf, calculations are more complex in that one has to take account of its angle relative to the solar beam, and of the different distribution of diffuse radiation from the ground and the sky.

Net radiation is measured using net radiometers that detect the difference between upward and downward radiation fluxes. For a surface such as a lawn, $\mathbf{\Phi}_n$, measured above the surface, equals the net radiation absorbed by the surface. The net radiation absorbed by a layer of leaves in a plant canopy, however, must be measured as the difference between $\mathbf{\Phi}_n$ above and below that layer. For more complex objects the net radiation absorbed can be obtained from net radiometers arranged around the object so as to integrate over all directions the net flux perpendicular to the surface.

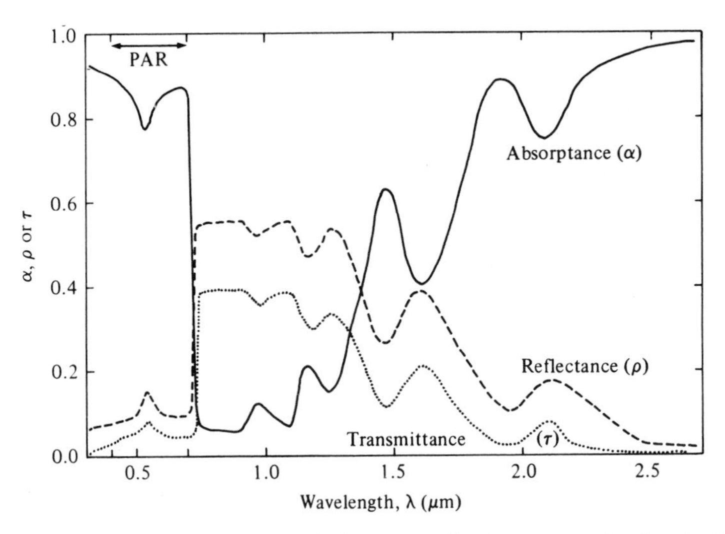

Fig. 2.12. Absorption, transmission and reflection spectra for 'typical' leaves (mean from various sources).

Radiation in plant communities

Radiative properties of plants

Absorption, reflection and transmission spectra for typical plant leaves are shown in Fig. 2.12. Although this general picture is true for most species, details vary with thickness, age, water content, surface morphology and orientation. The main features of the spectra are the high absorptance in most of the PAR except the green (hence the green colour of leaves) and the low absorptance in the near-IR. Leaves are good absorbers in the far-IR so that they behave approximately as black bodies in the longwave, with ε being between 0.94 and 0.99 for most species (Idso *et al.* 1969).

Table 2.4 presents some typical values for absorption and reflection coefficients for leaves, vegetation and other natural surfaces. These results show that the solar reflection coefficient for leaves of temperature crop species is usually close to 0.30 with relatively little variation between species. Reflectance can be higher for white pubescent leaves, waxy leaves or for leaves with a low moisture content. For example, as shown in Fig. 2.13, the reflectivity at 550 nm (approximately proportional to ρ_S) for *Atriplex hymenelytra* leaves can vary between 0.35 in winter and 0.6 in summer as a function of leaf moisture content. Not only does the presence of leaf hairs tend to increase the reflectance of that surface, but there is evidence that hairs on the lower surface of a leaf can increase reflection from the upper

Table 2.4. *Reflection* (ρ_S) *and absorption* (α_S) *coefficients for leaves, vegetation and other surfaces*

All values are for typical shortwave radiation unless otherwise stated (see Linacre 1969; Gates 1980; Stanhill 1981; Monteith & Unsworth 1990).

		ρ_S(%)		α_S(%)
Single leaves				
Crop species		29–33		40–60
Deciduous broad leaves (low sun)		26–32		34–44
Deciduous broad leaves (high sun)		20–26		48–56
Artemisia sp. } white, pubescent		39		55
Verbascum } (high sun)		36		52
Conifers		12		88
Typical mean values for total shortwave	ρ_S	~ 30	α_S	~ 50
Typical mean values for PAR	ρ_{PAR}	~ 9	α_{PAR}	~ 85
Vegetation				
Grass		24		
Crops		15–26		
Forests		12–18		
Typical mean values for total shortwave	ρ_S	~ 20		
Typical mean values for PAR	ρ_{PAR}	~ 5		
Other surfaces				
Snow		75–95		
Wet soil		9±4		
Dry soil		19±6		
Water		5–> 20		

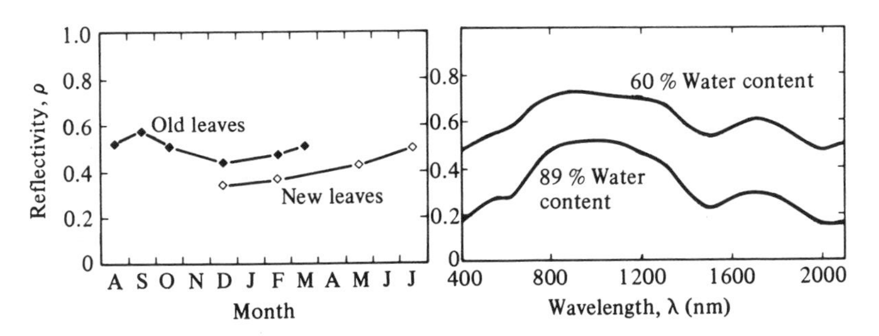

Fig. 2.13. (*a*) Seasonal course of leaf reflectivity at 550 nm for leaves of *Atriplex hymenelytra* growing in Death Valley, Californa. (*b*) Reflection spectra for *A. hymenelytra* leaves at high (89 %) and low (60 %) moisture contents (expressed as % fresh weight). (After Mooney *et al.* 1977.)

surface (e.g. Eller 1977). Reflectance in the PAR tends to be rather less than in the total shortwave.

As shown in Table 2.4, solar absorptance for many leaves is about 0.5, though this is quite variable, reaching as high as 0.88 for conifer needles. The absorptance in the PAR is rather higher, averaging about 0.85. Absorptance is strongly dependent on water content and pubescence, largely as a result of their effects on reflection. The effect of leaf hairs on α_{PAR} in closely related *Encelia* species is illustrated in Fig. 2.14, where the extremely pubescent desert species *E. farinosa* has a reflectance in the PAR 50% greater than that of the glabrous coastal species *E. californica.* Leaf thickness is another factor determining absorptance, because thick leaves (e.g. succulent species) have a very low transmittance.

The reflectance of plant canopies tends to be rather lower than that of the component single leaves because the multiple reflections between adjacent leaves and between leaves and stems leads to trapping of radiation. This effect is particularly marked for taller crops such as forests where ρ_S may be as low as 0.10, while ρ_S for some dense short canopies may approach that of individual leaves (some typical values are given in Table 2.4). As with other surfaces, the crop reflection coefficient depends on solar elevation, increasing by a factor of about two as the solar elevation decreases from 60° to 10° (Ross 1975).

Table 2.4 emphasises the different reflection and absorption behaviour in the PAR and in the near-IR. Although only about 50% of the incident shortwave radiation is in the PAR, about 80–85% of all absorbed shortwave radiation is in this region. Therefore spectral properties in the visible dominate the leaf radiation balance. Transmission of different wavelengths in plant canopies is discussed in the next section, while the implications of leaf spectral properties to the energy balance and thermal regime of plant leaves are discussed in Chapter 9.

Radiation distribution within plant canopies

A precise description of the pattern of radiation distribution within any plant canopy is difficult because of the necessity of taking account of detailed canopy architecture, the angular distribution of the incident radiation and the spectral properties of the leaves. There are, however, useful simplifications that give adequate precision for most purposes, including the modelling of photosynthesis and productivity (Lemeur & Blad 1974; Ross 1975; Campbell 1986; Campbell & Norman 1989; Monteith & Unsworth 1990). Only some of the most useful approximations are reviewed below.

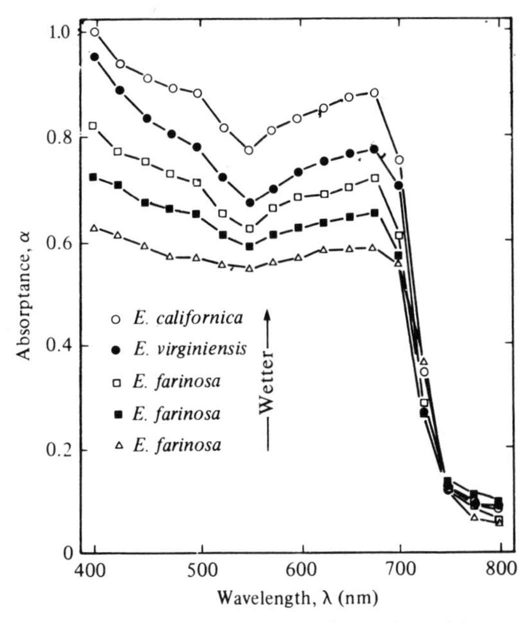

Fig. 2.14. Absorption spectra of *Encelia californica*, *E. virginiensis* and *E. farinosa* along an aridity gradient during April. (After Ehleringer 1980.)

Horizontal leaves

A common simplifying assumption is that the stand is horizontally uniform so that radiation is constant in any horizontal layer, only changing with height. In general the average irradiance at any level tends to decrease exponentially with increasing depth in a way similar to that predicted by Beer's Law (equation 2.4) assuming that the canopy is an homogeneous absorber.

One simple derivation assumes that the canopy consists of randomly arranged horizontal leaves with a canopy **leaf area index** (L, that is the total projected leaf area per unit area of ground) that can be divided into a number of horizontal layers each containing equal areas and within which no leaves overlap. If one considers a layer of canopy containing a small leaf area index, $\mathrm{d}L$, this will intercept an amount of radiation equal to $\mathbf{I}_0 \mathrm{d}L$, where $\mathbf{I}_0$ is the irradiance at the top of the layer. If the leaves are opaque, the change in irradiance ($\mathrm{d}\mathbf{I}$) on passing through this layer of canopy is equal to $-\mathbf{I}_0 \mathrm{d}L$. Integration downwards through a total leaf area index L, gives the average irradiance on a horizontal surface below that leaf area index as

$$\mathbf{I} = \mathbf{I}_0 \mathrm{e}^{-L} \qquad (2.15)$$

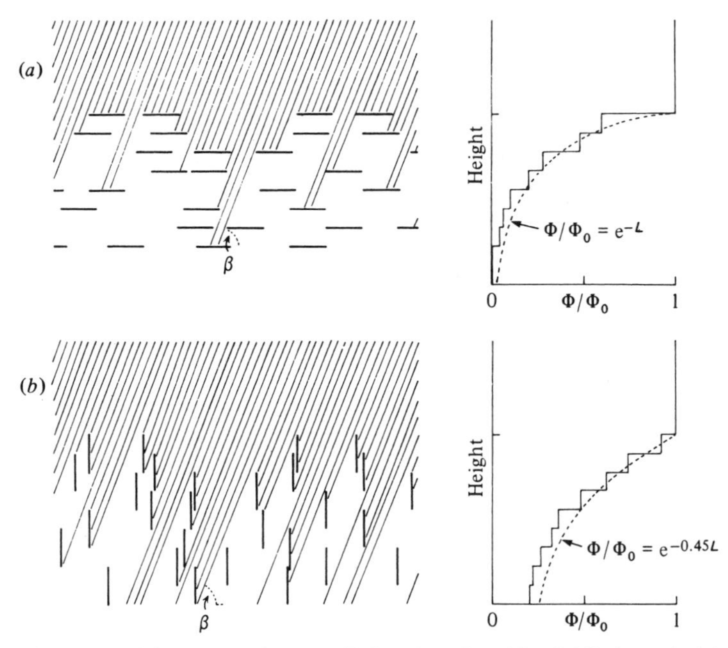

Fig. 2.15. Light penetration at a solar elevation (β) of 66° through (*a*) a horizontal-leaved canopy (where k in equation 2.15 = 1) and (*b*) a vertical-leaved canopy (where $k = \cot 66° = 0.45$). This illustrates the more rapid extinction in the horizontal-leaved canopy, though the converse would be true for $\beta < 45°$

Although this is similar to Beer's Law, in this case the irradiance at any level is in fact the average of some areas with unattenuated light (sunflecks) and some completely shaded areas. This pattern of radiation attenuation is illustrated for a hypothetical horizontal-leaved canopy in Fig. 2.15*a*.

It follows from equation 2.15 that, for opaque leaves, $\mathbf{I}/\mathbf{I}_o$ ($= e^{-L}$) is the fraction of the area on a plane below a leaf area index of L that is sunlit. Conversely, the sunlit leaf area index (L_{sunlit}) in the canopy is given by

$$L_{sunlit} = 1 - \mathbf{I}/\mathbf{I}_o = 1 - e^{-L} \tag{2.16}$$

This function approaches 1 at high leaf area indices, and indicates that the maximum leaf area index that can be in full sunlight, in a horizontal-leaved canopy, is equal to 1.

Other leaf angle distributions

The simple model for horizontal-leaved canopies may be extended to other patterns of leaf distribution. The general principle is to project the shadow of the leaves onto a horizontal plane, and to use this area as the exponent in equation 2.15. If k is the ratio of shadow area to actual leaf area, equation 2.15 becomes

$$\mathbf{I} = \mathbf{I}_o e^{-kL} \tag{2.17}$$

where k is an extinction coefficient. The situation for vertical leaves oriented towards the sun is illustrated in Fig. 2.15*b*. In this case the ratio of shadow area to actual area is equal to $\cot \beta$ (where β is the solar elevation). In contrast to the horizontal-leaved situation, the radiation profile is dependent on β. With high sun, k is less than 1 and light penetrates further into the canopy than with a corresponding horizontal-leaved canopy. With low sun, however, the converse is true.

Using an argument similar to that above, it is possible to estimate the value of L_{sunlit}. In this case the maximum value of L_{sunlit} is equal to $1/k$, so

$$L_{sunlit} = (1 - e^{-kL})/k \tag{2.18}$$

though the actual irradiance per unit illuminated leaf area will be given by $k\mathbf{I}_o$, that is lower for high sun and vertical leaves. This has important implications for both leaf temperature regulation (Chapter 9) and canopy photosynthesis (Chapter 7).

In most real canopies leaves assume a range of orientations, with some canopies (**planophile**) having predominantly, but not exclusively, horizontal leaves, others (**erectophile**) having predominantly vertically oriented leaves, but many other distributions are also found. Some typical distribution functions of leaf inclination (the angle between the leaf and the horizontal) for different canopies are illustrated in Fig. 2.16. It is often possible to approximate actual leaf angle distribution functions by simplified geometrical treatments that can be used to model light penetration through canopies.

One erectophile distribution that is of particular interest is the spherical leaf distribution. In this it is assumed that the leaves have an equal probability of any orientation so that they could be thought of as being capable of being rearranged on the surface of a sphere. Although on this assumption leaves have an equal probability of any azimuth (that is direction N, S, E, W, etc.), erect leaves (all around the equator) are more common than horizontal ones (only at the top or bottom). It follows that the extinction coefficient is related to the ratio of the projection of a sphere onto a horizontal plane to its surface area, which is $\pi r^2/(4\pi r^2 \sin \beta)$ ($= 0.25 \operatorname{cosec} \beta$), but because either side of the leaf can intercept radiation, the appropriate value for k is twice this value or $0.5 \operatorname{cosec} \beta$.

A more general function that is continuous over a range of leaf angles (like the spherical distribution) but can accommodate canopies with tendencies towards the erect or to the horizontal, as necessary, is the

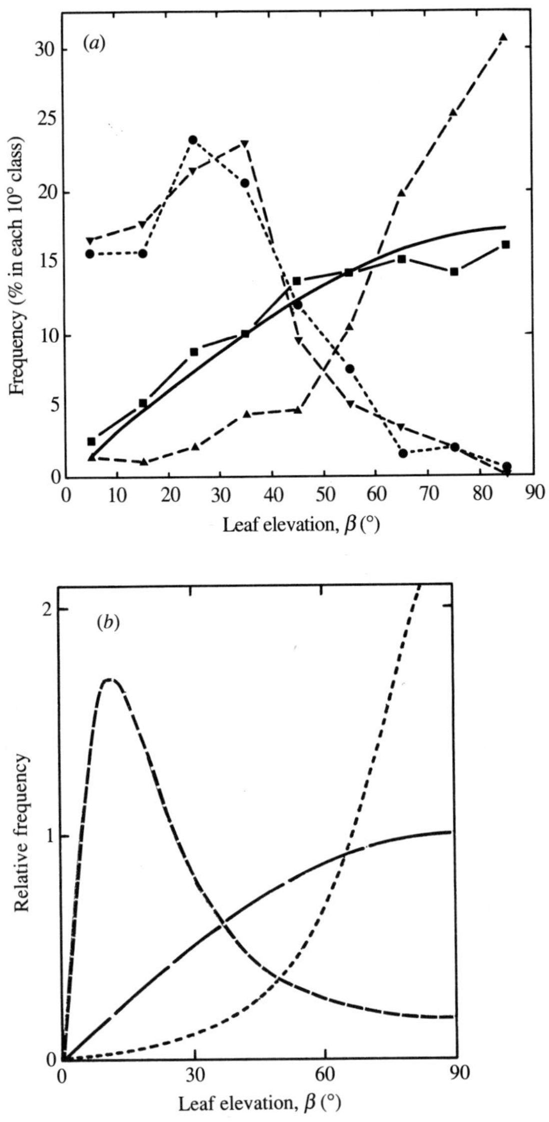

Fig. 2.16 (*a*) Some examples of leaf inclination distribution functions (where β is the leaf elevation above the horizontal) for different canopies: Shamouti orange (—■—), perennial ryegrass (May 6 ‒‒‒▲‒‒‒, August 21 ‒‒‒▼‒‒‒) and flowering white clover (·····●·····). (Data from de Wit 1965; Cohen & Fuchs 1987.) (*b*) Theoretical leaf inclination distribution functions, for ellipsoidal distributions with $x = 0.5$ (‒ ‒ ‒ ‒), $x = 1.0$ (spherical, ——— ———), and $x = 3.0$ (———).

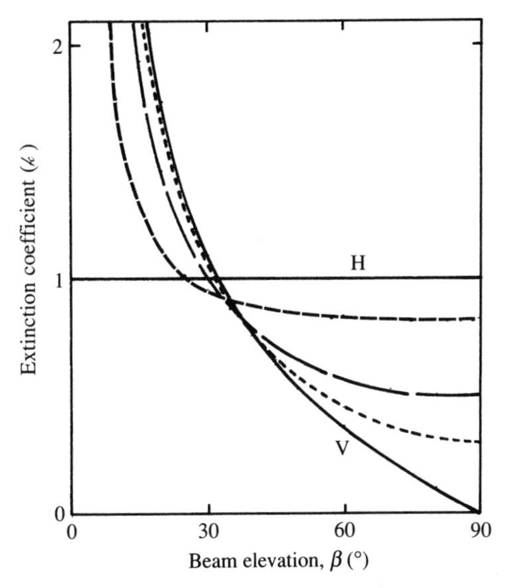

Fig. 2.17. Dependence of extinction coefficients (ℓ) for direct radiation as a function of beam elevation (β) for the ellipsoidal leaf angle distribution functions with $x = 0.5$ (-----), $x = 1.0$ (spherical, —— ——), and $x = 3.0$ (– – – –), together with curves for horizontal (H) and vertical (V) distributions.

ellipsoidal distribution. A single parameter, x (the ratio of the horizontal semi-axis of the ellipsoid to the vertical semi-axis) is used to describe the shape of the distribution. The spherical distribution is a special case of the ellipsoidal distribution when $x = 1$. The leaf inclination distributions for several models are illustrated in Fig. 2.16*b*.

Extinction coefficients are more generally useful in modelling light climates than are leaf inclination distributions. The variation of extinction coefficients with angle of elevation for different canopy models are illustrated in Fig. 2.17, and the functions tabulated in Table 2.5. Further information on radiation modelling in canopies may be found in Monteith & Unsworth (1990) and Campbell & Norman (1989).

Use in practice

In practice many real canopies cannot be approximated by one of these simple geometrical models and, furthermore, a proportion of intercepted radiation is either transmitted by the leaves or scattered downwards through the canopy. Therefore it is convenient for many purposes to replace the geometrically derived value of the extinction coefficient ℓ by an empirically determined value. Equation 2.17, with an empirical ℓ, was

Table 2.5. *Dependence of the extinction coefficient* (k) *on beam elevation* (β) *for different leaf angle distribution functions that are commonly used in modelling canopy light climates, together with corresponding leaf angle distribution functions*

Leaf angle distribution	Extinction coefficient (k)
Horizontal	$k = 1$
Vertical	$k = (2\cot\beta)/\pi$
Spherical	$k = 1/(2\sin\beta)$
Ellipsoidal[a]	$k = (x^2+\cot^2\beta)^{\frac{1}{2}}/(\mathrm{A}x)$
Diaheliotropic	$k = 1/\sin\beta$

[a] where x is the ratio of horizontal to the vertical axis of the ellipsoid and $\mathrm{A} \simeq (x+1.774(x+1.182)^{-0.733})/x$

originally proposed by Monsi & Saeki (1953) in their classical work. Observed values of k have been found to vary from about 0.3 to 1.5 depending on species, stand type and so on. Values less than 1.0 are obtained for non-horizontal leaves or clumped-leaf distributions, while values greater than 1.0 occur with horizontal leaves or more regular arrangement in space.

There are several other complications that it is necessary to consider when modelling radiation in plant canopies. These include the following:

1. *Spectral distribution.* Because leaves are relatively transparent in the IR, the shortwave radiation deep in plant canopies is relatively enriched in the near-IR. Figure 2.18 illustrates the relatively greater attenuation of PAR than near-IR within a wheat crop. For this particular example, k for PAR was more than twice that for near-IR. The extinction profile for total shortwave radiation is between those for PAR and IR, being very close to the observed profile for net radiation (because the net radiation profile is dominated by the shortwave component).

2. *Heliotropism.* Radiation penetration models are further complicated in the many species, particularly among the Leguminoseae, that show heliotropic leaf movements with the leaves tracking the sun during the day (see e.g. Ehleringer & Forseth 1980). Such movements can have large effects on the radiation received by the leaves, as illustrated in Fig. 2.19 for isolated leaves. This figure compares the diurnal trend of direct beam irradiance on a clear day, for a horizontal leaf, a vertical leaf oriented north–south, and a diaheliotropic leaf that is continually oriented perpendicular to the solar beam. For these conditions the diaheliotropic leaf can receive nearly 50% more radiation over a day than the horizontal leaf. Conversely, para-

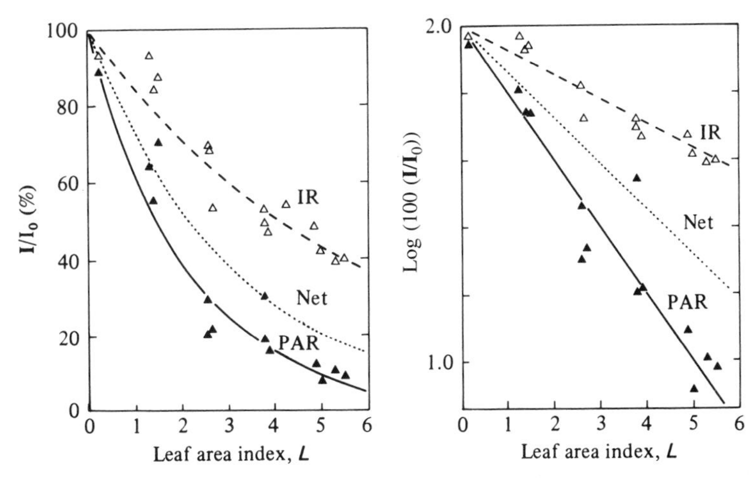

Fig. 2.18. Daily means of transmission of near-IR (--△--), PAR (—▲—) and net (·······) radiation in a wheat crop in early June (data from Szeicz 1974). In (*a*), irradiance in each waveband is expressed as a percentage of that incident on the top of the crop, and in (*b*) the data are transformed to logarithms.

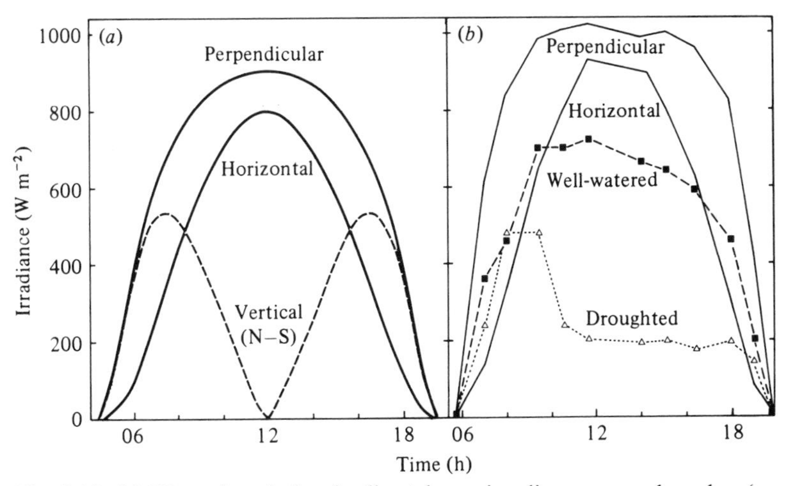

Fig. 2.19. (*a*) Diurnal variation in direct beam irradiance on a clear day ($\tau = 0.7$) in early June at 50°N latitude for a horizontal leaf, a leaf normal to the beam (perpendicular) and a vertical leaf oriented north–south. (*b*) Actual measurements for irradiance received by the upper surface of leaflets of water-stressed (·······△·······) and well-watered (- -■- -) *Vigna* plants at Davis, California (after Shackel & Hall 1979).

heliotropic movements (orientation parallel to the solar beam) minimise interception of solar radiation, as shown for water-stressed *Vigna* (cowpea) leaflets in Fig. 2.19*b*.

3. *Diffuse radiation.* Diffuse radiation is often an important component of incident shortwave radiation, but its penetration into canopies is not identical to that of direct radiation. In fact the leaf area index irradiated by diffuse radiation is greater than the corresponding L_{sunlit} for direct radiation. This is because a leaf that is shaded from the sky in one direction may be exposed to the sky in another. It can be shown that for horizontal leaves the irradiated leaf area is $\pi/2$ times L_{sunlit}. There is also a tendency for radiation to become more concentrated near the zenith as one goes down through a canopy. As the leaf area index increases, an increasing proportion of the leaf area is illuminated at low irradiances by diffuse light alone. Some data on the proportion of leaves in a mature sorghum canopy illuminated at different irradiances are presented in Fig. 2.20. It is apparent from this figure, which is fairly typical, that much of the leaf area in this canopy receives only weak diffuse irradiation yet a large proportion of the total energy received is at relatively high irradiance.

4. *Discontinuous canopies.* An approach that is useful for calculating light interception or transmission ($\mathcal{T}$) in discontinuous canopies such as orchards and widely spaced row crops is to divide the incident radiation into two components (see Jackson & Palmer 1979). These are a fraction $\mathcal{T}_f$ which would have reached the ground even if the plants were completely opaque solid bodies, and a fraction $(1-\mathcal{T}_f)$ that obeys the usual extinction law (equation 2.17). Therefore one can write for the total transmission

$$\mathcal{T} = \mathcal{T}_f + (1-\mathcal{T}_f)\,\mathrm{e}^{-\ell L'} \qquad (2.19)$$

where L' is the leaf area index expressed per unit area of ground that would be shaded by non-transmitting structures of the same three-dimensional outline as the plants (i.e. L' = 'orchard' leaf area index divided by $(1-\mathcal{T}_f)$).

5. *Penumbral effects.* When considering light penetration through canopies and the effects of sunflecks it can be necessary to take account of the apparent size of the sun's disk (approximately 0.5° near the zenith). As indicated in Fig. 2.21, the edges of shadows cast by leaves are not sharp and there is a region, called the penumbra, where the irradiance is less than that of full sunlight because direct light is coming from only part of the sun's disk. This effect commonly reduces the peak irradiance in sunflecks arising from small holes in canopies (see the photograph opposite p. 68 of Bainbridge *et al.* 1966 for an illustration of this effect). From simple trigonometry it is apparent that for holes in a 10 m tall tree canopy, the sunflecks projected at ground level will be entirely penumbral if the holes

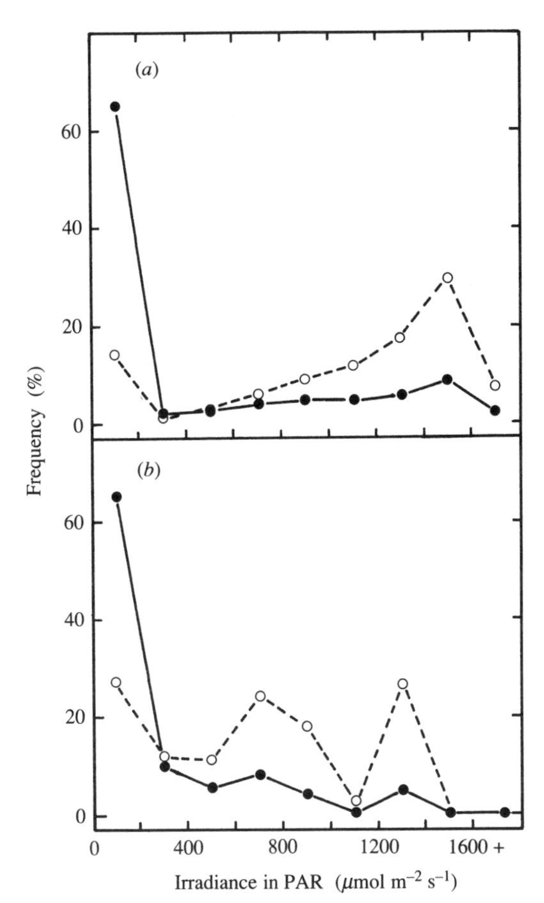

Fig. 2.20. Frequency distributions of leaf area having different photon irradiances (in the PAR, in 200 μmol m^{-2} s^{-1} intervals) (———) for a sorghum leaf canopy under clear conditions in Rome, Italy, together with the corresponding distribution of total energy received in each irradiance interval (- - - - - -), (*a*) for a solar elevation of 25°, and (*b*) for a solar elevation of 60°. The leaf area was approximately 6. Measurements were obtained by a canopy survey technique using a cosine-corrected quantum sensor and represent the higher of the abaxial or adaxial irradiances. (H. G. Jones, D. O. Hall & J. E. Corlett, unpublished data.)

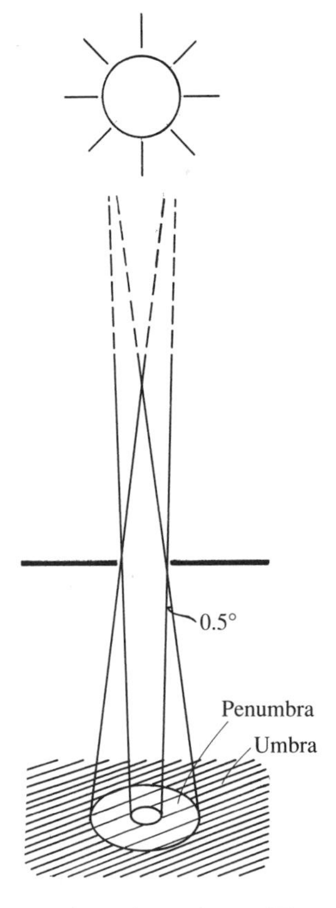

Fig. 2.21. An illustration of the penumbral effect, showing the umbral and penumbral regions for a gap in the canopy slightly larger than the apparent diameter of the sun.

are less than about 87 mm diameter ($10 \times \tan 0.5 = 0.0873$ m). Penumbral effects are particularly important in deep fine-leaved canopies.

Indirect methods for determining canopy structure

In addition to the obvious direct approaches for estimation of canopy structure, many of which involve very tedious measurements of leaf angles, or destructive measurements of leaf area, there are a number of indirect methods that are based on relatively simple measures of light penetration through canopies that can be used by a process of inversion to estimate leaf area and its angular distribution. A useful discussion of these approaches and their limitations is given by Norman & Campbell (1989). Two approaches are worth discussing in some detail as they illustrate important principles.

1. *Gap-fraction methods*. The basic approach is to determine the transmitted fraction ($\mathscr{T}$) of a beam of radiation for a range of zenith

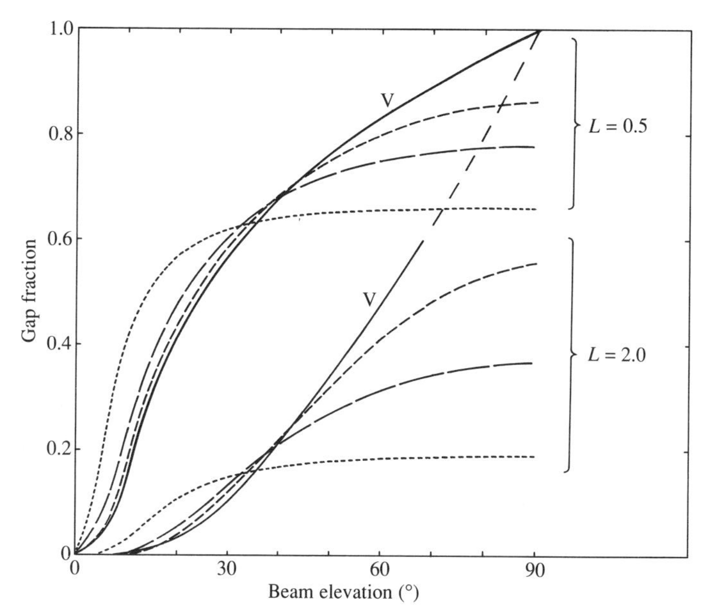

Fig. 2.22. Variation of the gap fraction with beam inclination above the horizontal for different leaf angle distributions including ellipsoidal with $x = 0.5$ (------), $x = 1.0$ (spherical, ——), or $x = 3.0$ (------), and vertical (V) distributions for canopies with leaf area indices (L) of 0.5 or 2.0.

angles. This information can then be compared with theoretical values for different canopy types and leaf area indices so as to estimate both the leaf angle distribution and L, usually by a statistical fitting procedure. The value of $\mathscr{T}(= \mathbf{I}/\mathbf{I}_0)$ is given by equation 2.17, so by substituting the appropriate function for the dependence of $\mathscr{k}$ on beam elevation, one can calculate the expected variation of $\mathscr{T}$, as shown for several canopy geometries in Fig. 2.22. Several algorithms to accomplish the inversion are available, but they all make some assumptions about the expected canopy geometry. It is worth noting from Fig. 2.17 and from Fig. 2.22 that the value of $\mathscr{k}$ (and hence $\mathscr{T}$) is nearly independent of canopy structure for an angle of elevation of about 32°, so that measurements at this angle give the most robust estimates of L.

The necessary information on gap-fraction can be obtained by a number of approaches including: (a) the use of hemispherical photographs taken from the ground under overcast conditions to estimate the fraction of sky visible at different angles of elevation; (b) use of an automated version of this approach (marketed by Li-Cor as the LAI-2000 Plant Canopy Analyzer). This instrument has an optical system that collects light from a whole hemisphere and splits it into fractions coming from different elevation classes. The contribution of radiation scattered by foliage is minimised by using filtered silicon light sensors that are sensitive only to radiation less than 490 nm. (c) Sunfleck area can be estimated for different solar elevations using instruments such as the Sunfleck Ceptometer (see above) set up with an appropriate threshold to avoid detection of diffuse radiation or scattered radiation.

2. *Remote sensing*. Spectral observations of plant canopies can be made using hand-held or boom-mounted instruments or with aircraft or satellite-based sensors. There is much interest in remote sensing for determining vegetation cover, for vegetation classification, as well as in estimation of crop health, productivity and even water use. One approach is to make use of the different spectral properties of leaves and soil. Green leaves absorb about 85% of visible radiation (less than 700 nm), but they scatter at least 85% of the near infra-red (above 700 nm) (see Fig. 2.12). In contrast, soils reflect approximately similar amounts of PAR and near-IR (NIR). If sensors in the red (R) and NIR are used problems of atmospheric interference are minimised because there is little difference in wavelength between the bands employed. The ratio of the spectral reflectances of a surface for red and near-IR therefore can be used as a measure of vegetation cover or leaf area index, often expressed as the **normalised difference vegetation index** (NDVI) defined as

$$\mathrm{NDVI} = \frac{\rho_{\mathrm{NIR}} - \rho_{\mathrm{R}}}{\rho_{\mathrm{NIR}} + \rho_{\mathrm{R}}} \qquad (2.20)$$

where ρ_{NIR} and ρ_{R} respectively are the reflectances in the near-IR (e.g. at 800 nm) and in the red (e.g. at about 650 nm). Because the remote sensors normally measure only the upward radiation flux in each waveband, it is useful to make the substitution

$$\rho_{\mathrm{NIR}}/\rho_{\mathrm{R}} = r_1/r_2 \qquad (2.21)$$

where r_1 is the ratio of reflected NIR to reflected R and r_2 is the ratio of incident NIR to incident R. This gives

$$\mathrm{NDVI} = \frac{r_1 - r_2}{r_1 + r_2} \qquad (2.22)$$

The necessary information on reflected radiation can be obtained from the multispectral scanners in satellites such as SPOT, and can then be used, after calibration, to estimate L.

Sample problems

2.1 Given the spectral properties of a leaf and the incoming solar radiation:

Wavelength interval	Average leaf absorptance	Total incident energy
0.3–0.7 μm	0.85	450 W m^{-2}
0.7–1.5 μm	0.20	380 W m^{-2}
1.5–3.0 μm	0.65	70 W m^{-2}

(i) Calculate (*a*) the shortwave energy absorbed by the leaf, (*b*) the shortwave absorption coefficient, (*c*) the leaf temperature (assuming that the environment is at 20 °C, there is no latent or sensible heat exchange and $\varepsilon = 1.0$ for wavelengths greater that 3 μm). (ii) Why do leaves not usually reach this temperature?

2.2 (i) Calculate net radiation absorbed by an isolated horizontal leaf ($\alpha = 0.5$) exposed above bare soil ($\rho_S = 0.3$), given that the total shortwave irradiance is 500 W m^{-2}, effective sky temperature is −5 °C, soil temperature is 24 °C and leaf temperature 20 °C. (ii) What extra assumptions are required?

2.3 (i) What is the average energy per photon of green light ($\lambda = 500$ nm) and infra-red ($\lambda = 2000$ nm)? (ii) What are the wavenumbers that correspond to these wavelengths? (iii) For a source that produces equal energy per nm centred on each of these wavelengths, calculate the ratio of the photon flux densities per nm at these wavelengths.

2.4 For a horizontal-leaved canopy with randomly distributed leaves: (i) What is the fraction of ground area sunlit if (*a*) $L = 1$ or (*b*) $L = 5$? (ii) What are the corresponding values of sunlit leaf area index? (iii) If instead, the leaves had random orientation and random inclination, what would be the corresponding values of sunlit leaf area index for a solar elevation of 40°?

2.5 Estimate the leaf area index for homogeneous canopies with either (i) a spherical leaf angle distribution, or (ii) a horizontal leaf distribution, given that the solar elevation is 60° and the transmitted fraction of direct radiation is 0.25.

3

Heat, mass and momentum transfer

Chapter 2 considered radiative energy exchange between plants and their environment. Other ways in which plants interact with their aerial environment include the transfer of matter, heat and momentum. The mechanisms involved in mass transfer processes, such as the exchanges of CO_2 and water vapour between plant leaves and the atmosphere, and in heat transfer are very closely related so will be treated together. These can be broadly divided into those operating at a molecular level which do not involve mass movement of the medium (i.e. diffusion of matter and conduction of heat) and those processes, generally termed convection, where the entity is transported by mass movement of the fluid. The forces exerted on plants by the wind are a manifestation of momentum transfer.

Clear discussion of heat and mass transfer processes may be found in Campbell (1977) and in Monteith & Unsworth (1990) with more advanced treatments in Monteith (1975) and Edwards *et al.* (1979). The physical principles underlying these transfer processes and the analogies between them are outlined in this chapter, and this information is used to analyse transfer between the atmosphere and both single leaves and whole canopies. Although the principles described are applicable to transfer in any fluid, the examples in this chapter will be confined to transfer in air.

Measures of concentration

Before going into details of the different mechanisms of heat and mass transfer it is necessary to define what is meant by concentration. In general the spontaneous transfer of mass, or other entities such as heat or momentum, occurs from a region of high 'concentration' to one of low 'concentration'. There are, however, many alternative ways in which one can specify the amount or concentration of an entity 'i' in a mixture, each of which may be appropriate for certain purposes, as can be seen in the following discussion.

1. *Concentration*. A widely used measure of composition is the (mass) concentration (c_i) or density (ρ_i) where

$$c_i = \rho_i = \text{mass of i per unit volume of mixture} \tag{3.1}$$

Alternatively one can use the molar concentration (c_i^m)

$$c_i^m = \text{number of moles of i per unit volume of mixture} = c_i/M_i \tag{3.2}$$

where M_i is the molecular weight. Although concentration is often used as a fundamental measure of gas composition, in a closed system concentration changes with temperature or pressure as these factors alter the volume according to the ideal gas laws:

$$PV = n\mathcal{R}T \tag{3.3}$$

where n is the number of moles present, T is the absolute temperature, P is the pressure, V is the volume and $\mathcal{R}$ is the universal gas constant. Because liquids are not as compressible as gases, concentration is much less sensitive to pressure or temperature in solutions.

2. *Mole fraction*. A more conservative measure of composition is the mole fraction (x_i), which is the number of moles of i (n_i) as a fraction of the total number of moles present in the mixture (Σn):

$$x_i = n_i/\Sigma n \tag{3.4}$$

In this case alterations in temperature, pressure or volume do not affect the mole fraction as they affect all components equally. A related measure appropriate for gases is partial pressure (p_i), which for any component is the pressure that it would exert if allowed to occupy the whole volume available. The equivalence with mole fraction follows from combining the ideal gas law with *Dalton's Law of Partial Pressures* which states that, in a gas mixture of several components, the total pressure equals the sum of the partial pressures of the components, therefore

$$x_i = p_i/P \tag{3.5}$$

Using the above relationships it may easily be shown that gas concentration is related to partial pressure by

$$c_i = \frac{\text{mass}_i}{V} = \frac{n_i M_i}{V} = \frac{p_i M_i}{\mathcal{R}T} \tag{3.6}$$

3. *Mass fraction*. Another useful term is the mass fraction (m_i):

$$m_i = \text{mass of i per unit total mass of mixture} = c_i/\rho \tag{3.7}$$

where ρ is the density of the mixture. This is also independent of temperature and pressure. The mass fraction is related to mole fraction by

$$m_i = x_i M_i/M \tag{3.8}$$

where M is the average molecular weight of the mixture.

4. *Volume fraction*. For a gas the volume fraction (the volume of i per unit total volume of mixture) is identical to the mole fraction.

5. *Mixing ratio*. A term common in the meteorological literature to describe the composition of air is the mixing ratio (w_i) where

$$w_i = \text{mass of i per (total mass} - \text{mass of i)} \tag{3.9}$$

Molecular transport processes

Diffusion – Fick's First Law

The rapid thermal motions of the individual molecules in a fluid lead to random rearrangement of molecular position and, in an inhomogeneous fluid, to transfer of mass and heat. This process is called diffusion. For example, in a motionless fluid, mass transfer occurs as a result of the net movement of molecules of one species from any area of high concentration to one of lower concentration. In a one-dimensional system, the flux density or rate of mass transfer ($\mathbf{J}_i$) of an entity i per unit area through a plane is directly related to the concentration gradient ($\partial c_i/\partial x$) of i across the plane by a constant called the diffusion coefficient ($\boldsymbol{D}_i$). This can be written mathematically as

$$\mathbf{J}_i = -\mathbf{D}_i \frac{\partial c_i}{\partial x} \tag{3.10}$$

This is the one-dimensional form of *Fick's First Law of Diffusion*. The minus sign is a mathematical convention to show that the flux is in the direction of decreasing concentration. Corresponding equations can be written for transfer in more than one dimension, but in what follows only the one-dimensional case will be treated.

Although it is common to use the concentration gradient as the driving force for diffusion as in equation 3.10, and this will be done in much of what follows, it can be inadequate for precise work when other factors are varying. For example, in solutions that depart significantly from ideal behaviour one needs to replace concentration by activity (see physical chemistry texts, e.g. Atkins 1990). Similarly in gases where there is a temperature gradient between the source and the sink, use of concentration can lead to significant errors (Cowan 1977). This is because the rate of diffusion depends on the rate at which the individual molecules move (a function of temperature) as well as on their concentration. The use of mole fraction, partial pressure or mass fraction takes this effect into account. By using the appropriate substitutions for c_i (equations 3.7, 3.6 and 3.5) equation 3.10 may be rewritten in any of the following forms which are more appropriate for non-isothermal gases:

$$\mathbf{J}_i = -\boldsymbol{D}_i \rho \frac{\partial m_i}{\partial x} \tag{3.11}$$

$$\mathbf{J}_i = -\boldsymbol{D}_i \frac{M_i}{\mathscr{R}T} \frac{\partial p_i}{\partial x} \tag{3.12}$$

$$\mathbf{J}_i = -\boldsymbol{D}_i \frac{PM_i}{\mathscr{R}T} \frac{\partial x_i}{\partial x} \tag{3.13}$$

These equations may appear similar to equation 3.10 since, for example, $\rho m_i = c_i$. However, $\rho \Delta m_i$ is not necessarily identical to Δc_i (where Δ represents a finite difference) so they can provide a significant improvement in non-isothermal systems. Unfortunately even these equations still involve some simplification.

Heat conduction

Heat transfer by conduction is analogous to diffusion. Conduction is the transfer of heat along a temperature gradient from a region of higher temperature (or kinetic energy) to one of lower temperature, without mass movement of the medium. In solids this energy transfer occurs as a result of molecular collisons transferring kinetic energy between molecules that are not themselves displaced, while in fluids the higher energy molecules themselves may diffuse.

Conductive heat transfer is described by *Fourier's Law*, where the rate of sensible heat transfer per unit area (**C**, with units of W m^{-2} = J m^{-2} s^{-1}) is given by

$$\mathbf{C} = -k \frac{\partial T}{\partial x} \tag{3.14}$$

where k is the **thermal conductivity** (W m^{-1} K^{-1}). Although the driving force for heat transfer is the temperature gradient, it is convenient to make a simple mathematical manipulation so as to obtain the proportionality constant in the same units as were used for mass transfer (Monteith & Unsworth 1990). If T is replaced by a 'heat concentration' $c_H = \rho c_p T$, where c_p is the specific heat capacity of the fluid (J kg^{-1}), one obtains an equation analogous to 3.10:

$$\mathbf{C} = -\boldsymbol{D}_H \rho c_p \frac{\partial T}{\partial x} \tag{3.15}$$

where $\boldsymbol{D}_{\mathrm{H}}$ is a thermal diffusion coefficient (often called a thermal diffusivity). Values of $\boldsymbol{D}_{\mathrm{H}}$ and k for various fluids and solids are given in Appendixes 2 and 5.

Momentum transfer

When a force is applied tangentially to a surface it tends to cause the surface layer to slide or shear in relation to the underlying material. A rigid solid transmits such a shearing stress, which is given the symbol τ and has units of force per unit area (kg m^{-1} s^{-2}), without undergoing deformation. In a fluid, however, adjacent layers slide relative to each other with any one layer being relatively ineffective at transmitting a shearing stress to the next layer of fluid, so that a velocity gradient develops when a fluid flows over a surface. The viscosity of a fluid is a measure of the internal frictional forces that arise from molecular interactions between adjacent layers, with viscous fluids being more effective at transmitting a shearing stress than non-viscous fluids. This process is described by *Newton's Law of Viscosity*, which states that the shearing stress at a plane in a fluid is directly proportional to the velocity gradient ($\partial u/\partial x$):

$$\tau = \eta \frac{\partial u}{\partial x} \tag{3.16}$$

where η is called the dynamic viscosity (kg m^{-1} s^{-1}).

This equation is similar in form to those already introduced for heat and mass transfer. In this case the shearing stress has the dimensions of a momentum flux density, where momentum is mass × velocity (i.e. $\mathrm{ML^{-1}\ T^{-2} = MLT^{-1} \cdot L^{-2} \cdot T^{-1}}$). As for heat transfer, the velocity gradient can be replaced by a gradient of momentum 'concentration' (c_{M} = mass × velocity/volume = ρu) thus giving a proportionality constant with the dimensions of a diffusion coefficient ($\mathrm{L^2\ T^{-1}}$). This diffusion coefficient for momentum ($\boldsymbol{D}_{\mathrm{M}}$) is also called the **kinematic viscosity** ($\boldsymbol{\nu}$).

The tangential force exerted on a surface by a fluid flowing over it is termed skin friction. In addition to this transfer of momentum to a body across the streamlines of flow, a moving fluid can also exert a force as a consequence of form drag (τ_{f}) where the pressure exerted on the front of an object is greater than that on the rear. This is the main force that causes the bending of trees and other plants in the wind. The magnitude of the form drag, that is the force per unit cross-sectional area normal to the flow (A) for any object is given by

$$\tau_{\mathrm{f}} = c_D \tfrac{1}{2} \rho u^2 \tag{3.17}$$

where c_D is a drag coefficient that relates the actual drag to the maximum potential force that could be exerted if all the air movement was completely stopped ($\frac{1}{2}\rho u^2$). 'Streamlined' objects, such as aircraft, will have much lower drag coefficients than objects without streamlining (e.g. buildings). The value of c_D decreases dramatically if the airflow is turbulent. Further discussion of turbulence and its importance may be found in the section on convective and turbulent transfer (see p. 57), while drag and its significance is considered further in Chapter 11.

Diffusion coefficients

Choice of an appropriate 'concentration' gradient enables us to express the proportionality constant $\boldsymbol{D}$ for a wide range of transfer processes in common units. For example, dimensional analyses of equations 3.10 to 3.13 for mass transfer and of 3.15 for heat transfer all give $\boldsymbol{D}$ with dimensions $L^2\ T^{-1}$, while the same is true for momentum transfer.

The value of $\boldsymbol{D}$ in a binary mixture (e.g. CO_2 and air) is called a mutual diffusion coefficient, where $\boldsymbol{D}_{CA}$ for CO_2 diffusing into air is the same as $\boldsymbol{D}_{AC}$ for air diffusing into CO_2, with very little effect of altering the proportions of air and CO_2. Values for diffusion coefficients for quantities including various gases, heat and momentum in both air and in water are listed in Appendix 2. When a substance is diffusing within itself, $\boldsymbol{D}$ is called a self-diffusion coefficient; this can be very different from the mutual diffusion coefficient. For example the self-diffusion coefficient for CO_2 ($\boldsymbol{D}_C$) is $5.8\ mm^2\ s^{-1}$ compared with $\boldsymbol{D}_{CA}$ of $14.7\ mm^2\ s^{-1}$ at 20 °C. Plant physiology is often concerned with ternary systems of air, CO_2 and H_2O, where CO_2 and H_2O fluxes may interfere; rigorous treatment of this effect can modify equation 3.10 (Jarman 1974).

The relative values of $\boldsymbol{D}$ for different gases are approximately as predicted by *Graham's Law*, which states that the diffusion coefficients of gases are inversely proportional to the square roots of their densities when pure (i.e. $\boldsymbol{D}_i \propto (M_i)^{-\frac{1}{2}}$, since density is proportional to M_i). Effects of temperature and pressure on $\boldsymbol{D}$ are given by

$$\boldsymbol{D} = \boldsymbol{D}^o(T/T^o)^n\ (P^o/P) \tag{3.18}$$

where the superscript 'o' refers to a reference value which can be taken as the value at 20 °C (293.15 K) and a pressure of 101.3 kPa (1013 mbar). Although the exponent n depends on the gas, a value of 1.75 predicts $\boldsymbol{D}$ over the normal range of environmental temperatures with less than 1 % error. In addition $\boldsymbol{D}$ is modified when diffusion occurs in a confined physical system where the average distance that a molecule travels between collisions (the mean free path, which for CO_2 in air at 20 °C is approximately 54 nm) is of the same order as the size of the system. An example of a situation where this effect may be relevant is gaseous diffusion through nearly closed stomata.

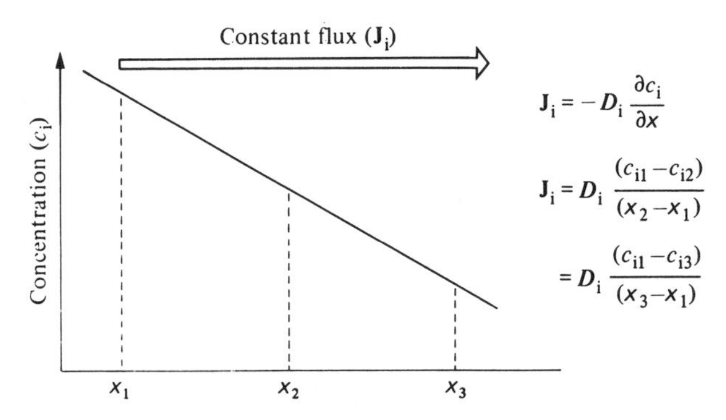

Fig. 3.1. Diffusion down a concentration gradient showing that where $\boldsymbol{D}_i$ is constant, the flux is constant with distance and concentration drop is proportional to distance.

Integrated form of the transport equation

The close similarity between the equations describing transport of a wide range of different entities including water vapour, carbon dioxide, electric charge, heat and momentum, has led to them being sometimes referred to as particular examples of the **general transport equation**

$$\text{Flux density (or flux)} = \text{proportionality constant} \times \text{driving force} \quad (3.19)$$

In many practical situations it is more convenient to measure concentrations at two positions in a system, rather than to determine the concentration gradient at a point: therefore the transport equation is commonly applied in an integrated form. In the simple case where the flux is constant over the path being considered (i.e. there is no absorption or evolution of the transported species in that region) and where $\boldsymbol{D}$ does not change with position (generally true for molecular diffusion), integration of equation 3.10 between planes at x_1 and x_2, a distance ℓ apart, gives

$$\mathbf{J}_i = \boldsymbol{D}_i(c_{i1} - c_{i2})/\ell \quad (3.20)$$

where c_{i1} and c_{i2} are the concentrations at x_1 and x_2 (see Fig. 3.1). In this equation, the driving force is the concentration difference across the path. Therefore the proportionality constant relating any flux density to the appropriate concentration difference is equal to $\boldsymbol{D}_i/\ell$. In plant physiology this constant is conventionally called a **conductance** (and given the symbol g for diffusive transfer and for heat transfer). For many purposes, as will be seen later, it is more convenient to replace the conductance by its reciprocal, termed a **resistance** (given the symbol r), so that

$$r_i = 1/g_i = \ell/\boldsymbol{D}_i \quad (3.21)$$

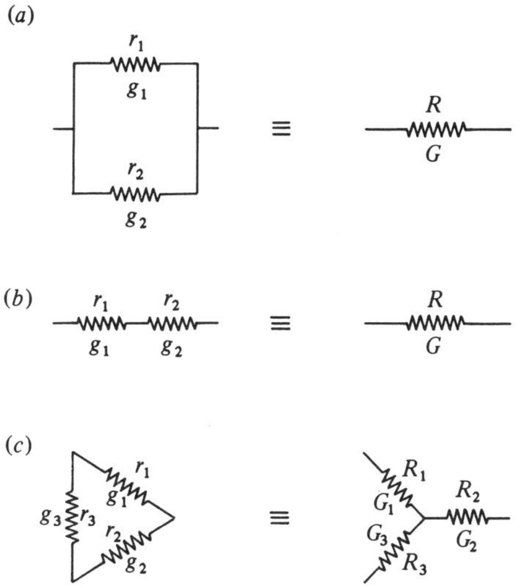

(*a*)

$$G = g_1 + g_2 \qquad \frac{1}{R} = \frac{1}{r_1} + \frac{1}{r_2} \qquad R = r_1 r_2/(r_1 + r_2)$$

(*b*)

$$R = r_1 + r_2 \qquad \frac{1}{G} = \frac{1}{g_1} + \frac{1}{g_2} \qquad G = g_1 g_2/(g_1 + g_2)$$

(*c*)

$$R_1 = r_1 r_3/(r_1 + r_2 + r_3) \qquad G_1 = g_1 + g_3 + (g_1 g_3/g_2)$$

etc.

Fig. 3.2. The rules for simplifying complex networks of resistors: (*a*) resistors in parallel; (*b*) resistors in series and (*c*) the Delta–Wye transform.

Another transport process that fits the general transport equation is the transfer of electric charge as described by *Ohm's Law* (when a steady current (I) is flowing through a conductor, the potential difference between its ends (V) is directly proportional to the current, with the constant of proportionality being called a resistance (R) – i.e. $V = IR$). The close analogy between Ohm's Law and other transport processes has proved extremely useful in the analysis of transfer processes in plants. This is because electrical circuit theory is well developed and is directly applicable to the analysis of the complex networks that occur in plant systems. As a simple example, for a leaf losing water by evaporation from both surfaces, the analogous electrical system is two resistors in parallel with the same potential difference across them. The rules for simplifying complex electrical networks are summarised in Fig. 3.2.

It is often preferable to use conductances rather than resistances in transport studies because the flux across a path with a given driving force is directly proportional to its conductance, but inversely related to the resistance. This inverse relationship to resistance can be misleading in simple systems with only one dominant resistance (see e.g. Chapter 6). It is readily apparent from Fig. 3.2, however, that when a system is predominantly composed of resistors in series, it is more convenient to work with resistances, particularly if one is concerned with the relative limitation imposed by each component (see Chapter 7). A system of

Table 3.1. *Analogies between different molecular transfer processes*

General transport equation	Flux density	= (apparent) driving force	× conductance
Fick's Law (mass transfer)	$\mathbf{J}_i$ (kg m^{-2} s^{-1})	$= \Delta c_i$ (kg m^{-3})	$\times \boldsymbol{D}_i/\ell\ (= g_i)$ (m s^{-1})
	$\mathbf{J}_i^m$ (mol m^{-2} s^{-1})	$= \Delta x_i$ (dimensionless)	$\times P\boldsymbol{D}_i/\ell \mathscr{R} T\ (= g_i^m)$ (mol m^{-2} s^{-1})
	$\mathbf{J}_i^m$ (mol m^{-2} s^{-1})	$= (P/\mathscr{R}T)\,\Delta x_i$ (mol m^{-3})	$\times \boldsymbol{D}_i/\ell\ (= g_i)$ (m s^{-1})
Fourier's Law (heat conduction)	$\mathbf{C}$ (J m^{-2} s^{-1})	$= \Delta T$ (K)	$\times k/\ell$ (W m^{-2} K^{-1})
	$\mathbf{C}$ (J m^{-2} s^{-1})	$= \rho c_p \Delta T (= \Delta c_H)$ (J m^{-3})	$\times \boldsymbol{D}_H/\ell\ (= g_H)$ (m s^{-1})
Newton's Law of Viscosity (momentum transfer)	τ (kg m^{-1} s^{-2})	$= \Delta u$ (m s^{-1})	$\times \eta/\ell$ (kg m^{-2} s^{-1})
	τ (kg m^{-1} s^{-2})	$= \rho \Delta u (= \Delta c_M)$ (kg m^{-2} s^{-1})	$\times \boldsymbol{D}_M/\ell\ (= g_M)$ (m s^{-1})
Poiseuille's Law[a] (flow in pipes)	$\mathbf{J}_v$ (m^3 m^{-2} s^{-1})	$= \Delta P$ (kg m^{-1} s^{-2})	$\times r^2/8\ell\eta\ (= L_p)$ (m^2 s kg^{-1})
Ohm's Law (electric charge)	I (flux) (A)	$= V$ (W A^{-1})	$\times 1/R$ (A^2 W^{-1})

[a] For details of Poiseuille's Law see Chapter 4.

resistors in parallel, on the other hand, is most easily treated using conductances. Either form will be used as appropriate in the following chapters so it is necessary to be familiar with both type of expression and their conversion.

The analogies between different transfer processes are summarised in Table 3.1. It is clear from this that the units for conductance depend on what is chosen as the driving force, it being to some extent arbitrary which factors are included in which term. In each case it is possible to manipulate units to give a conductance in m s^{-1} (or mm s^{-1}). Note that for electricity the current is a flux rather than a flux density, so that the analogy is not complete.

Diffusion coefficients are fundamental properties of the medium and of the material diffusing rather than of the particular system geometry. This is in direct contrast to conductances or resistances that are basically a property of the whole system in that they vary with geometry (e.g. the distance over which the transport occurs) as well as with the mechanism of transport (e.g. molecular diffusion or the rather more rapid turbulent transport).

Units for resistance and conductance

Until recently it was normal practice among plant physiologists to express mass and heat transfer resistances in units of $s\,m^{-1}$ (or $s\,cm^{-1}$) and conductances in $mm\,s^{-1}$ (or $m\,s^{-1}$). These units arise (see equation 3.20) if the flux is expressed as a mass flux density (e.g. $kg\,m^{-2}\,s^{-1}$) and the driving force is a concentration difference ($kg\,m^{-3}$). The same units arise for heat transfer when treated according to equation 3.15, and for momentum transfer (see Table 3.1).

It is, however, becoming increasingly common, particularly in the biochemical literature, to express the flux as a molar flux density ($\mathbf{J}^m$, $mol\,m^{-2}\,s^{-1}$), because biochemical reactions concern numbers of molecules rather than the mass of material. [More correctly $\mathbf{J}^m$ should be termed a mole flux density because use of the term molar should strictly be limited to the meaning 'divided by moles'.] Similarly, concentrations (e.g. of water vapour and CO_2) are usually measured as partial pressures (or the related volume fraction), and as the appropriate driving force for diffusion is the gradient of partial pressure (p_i) or mole fraction (x_i) (rather than concentration), one can write the integrated form of the transport equation for a molar flux in either of the equivalent forms:

$$\mathbf{J}_i^m = \frac{\boldsymbol{D}_i P}{\ell \mathcal{R} T}(x_{i1} - x_{i2}) \qquad (3.22)$$

$$= \frac{\boldsymbol{D}_i}{\ell \mathcal{R} T}(p_{i1} - p_{i2})$$

If one now follows general usage and defines a molar conductance (g^m) as $P\boldsymbol{D}/\ell\mathcal{R}T$, it has dimensions $mol\,L^{-2}\,T^{-1}$, giving as appropriate units ($mol\,m^{-2}\,s^{-1}$) and the corresponding molar resistance (r^m) has units ($m^2\,s\,mol^{-1}$). These molar units will be used frequently in much of the rest of this book, especially when considering gas exchange through stomata, so the superscript 'm' will often be omitted in what follows so as to simplify presentation of equations, and the type of units will be indicated by the choice of font (i.e. $g^m = \mathsf{g}$ and $r^m = \mathsf{r}$).

This alternative definition of conductance has some advantages. In the more usual definition where $g = \boldsymbol{D}/\ell$, conductance is approximately proportional to the square of the temperature and inversely proportional to P (see equation 3.18). Where, however, $\mathsf{g} = P\boldsymbol{D}/\ell\mathcal{R}T$, it is relatively independent of the properties of the air, being independent of P and approximately proportional to absolute temperature. The usual formulation is clearly less appropriate if one is considering effects of altitude (and hence total pressure) on gas exchange. A further advantage of using partial pressure is that it obviates the need for correcting for changing temperature and pressure that arises when using concentration. It is particularly important to use partial pressures rather than concentration gradients where the system is non-isothermal (Cowan 1977).

It follows from equations 3.20 and 3.22 that conversion between the two types of units is by means of

$$g = g(P/\mathcal{R}T) \qquad (3.23a)$$

and

$$r = r(\mathcal{R}T/P) \qquad (3.23b)$$

At sea level and 25 °C approximate conversions are, for resistance

$$r(\mathrm{m^2\,s\,mol^{-1}}) = 2.5\,r\ (\mathrm{s\,cm^{-1}}) = 0.025\,r\ (\mathrm{s\,m^{-1}}) \qquad (3.24a)$$

and for conductance

$$g(\mathrm{mol\,m^{-2}\,s^{-1}}) = 0.04\,g\ (\mathrm{mm\,s^{-1}}) \qquad (3.24b)$$

Conversions at other temperatures are given in Appendix 3.

In spite of the advantages of using a molar basis for expression of mass transfer resistances and conductances, the units s $\mathrm{m^{-1}}$ and mm $\mathrm{s^{-1}}$ will be retained for some purposes in the following treatment and especially for the analysis of heat and momentum transfer where there are no obvious analogies to molar fluxes. In addition, these are still the most commonly used and appropriate units for most canopy-level studies, especially when considering evaporation, and will normally be used in that context. Of course it should be recognised that for any given conditions of temperature and pressure the two sets of units are directly interconvertible using equations 3.23 and 3.24 as appropriate.

Fick's Second Law of Diffusion

In many situations where diffusion takes place the flux is not constant with distance because some of the material diffusing goes into changing the concentration at any position. Using the principle of conservation, that is that matter cannot normally be created or destroyed, it is easy to show that, in a one-dimensional system where $\mathbf{J}_i$ is increasing with distance in the x direction, the extra material required must be obtained by decreasing c_i, so that

$$\frac{\partial \mathbf{J}_i}{\partial x} = -\frac{\partial c_i}{\partial t} \qquad (3.25)$$

This is known as the continuity equation. Substituting for $\mathbf{J}_i$ using Fick's First Law (equation 3.10) leads to

$$\frac{\partial c_i}{\partial t} = -\frac{\partial}{\partial x}\left(-\boldsymbol{D}_i \frac{\partial c_i}{\partial x}\right) = \boldsymbol{D}_i \frac{\partial^2 c_i}{\partial x^2} \qquad (3.26)$$

which is known as *Fick's Second Law*. This equation describes the time-distance relationships of concentration when diffusion occurs. The solution of this equation that is appropriate for any particular problem depends on the initial conditions and on the details of the system geometry. Solutions of this equation for a wide range of systems and boundary conditions are presented by Crank (1975). Here I will discuss only one example that can be used to illustrate the scale of diffusive transport in plant systems. This is the case where a finite amount of material is released at time zero in a plane at the origin and allowed to spread by diffusion in one dimension. The shape of the resulting curve relating concentration to distance is that of the Gaussian or normal distribution. For this curve the distance from the origin at which the concentration drops to 37% (= 1/e) of that at the origin is given by

$$x = \sqrt{(4\boldsymbol{D}t)} \qquad (3.27)$$

An alternative explanation of x is that 16% of the material initially placed at the origin will diffuse at least as far as x in time t. The distance over which diffusive transport occurs increases with the square root of time. Substituting a typical value for $\boldsymbol{D}$ in air of 20 mm s^{-1}, gives $x = 9$ mm for $t = 1$ s (i.e. $\sqrt{(4 \times 20 \times 10^{-6} \times 1)} \simeq 9 \times 10^{-3}$ m). This illustrates the sort of distance over which gaseous diffusion is an effective transport mechanism.

Convective and turbulent transfer

The transfer of mass or heat by diffusion is a consequence of the thermal movements of individual molecules and is the dominant mechanism in still fluids such as the air within the intercellular spaces of plant leaves. For surfaces that are exposed to the atmosphere, such as leaves, air movement over the surface can speed up heat and mass transfer considerably. There are two processes involved.

In the first, the air movement continuously replenishes the air close to the surface with unmodified air, thus maintaining a steep gradient of concentration (the driving force for diffusion) and therefore more rapid tranport than obtains in still air. This is only important for isolated surfaces (such as isolated leaves or plants – see Chapter 5). Where there is an extensive homogeneous surface an equilibrium is achieved such that the air flowing close to the surface has already been modified by passage over an

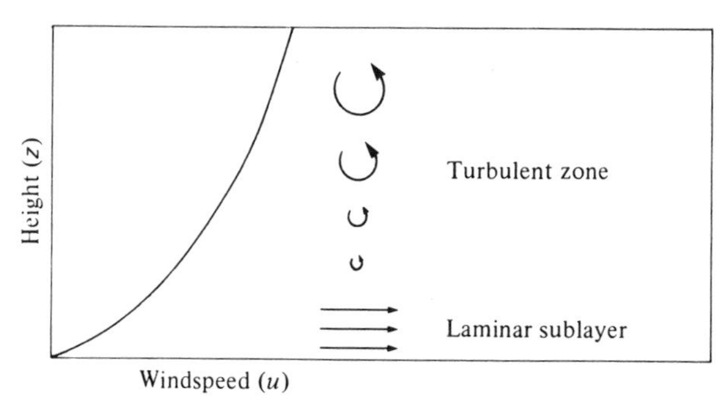

Fig. 3.3. Profile of windspeed moving over a surface showing the laminar sublayer where windspeed changes rapidly with height and a turbulent zone with eddy size increasing with distance.

identical surface upwind. In this case the concentration gradient is the same as would occur in still air unless the airstream is turbulent. Turbulence or random eddies in the airstream provides the second way in which air movement can speed up transfer processes; in this case, materials are transported directly in the moving air currents.

The air in the lower atmosphere is never completely still. Not only is there usually a net horizontal motion or wind, but there are also many random movements of small packets of air. The actual pattern of air movement depends on the type of convection regime that exists. This may be **free convection**, where the air movements are caused by changes in air density, as occur where the air adjacent to a heated surface expands and therefore rises, or where cold air sinks below a cool surface. Or it may be **forced convection**, where the air movement is determined by an external pressure gradient causing wind. Room heating with conventional 'radiators' relies largely on free convection, while fan-assisted radiators use forced convection to transfer heat to the room.

Forced convection may lead to the generation of eddies or turbulence as a result of the frictional forces acting between the wind and the surfaces over which it flows. The size and velocity of the individual eddies depend on a number of factors but they tend to decrease in magnitude as the surface is approached (Fig. 3.3). Evidence for the random spatial distribution and the persistence of these eddies may easily be seen if one looks at the patterns on a field of waving barley. On a smaller scale they can be detected by instruments such as hot-wire anemometers that respond rapidly to changes in air velocity (see Chapter 11). Because the size of eddies in an airstream tends to be similar to the scale of surface irregularities, they are therefore several orders of magnitude larger than the average molecular movements giving rise to diffusion. For this reason turbulent transfer tends to be much faster than diffusion, typically by between three and seven orders of magnitude.

The relative importance of free and forced convection in heat and mass transfer depends on the balance between the buoyancy forces arising from temperature gradients and the inertial forces arising from air movements that cause turbulence. In most plant environments, heat and mass transfer are rarely determined by free convection alone, though it may be an important component of the transfer mechanism in very light wind.

Boundary layers

It was pointed out earlier that when a fluid flows over a surface the flow velocity decreases towards that surface as a consequence of the friction between the surface and the fluid and of the viscous forces within the fluid. The zone adjacent to a surface, where the mean velocity is reduced significantly below that of the free stream, is termed the boundary layer. In what follows, the transfer conductances and resistances in the boundary layer will be distinguished by the use of the subscript 'a', so that the boundary layer conductance for heat transfer would be g_{aH}. One common arbitrary definition of the boundary layer defines its limit as that streamline where the velocity reaches 99 % of that in the free airstream. Because the depth of the boundary layer in air tends to be about two orders of magnitude less than the size of the object, mass and heat transfer can be regarded as one-dimensional processes at right angles to the surface.

The pattern of fluid movement within a boundary layer may be either laminar, where all the fluid movement is parallel to the surface, or it may be turbulent. In a turbulent boundary layer the movements of individual molecules rather resemble the movements of commuters going to or from work in a large city: although the individual particles may be moving in a very irregular pattern, the overall motion is regular and predictable. Whether or not a particular boundary layer is laminar or turbulent depends on the balance between inertial forces in the fluid (because of its velocity) and the viscous forces that tend to produce stability and a laminar flow pattern.

Experimentally it has been found that, for a smooth plate, the transition from a laminar to a turbulent boundary layer generally occurs when the value of a group of terms called the **Reynolds number** exceeds a value between 10^4 and 10^5. The Reynolds number is a dimensionless group given by $ud/\mathbf{v}$, where u is the free fluid velocity, d is a characteristic dimension of the object and $\mathbf{v}$ is the kinematic viscosity ($= \boldsymbol{D}_M$). For parallel-sided flat plates (approximately equivalent to grass leaves), d is the downwind width,

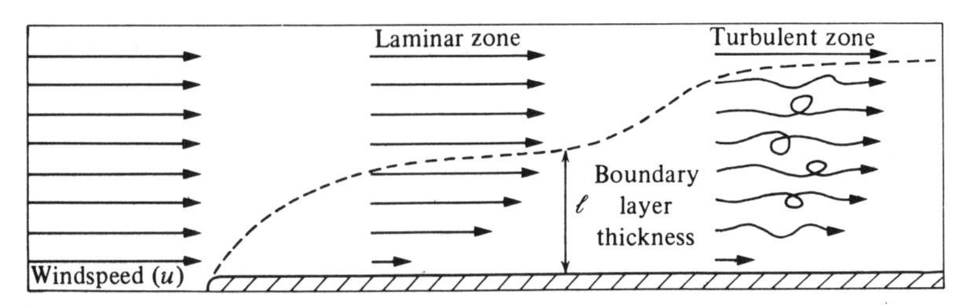

Fig. 3.4. Diagrammatic representation (with much exaggerated vertical scale) of the development of the boundary layer over a smooth flat plate in a laminar airstream, showing the windspeed profiles, the initial laminar zone and the onset of turbulence.

while for circular plates (appropriate for certain other leaves) d is $0.9 \times$ diameter. For irregular plates d is the average downwind width, while for spheres or cylinders with their long axis normal to the flow d is equal to the diameter.

The build-up of a boundary layer in an airstream flowing over a flat plate such as a leaf is illustrated in Fig. 3.4. Initially there is a laminar zone that gradually increases in thickness with increasing distance from the leading edge. The laminar layer may then break down to form a turbulent zone when d increases enough to make the local Reynolds number larger than the critical value. There is good evidence that this critical Reynolds number is achieved at values well below 10^4 (i.e. 400–3000: Grace 1981) for plant leaves, because of the tendency of their surface irregularities, such as veins and hairs, to induce turbulence. Turbulence in the boundary layer is also encouraged by any turbulence in the free stream that might be caused by objects such as leaves and stems upwind (see Haseba 1973). Even where the majority of the boundary layer is turbulent there remains a thin zone close to the surface called the laminar sublayer where the flow is laminar, though this may be only a few tens of micrometres thick. As an indication of the sort of conditions under which turbulence may occur with real leaves, a leaf only 1 cm wide would achieve a possibly critical Reynolds number of 500 at a windspeed of only $0.76\ \mathrm{m\ s^{-1}}$ (i.e. $500 \times 15.1 \times 10^{-6}/0.01$).

Mass and heat transfer through a boundary layer can be described by the general transport equation in the form already used for molecular diffusion in still air:

$$\mathbf{J}_i = g_i(c_{i1} - c_{i2}) \tag{3.28}$$

or for heat:

$$\mathbf{C} = g_H \rho c_p (T_1 - T_2) \tag{3.29}$$

As transport within a laminar boundary layer is by diffusion, the

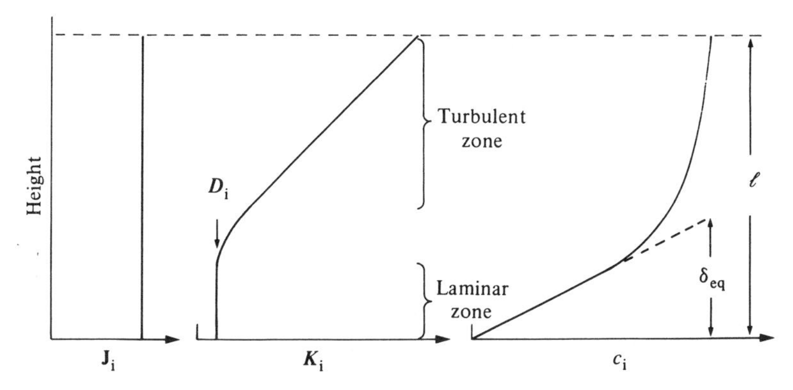

Fig. 3.5. Profiles of $\mathbf{J}_i$, $\boldsymbol{K}_i$ and c_i across a mixed boundary layer of total depth ℓ, illustrating the equivalent boundary layer thickness δ_{eq}.

conductance of a laminar layer with a mean thickness δ is given by $\boldsymbol{D}_i/\delta$ (see equation 3.21). The thickness of a laminar boundary layer over a flat surface increases in proportion to the square root of the distance from the leading edge and in proportion to the reciprocal of the square root of the free fluid velocity. The thickness, δ, is also weakly dependent on $\boldsymbol{D}_i$, being approximately proportional to $\boldsymbol{D}_i^{0.33}$, so that the boundary layer thickness is different for heat, mass and momentum.

Where the flow regime in the boundary layer is turbulent or mixed, the same form of equation applies but mass transfer is more rapid because of the eddies. In this case the boundary layer conductance is increased because $\boldsymbol{D}$, the molecular diffusion coefficient, is replaced by a larger eddy transfer coefficient, $\boldsymbol{K}$. The value of this transfer coefficient varies with the size of the eddies and tends to increase with distance from the surface (Fig. 3.3). The value of $\boldsymbol{K}$ may increase from around 10^{-5} m^2 s^{-1} near the leaves where the eddies are small to about 10^{-1} m^2 s^{-1} at the top of a plant canopy, reaching as much as 10^2 m^2 s^{-1} well above the canopy.

It has already been noted that the integration of Fick's First Law (equation 3.10) to obtain the integrated form (e.g. equation 3.29) is easiest where the transfer coefficient does not alter with distance. Where it does vary, as in a turbulent boundary layer, the definitions of conductance and resistance in equation 3.21 must be replaced by

$$r_i = g_i^{-1} = \int_{x1}^{x2} \frac{dx}{\boldsymbol{K}_i} \tag{3.30}$$

Figure 3.5 illustrates how the transfer coefficient and concentration gradient might vary across a typical mixed boundary layer having a laminar sublayer and a turbulent zone.

Because of the difficulties in integrating the transport equation in situations where the transfer coefficient varies, it is convenient to define an equivalent boundary layer of thickness δ_{eq}. This is the thickness of still air that would have the same conductance or resistance as the turbulent boundary layer of thickness ℓ (see Fig. 3.5). Thus for a turbulent boundary layer where the value of the transfer coefficient, $\boldsymbol{K}$, is say $10^3 \times \boldsymbol{D}$, the thickness of the equivalent boundary layer (δ_{eq}) is $10^{-3} \times \ell$. Note that both δ and δ_{eq} are average thicknesses, since the actual thickness of the boundary layer is less near the leading than the trailing edge.

As it is often difficult to determine the boundary layer thickness it has been found convenient to express heat and mass transfer in terms of the characteristic dimension (d). The ratio d/δ_{eq} is often called the **Nusselt number** when studying heat transfer or the **Sherwood number** when referring to mass transfer. These two dimensionless groups are among those widely used in the fluid dynamics literature to summarise information on heat and mass transfer. For our purposes it is more convenient to express this information directly in terms of the dependence of boundary layer conductance or resistance on windspeed and leaf dimensions. The application of dimensionless groups is discussed by Monteith & Unsworth (1990) and in textbooks on heat and mass transfer (e.g. Kreith 1973).

Conductance of leaf boundary layers

The conversions between conductances for different entities depend on the nature of the boundary layer. Both when the air surrounding a plant organ is still and within the intercellular spaces of leaves transfer of heat or mass depends on molecular diffusion. Conductances for different entities (e.g. CO_2, water vapour or heat) through such a layer of still air would be in the ratio of their molecular diffusion coefficients (Appendix 2). In a laminar boundary layer transport is still by diffusion so that one might expect the conductances to be in the same ratio, but as the effective boundary layer thicknesses for mass and heat transfer are proportional to $\boldsymbol{D}^{\frac{1}{3}}$, it follows that conductances are approximately in the ratio of the $\frac{2}{3}$ power of the diffusion coefficients. As turbulence increases, transport in eddies becomes rapid in relation to molecular diffusion, so that in a fully turbulent boundary layer above a canopy, heat, water vapour and carbon dioxide are all transported equally efficiently and therefore the conductances approach equality. Appropriate factors for converting between conductances for other entities are given in Table 3.2.

The value of the boundary layer conductance for a leaf or other object depends mainly on its shape and size and on the windspeed. It is best determined empirically for leaves of any given dimensions by measuring

Table 3.2. *Factors for converting conductances for different entities in different boundary layers relative to the heat transfer conductance* (g_{aH})

	Relationship	g_{aH}	g_{aW}	g_{aC}	g_{aM}
Still air	$(\boldsymbol{D}_i/\boldsymbol{D}_H)$	1.0	1.12	0.68	0.73
Laminar	$(\boldsymbol{D}_i/\boldsymbol{D}_H)^{0.67}$	1.0	1.08	0.76	0.80
Turbulent	$(\boldsymbol{D}_H)$	1.0	1.0	1.0	1.0

water loss from wet surfaces (e.g. blotting paper) of the same size with the same external conditions, or by energy balance measurements. These methods are outlined in Appendix 8. It has been found that conductance may be estimated with adequate precision for many purposes from the wind velocity (u) and the characteristic dimension (d), by making use of relationships that have been derived from a range of experiments and from heat transfer theory (see Monteith 1981*b*; Monteith & Unsworth 1990). For flat plates in laminar forced convection conditions, the value of the boundary layer conductance to heat transfer (mm s^{-1}) is given by

$$g_{aH} = r_{aH}^{-1} = 6.62\,(u/d)^{0.5} \tag{3.31}$$

where d is the characteristic dimension (m) and u is the wind velocity (m s^{-1}). Note that this conductance refers to unit projected area of leaf (that is the area of *one* side) but includes heat transfer from both surfaces in parallel. Since mean boundary layer thickness is inversely related to g (i.e. $\delta = \boldsymbol{D}_H/g_H$) it is easy to calculate the corresponding boundary layer thickness as $\delta = 2\ \boldsymbol{D}_H(u/d)^{-0.5}/6.62$, where the factor 2 converts the conductance to that appropriate for exchange from one side of a leaf. For a 1 cm leaf in a wind of 1 m s^{-1}, therefore, $\delta = 0.65$ mm (i.e. $2 \times 0.215 \times 10^{-4} \times (1/0.01)^{-0.5}/6.62 \times 10^{-3}$ m).

Corresponding expressions for conductances of other shaped objects are: for cylinders with their long axis normal to the flow

$$g_{aH} = r_{aH}^{-1} = 4.03\,(u^{0.6}/d^{0.4}) \tag{3.32}$$

and for spheres

$$g_{aH} = r_{aH}^{-1} = 5.71\,(u^{0.6}/d^{0.4}) \tag{3.33}$$

where d is the diameter of the cylinder or sphere. Equations 3.32 and 3.33 both refer to unit *surface* area.

Although equations 3.31–3.33 are really only applicable to smooth isothermal plates or other shapes in laminar flow, they are commonly used to estimate conductances for real leaves and other plant organs. In practice, however, surface temperatures are not uniform and some degree of turbulence in the leaf boundary layer is common. Where turbulence occurs, equations 3.31–3.33, which apply to laminar conditions, tend to underestimate the true conductance usually by a factor of between 1 and 2, though perhaps by as much as 3 in certain circumstances (see Monteith 1981*b* and Grace 1981). In addition to factors such as leaf size and windspeed, the turbulent regime in the airstream can be important: Haseba (1973), for example, has shown that the altered turbulence pattern within dense plant canopies can increase the conductance of rigid leaf models independently of windspeed.

The presence of leaf hairs also affects transfer in the leaf boundary layer (see Johnson 1975). Sparse hairs may increase surface roughness and the tendency for turbulence. On the other hand, a dense mat of hairs is likely to increase the effective depth of the boundary layer (certainly for water vapour or CO_2 transfer) by up to the depth of the hair mat. A layer of still air trapped by hairs 1 mm long would have a resistance to water vapour diffusion of $\ell/\boldsymbol{D}_W = 1 \times 10^{-3}/0.242 \times 10^{-4} = 41\ \mathrm{s\ m^{-1}}$. For momentum, however, the hairs would move the effective sink for momentum from the leaf surface to the surface of the hair mat, so hairs would affect the ratio between conductances for heat, mass and momentum.

Because of these complexities it is clear that it is difficult to estimate leaf conductance accurately. Perhaps the best available generalisation is to increase the conductances calculated from equations 3.31–3.33 by a factor of 1.5, giving the dependence of leaf conductance on windspeed and dimensions shown in Fig. 3.6. The characteristic dimensions used in these calculations range from 1 mm (as for narrow-leaved grasses and pine needles) to 30 cm corresponding to very large leaves such as bananas. Variation of leaf size over this range changes g_a by more than an order of magnitude. Windspeeds at the top of plant canopies can often exceed 1 m s^{-1}, but at times (e.g. at night) and deep in the canopy values may fall to 0.1 m s^{-1} or less.

Although forced convection is likely to dominate heat and mass transfer from leaves in natural environments, when large leaf-to-air temperature differentials occur, as with large leaves and high irradiances, there may be a significant contribution by free convection. With a 10 °C leaf–air temperature differential, the free convection conductance for heat is likely to be about 3.2 mm s^{-1} (Monteith 1981*b*), so that it is comparable to that arising from forced convection only for the largest leaves at windspeeds less than 0.3 m s^{-1}.

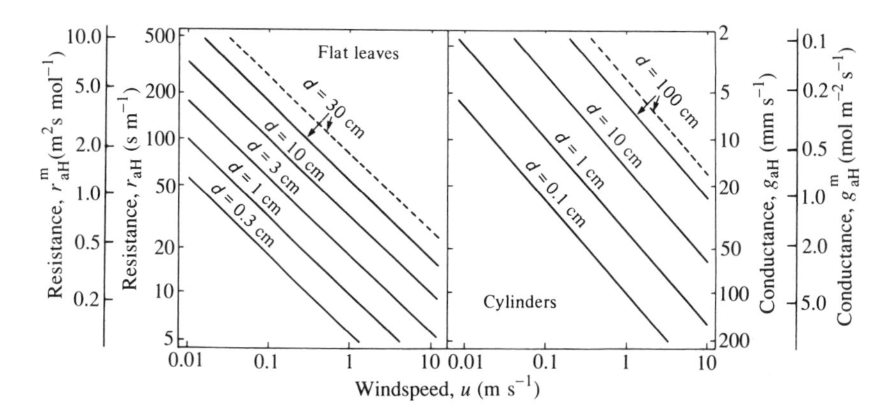

Fig. 3.6. Estimated dependence of g_H or r_{aH} on characteristic dimension (d) and windspeed for flat leaves or cylindrical leaves or stems in natural environments. The dashed lines give the value of g_{aH} predicted by equations 3.31 or 3.32 for laminar flow, the solid lines (g_{aH} = 1.5 × value predicted by these equations) are for more typical flow conditions.

Transfer processes within and above plant canopies

Many of the principles that have been applied to heat and mass transfer of individual leaves are also applicable to exchange by large areas of vegetation, but there are a number of important differences and complicating factors. First, the analysis is complicated by the fact that the 'surface' of a plant canopy (i.e. the sources or sinks of heat, water, CO_2 and momentum) is usually distributed over a significant depth of canopy and also the distribution with depth is different for each entity. A second feature that has been of particular value in the development of micrometeorological techniques for the study of transfer processes between vegetation and the atmosphere is the difference in scale, with the boundary layer above a canopy being much deeper than that for a single leaf, so that it is possible to make measurements within the boundary layer and these have been used to infer fluxes. The third feature of transfer within and above canopies is that the crop boundary layer is generally turbulent so that the transfer coefficients (***K***) for heat and mass transfer are usually assumed equal, though there can be great spatial and temporal heterogeneity. This similarity assumption forms the basis of several of the micrometeorological methods used to study canopy exchange processes.

Transfer above plant canopies

The theory of transfer processes above plant canopies is outlined by Thom (1975) and well summarised by Monteith & Unsworth (1990). Many examples are presented in Monteith (1976). The analysis of micrometeorological measurements within the crop boundary layer requires that there is no convergence or divergence of flux (i.e. there are no sources or sinks within the boundary layer for the entity being transported) and no advection. In other words, the conservation equation applies and a one-dimensional vertical flux is assumed, the flux being constant at different heights, but the transfer coefficient at any height z, $\boldsymbol{K}_i(z)$, varies. The flux of mass or momentum above the crop, therefore, can be described by the standard gradient–diffusion assumption where the flux is proportional to the transfer coefficient multiplied by the driving concentration gradient:

$$\mathbf{J}_i = -\boldsymbol{K}_i(z)\,(\partial c_i/\partial z) \tag{3.34}$$

For this equation to hold, measurements must be made entirely within the crop boundary layer that has developed from the 'leading edge' of the field or area of vegetation being studied. The depth of the boundary layer increases with distance or 'fetch' from the leading edge. In general it is assumed that measurements may be made with adequate precision up to a height above the canopy equal to about $0.01 \times$ fetch. It is also found that measurements need to be made well above the underlying canopy because the erratic turbulence structure near and within canopies leads to such great variability in $\boldsymbol{K}$ that equation 3.34 has little practical value in this zone (see Raupach, 1989). It follows that micrometeorological studies of fluxes through the crop boundary layer require large areas of homogeneous vegetation, the size of which depend on the height above the canopy at which sensors are placed.

Wind profiles and estimation of conductance

Windspeed increases with height above open ground or above plant communities, with the rate of increase being greatest near the ground, as shown in Fig. 3.7. The shape of this windspeed profile is such that over open ground the logarithm of height ($\ln z$) is linearly related to the windspeed at that height (u_z). Expressing u_z in terms of $\ln z$ gives

$$u_z = A\,(\ln z - \ln z_o) = A \ln (z/z_o) \tag{3.35}$$

The intercept on the $\ln z$ axis is $\ln z_o$, where z_o is called the **roughness length**, and is a measure of the aerodynamic roughness of the surface. The slope, A, is usually replaced by the term u_*/k, where u_* is called the **friction**

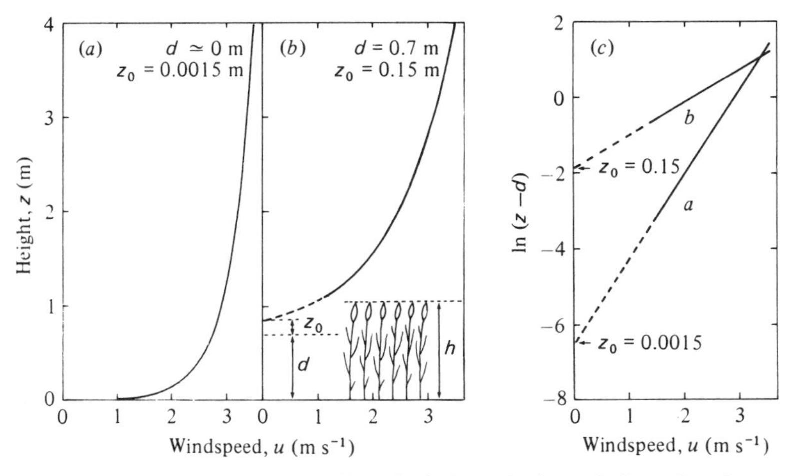

Fig. 3.7. Hypothetical mean profiles of windspeed when windspeed at 4 m equals 3.5 m s^{-1} for (*a*) bare ground and (*b*) a cereal crop together with (*c*) corresponding linearising logarithmic transformations.

velocity (having dimensions of velocity) and characterises the turbulent regime, and k is a dimensionless constant ($= 0.41$) named after von Karman.

Over vegetation, unlike over open ground, the windspeed profile is no longer linear when u is related to $\ln z$. Instead u is linearly related to $\ln(z-d)$, where d is an apparent reference height, the **zero plane displacement** (Fig. 3.7). As shown in Fig. 3.7 windspeed extrapolates to zero at a height of $d+z_o$ (though actual windspeed at this height is still finite). Substituting $(z-d)$ for z in equation 3.35 gives

$$u_z = (u_*/k)\ln[(z-d)/z_o] \quad (3.36)$$

as describing a windspeed profile above vegetation. The plane at a height $d+z_o$ may be regarded as an apparent sink for momentum.

It has been found that reasonable approximations to d and z_o for a range of relatively dense vegetation types are (Campbell 1977)

$$d = 0.64\,h \quad (3.37)$$

and

$$z_o = 0.13\,h \quad (3.38)$$

where h is the crop height. More appropriate values for coniferous forest are given by the following equations (Jarvis *et al.* 1976):

$$d = 0.78\,h \quad (3.37a)$$

$$z_o = 0.075\,h \quad (3.38a)$$

In practice d and z_o vary with windspeed and canopy structure in a fairly complex manner (see Monteith 1976; Monteith & Unsworth 1990).

It is possible to use the windspeed profile to estimate transfer coefficients and conductances in the crop boundary layer. As momentum transfer is analogous to other transport processes, it is possible to define a conductance for momentum transfer between height z and the reference plane ($z = d+z_o$) using the usual transport equation

$$\tau = g_{AM}\,\rho[u_z - u_{(d+z_o)}] = g_{AM}\,\rho u_z \tag{3.39}$$

where g_{AM} is the canopy boundary layer conductance for momentum ($= r_{AM}^{-1}$). It can also be shown (see Monteith & Unsworth 1990) that

$$\tau = \rho u_*^2 \tag{3.40}$$

Combining these two equations gives

$$g_{AM} = u_*^2/u_z \tag{3.41}$$

which can be expressed in terms of the parameters of the wind profile equation (equation 3.36) to give

$$g_{AM} = \frac{k^2 u_z}{\{\ln[(z-d)/z_o]\}^2} \tag{3.42}$$

Not only does this equation imply that g_{AM} increases with windspeed, but it also indicates that conductance tends to increase with crop height (as d and z_o both increase with height). Substituting, for example, values of u, d and z_o from Fig. 3.7 into equation 3.42 gives, for a windspeed of 3.5 m s^{-1} at 4 m, $g_{AM} = 9$ mm s^{-1} for the bare ground ($d \simeq 0$, $z_o = 0.0015$) and $g_{AM} = 62$ mm s^{-1} for the cereal crop ($d = 0.7$, $z_o = 0.15$).

Because the apparent sink for momentum in a canopy is above those for heat or mass exchange, there is a small extra resistance required when converting from r_{AM} to the corresponding resistances for heat or mass transfer. This extra resistance refers to transfer between the level of the momentum sink ($d+z_o$) and the alternate sink (Thom 1975).

Making use of the similarity assumption for the turbulent transfer of different entities in the boundary layer, equation 3.42 can be used as an estimate for the crop boundary layer conductances for other entities such as heat, CO_2 and water vapour. This forms the basis for an important method for estimating fluxes of these quantities. Once the conductance is known (from the wind profile), fluxes may be obtained from measurements of the appropriate concentration differences using equations 3.28 or 3.29. Alternatively fluxes may be estimated directly from equation 3.34 if the concentration gradient and $\boldsymbol{K}_i$ at any height are known.

A problem with equations 3.36 and 3.42 is that they hold only when the temperature profile in the atmosphere is close to neutrality. At neutrality the temperature decreases with height according to the dry adiabatic lapse rate (0.01 °C m^{-1} – see Chapter 11). If temperature decreases more rapidly with height there is a tendency for free convection to occur as a result of 'buoyancy' effects. This makes the atmosphere unstable and turbulent transfer is enhanced. Conversely, when temperature increases with height (a temperature inversion), the atmosphere is stable and transfer is suppressed because the less dense air is above the cooler denser air. In either case the normal profile equations need modification (see Thom 1975; Monteith & Unsworth 1990).

Transfer within plant canopies

The erratic turbulence structure within plant canopies and the complexities introduced by the distribution of sources and sinks for heat, mass and momentum make the application of the gradient–diffusion analogue to transfer processes within the canopy extremely difficult. Examples of the range of within-canopy wind profiles for different types of plant stand are shown in Fig. 3.8. In some canopies windspeed may be highest near the ground, particularly in forests that have little understorey vegetation.

The microclimate within a canopy depends on the source distributions and concentration fields of heat, water vapour and CO_2. The variation in source density for an entity i with height (the source density profile, $S_i(z)$, where a sink is a negative S), depends on physical and physiological processes particularly at the leaves, while the concentration profile $c_i(z)$ depends on the turbulent wind flow and the way this distributes the entity under consideration. Since the law of conservation must apply, the source density in any horizontal plane is related to the change in flux across that plane:

$$S_i(z) = \mathrm{d}\mathbf{J}_i/\mathrm{d}z \qquad (3.43)$$

similarly the flux across the plane at height z is given by the integral from the ground to height z:

$$\mathbf{J}_i(z) = \mathbf{J}_i(0) + \int_0^z S_i(z)\,\mathrm{d}z \qquad (3.44)$$

where $\mathbf{J}_i(0)$ is the flux density from the ground at $z = 0$.

Analysis of the turbulence structure within and above canopies has demonstrated that the strongest turbulent events in a wide range of canopy types are gusts: energetic, downward incursions of air into the canopy space from the faster moving air above (see Raupach, 1989). These gusts tend to be very intermittent but are responsible for most of the momentum transfer

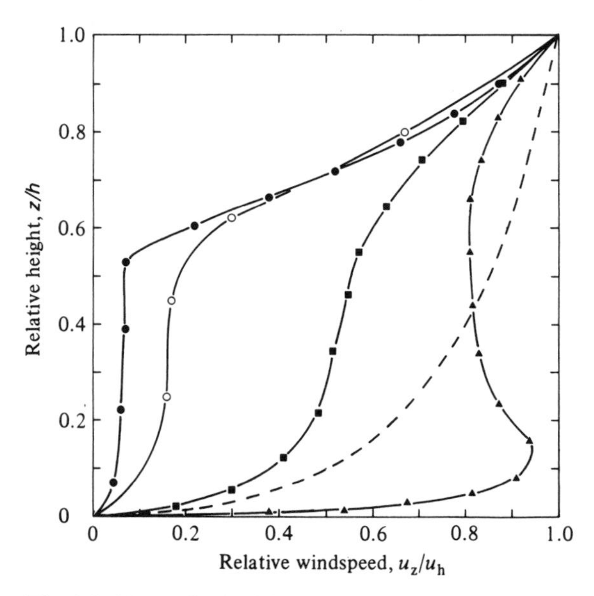

Fig. 3.8. Normalised within canopy windspeed profiles for (●) a dense stand of cotton, (■) dense hardwood jungle with understorey, (▲) isolated conifer stand with no understorey (see Businger 1975 for original references), (○) a corn crop (data from Lemon 1967) and (––––) a logarithmic profile (equation 3.35) with $z_o = 0.01$ h.

(more than 50 % of energy may be transferred in events occupying less than 5 % of the time). The resulting variability of $\boldsymbol{K}$ means that gradient analogies are not helpful, with counter gradient fluxes (and hence apparently negative $\boldsymbol{K}$ values) having been observed for heat, water vapour and CO_2 fluxes within a pine forest (Denmead & Bradley, 1987). The typical canopy eddies are coherent structures of similar dimensions to the canopy height that persist for long periods: the wind waves across cereal fields provide familiar visual evidence for the persistence of turbulent motions of this scale.

As a result of the common failure of the normal gradient diffusion analogue in plant canopies there has been considerable effort aimed at developing an approach to the analysis of turbulent dispersion in canopies which is applicable to the non-diffusive flow that is found in the presence of persistent, large-scale eddies. One approach is to consider all the individual canopy elements as independent point sources releasing material (for example water vapour) into small parcels of air as they pass and to estimate the statistical probability of independent parcels released into the airstream from all these sources reaching a specific point at a particular time (see Raupach 1989). It follows that transport depends on the turbulence structure of the airflow.

Sample problems

3.1 Water vapour is diffusing down a 10 cm isothermal tube at 20 °C from a wet surface ($c_W = 17.3$ g m^{-3}) to a sink consisting of saturated salt solution (equilibrium $c_W = 11$ g m^{-3}). Calculate (i) $\mathbf{J}_W$, (ii) g_W; and the equivalent molar values (iii) $\mathbf{J}_W^m$ and (iv) g_W^m.

3.2 For a 2 cm diameter circular leaf exposed in a laminar airstream moving at 1 m s^{-1}, (i) what are (*a*) g_{aH}, (*b*) g_{aW}, (*c*) g_{aM}, (*d*) the mean boundary layer thickness for momentum? (ii) What would be the values for these conductances if the leaf was covered in a mat of hairs 1 mm deep? (iii) Is the assumption of a laminar boundary layer likely to be valid?

3.3 If the windspeed at 2 m is 4 m s^{-1} when blowing over a wheat canopy 80 cm tall, what are (i) u_*, (ii) the windspeed at the top of the canopy, (iii) τ and (iv) g_{AM} between the reference plane and 2 m?

4

Plant water relations

Physical and chemical properties of water

Water is an essential plant component being a major constituent of plant cells, and ranging from about 10 % of fresh weight in many dried seeds to more than 95 % in some fruits and young leaves. Many of the morphological and physiological characteristics of land plants discussed in this book are adaptations permitting life on land by maintaining an adequate internal water status in spite of the typically rather dry aerial environment.

The unique properties of water (see Slatyer 1967, Eisenberg & Kauzmann 1969, and Nobel 1991, for details) form the basis of much environmental physiology. For example it is a liquid at normal temperatures and is a strong solvent, thus providing a good medium for biochemical reactions and for transport (both short-distance diffusion and long-distance movement in the xylem and phloem). Water is also involved as a reactant in processes such as photosynthesis and hydrolysis, while its thermal properties are important in temperature regulation and its incompressibility is important in support and growth.

The properties of water stem from its structure (Fig. 4.1), and from the fact that it dissociates into hydrogen and hydroxyl ions that are always present in solution. The angle between the two covalent O–H bonds and the asymmetry of the charge distribution along the bonds gives rise to a marked polarity of charge so that water is a dipole. This polarity allows the development of so called hydrogen bonds between adjacent water molecules, as illustrated in Fig. 4.1, or between water and other charged surfaces or molecules. This hydrogen bonding, though weak (only having a bond energy of $\sim$ 20 kJ mol^{-1}, compared with the bond energy of the covalent O–H bond of $\sim$ 450 kJ mol^{-1}), can give rise to significant 'structuring' even in liquid water.

It is a consequence of the intermolecular hydrogen bonds that water tends to be a liquid at much lower temperatures than other small molecules (e.g. NH_3, CH_4 or CO_2) which do not have the marked polarity and

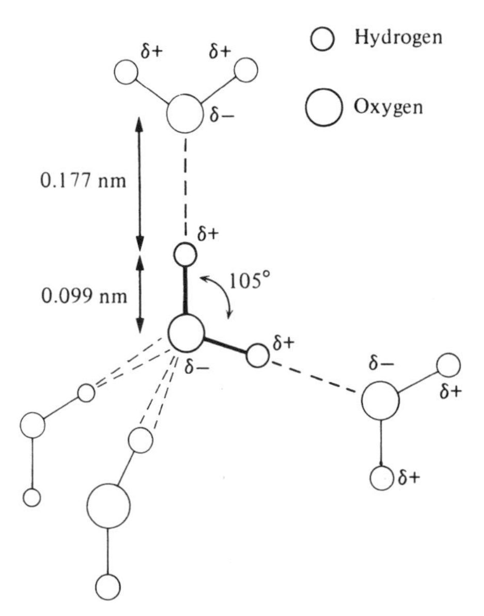

Fig. 4.1. Schematic structure of water molecules showing the hydrogen bonding that results from the electrostatic attraction between the net positive charge on hydrogen and net negative charge on oxygen atoms. In ice, the tetrahedral structure shown is fairly rigid, but there is ordering even in liquid water.

hydrogen bonding characteristic of water. The hydrogen bonds help to maintain water in a semi-ordered liquid form. A measure of the high degree of ordering in liquid water can be obtained from the fact that the amount of energy required to convert solid ice to liquid water (the latent heat of fusion) is 6.01 kJ mol^{-1}, only about 15% of the total energy required to break all the hydrogen bonds in the ice.

The strength of the hydrogen bonding between water molecules gives rise to strong cohesive forces in the water. A property of water that results from these cohesive forces is its extremely high surface tension. The surface tension (σ) is defined as the force transmitted across a line in the surface and equals 7.28×10^{-2} N m^{-1} for water in air at 20 °C. This property is particularly important in capillary rise phenomena including the retention of water within the interstices in soils and in the cellulose matrix of plant cell walls. Capillary rise involves both adhesive forces between the liquid water and the solid phase of the capillary wall and cohesive forces within the liquid water. The phenomenon is illustrated in Fig. 4.2, where it can be seen that as long as the contact angle (α) between the liquid and the surface is less than 90° there will be a vertical force acting on the liquid tending to draw it up a capillary. The magnitude of this force is given by the vertical

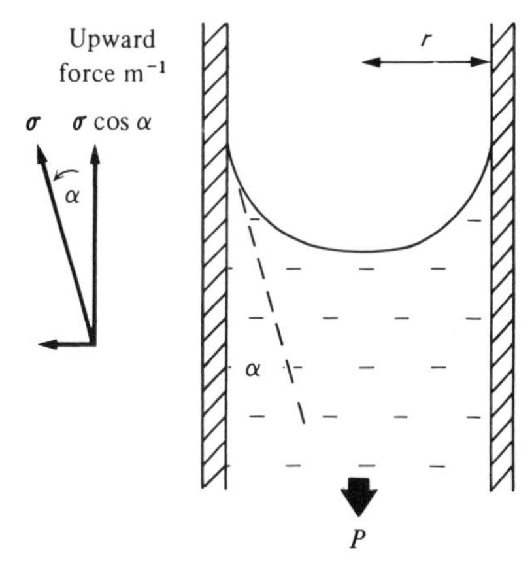

Fig. 4.2. Capillary rise in liquids. At equilibrium the tension (P) is balanced by the vertical component of the adhesive force divided by the area over which it acts (πr^2). The vertical component of the adhesive force is given by $\boldsymbol{\sigma} \cos \alpha \times$ the perimeter of the capillary ($2\pi r$), so that $P = (2\boldsymbol{\sigma} \cos \alpha)/r$.

component of the surface tension ($= \boldsymbol{\sigma} \cos \alpha$) multiplied by the distance over which it acts (the perimeter of the capillary which, for a circular cross-section $= 2\pi r$), so that the total upward force $= 2\pi r \boldsymbol{\sigma} \cos \alpha$. At equilibrium, this force is balanced by the gravitational force acting on the column of liquid. This gravitational force is approximately the product of the mass of liquid in the column ($\simeq \pi r^2 h \rho$, where h is the height of capillary rise, ρ is the liquid density $= 998$ kg m^{-3} for water at 20 °C) and the acceleration due to gravity ($\boldsymbol{g} = 9.8$ m s^{-2} at sea level). Equating these forces leads to

$$h = 2\boldsymbol{\sigma} \cos \alpha / r \rho \boldsymbol{g} \tag{4.1}$$

The contact angle is close to zero for wettable surfaces such as soils, cell walls and xylem vessels which have many polar groups on their surfaces. Equation 4.1 shows that capillary rise is inversely proportional to the capillary radius.

Instead of calculating the capillary rise, it is often more convenient to have a measure of the suction or tension in the liquid (P) required to drain a capillary of given radius. The tension (with units of pressure, N m^{-2} or Pa) is given by the force divided by the area (πr^2) over which it acts, so that

$$P = (2\pi r \boldsymbol{\sigma} \cos \alpha)/\pi r^2 = (2\boldsymbol{\sigma} \cos \alpha)/r \tag{4.2}$$

which reduces to

$$P = 2\boldsymbol{\sigma}/r \tag{4.3}$$

when α is zero. Applying this to the pores in the cell wall matrix of higher plants, which have a typical radius of approximately 5 nm, gives a suction of about 30 MPa ($(2 \times 7.28 \times 10^{-2})/(5 \times 10^{-9})$ Pa). This is the pressure required to force the air-water interface into the cell wall, i.e. to drain the cell wall. A suction of 30 MPa would support a column of water approximately 3 km tall! The following approximation of equation 4.3 is a useful 'rule-of-thumb' for calculating the pressure required to drain any pore: pressure (MPa) $\simeq$ 0.3/diameter (in μm).

The strong cohesive forces between the molecules of water are also directly involved in drawing water up in transpiring plants. One can calculate from the energy in the hydrogen bonds that the maximum theoretical tensile strength in an unbroken column of pure water would be greater than 1000 MPa. In practice, however, tensions greater than about 6.0 MPa can rarely be sustained in xylem vessels as is discussed below. Although the cohesive forces are normally strong enough to permit water to be drawn to the top of tall trees, on occasions the water columns in the xylem vessels rupture or cavitate. The role of capillarity and cohesion in water movement in plants and the significance of cavitation are discussed later in this chapter.

The high specific heat capacity of water ($c_p = 4182$ J kg^{-1} K^{-1}) (that is, the heat energy required to change the temperature of 1 kg by 1 K at constant pressure) contributes to temperature stability, both in aquatic environments and in plants with large water stores (e.g. cacti). Equally, the high latent heat of vaporisation ($\lambda = 2.454$ MJ kg^{-1} at 20 °C) (that is, the energy required to convert 1 kg of liquid water to vapour at constant temperature) has important consequences for temperature regulation of plant leaves (Chapter 10).

The large dielectric constant of water ($\mathscr{D} = 80.2$ at 20 °C) also results from the polar structure of water. This value is over 40 times that for a non-polar liquid like hexane. The dielectric constant is a measure of the impermeability of a medium to the attraction between electric charges, so that a material with high $\mathscr{D}$ reduces the strength of ionic attractions between different molecules. This effect is involved in making water an extremely powerful solvent and hence a good medium for many biochemical reactions. Other properties, such as its dissociation into H^+ and OH^-, its spectral absorptance, and the properties of its solid form, ice, also have important physiological and ecological consequences.

Water potential

The amount of water present in a system is a useful measure of plant or soil water status for some purposes. More commonly, however, the water status in plant systems is measured in terms of water potential (Ψ), which is a measure of free energy available to do work. **Water potential** is defined in terms of the chemical potential of water (μ_W), which in turn is the amount by which the Gibbs free energy (G) in the system changes as water is added or removed while temperature, pressure and other constituents remain constant. In mathematical notation

$$\mu_W = \left(\frac{\partial G}{\partial n_W}\right)_{T,P,n_i} \quad (4.4)$$

where n_W is the number of moles of water added. Water moves spontaneously only from a region of higher chemical potential to one of a lower chemical potential. As water moves down its chemical potential gradient, it releases free energy so that such a flow has the potential to do work.

The chemical potential has units of energy content (i.e. J mol^{-1}). It is, however, the practice in plant physiology to express water status as water potential using pressure units. This can be done by dividing the chemical potential by the partial molal volume of water ($\bar{V}_W = 18.05 \times 10^{-6}$ m^3 mol^{-1} at 20 °C), and using the following definition for water potential:

$$\Psi = \frac{\mu_W - \mu_W^o}{\bar{V}_W} \quad (4.5)$$

where μ_W^o is the chemical potential of water at a reference state consisting of pure free water at the same temperature, at atmospheric pressure and at a reference elevation. This definition has the consequence that Ψ is zero when water is freely available decreasing to negative values as water becomes scarce, so 'higher' water potentials, at least in plant systems, are generally less negative. Although the bar is often used as the unit of water potential, the appropriate SI unit is the pascal (1 Pa = 1 N m^{-2} = 10^{-5} bar) so water potentials will normally be expressed as MPa (1 MPa = 10 bar).

The total water potential may be partitioned into several components, one or more of which may be relevant in any particular system:

$$\Psi = \Psi_p + \Psi_\pi + \Psi_m + \Psi_g \quad (4.6)$$

where Ψ_p, Ψ_π, Ψ_m and Ψ_g respectively are components due to pressure, osmotic, matric and gravitational forces. The pressure component (Ψ_p) represents the difference in hydrostatic pressure from the reference and can be positive or negative. The osmotic component (Ψ_π) results from dissolved solutes lowering the free energy of the water and is always negative. Rather than referring to osmotic potential which is negative, many authors use the term osmotic pressure ($\Pi = -\Psi_\pi$). It can be shown that the osmotic potential is related to the mole fraction of water (x_W) or its activity (a_W) by

$$\Psi_\pi = \frac{\mathcal{R}T}{\bar{V}_W}\ln(\gamma_W x_W) = \frac{\mathcal{R}T}{\bar{V}_W}\ln a_W \quad (4.7)$$

where γ_W is an activity coefficient that measures departure from ideal behaviour by the solution. As the concentration of solutes increases, x_W and Ψ_π decrease. Although γ_W equals 1 in very dilute solutions, most plant systems show some departure from ideal behaviour (see Milburn 1979 for useful tabulations). A very useful approximation of equation 4.7 that is reasonably accurate for many biological solutions is the van't Hoff relation:

$$\Psi_\pi = -\mathscr{R}Tc_s \qquad (4.8)$$

where c_s is the concentration of solute expressed as mol m^{-3} solvent (or more precisely as mol per 10^3 kg solvent). Typical cell sap from many plants has an osmotic potential of about -1 MPa ($\Pi = 1$ MPa). Using equation 4.8 and substituting the value of $\mathscr{R}T$ at 20 °C (2437 J mol^{-1}), gives a total cell sap solute concentration of $-(-10^6/2437) \simeq 410$ osmol m^{-3}. (An osmole is analogous to a mole in that it contains Avogadro's number of osmotically active particles – i.e. 1 mol of NaCl contains 2 osmol.)

The matric potential (Ψ_m) is similar to Ψ_π, except that the reduction of a_W results from forces at the surfaces of solids. The distinction between Ψ_π and Ψ_m is to some extent arbitrary, since it is often difficult to decide whether the particles are solutes or solids, so that Ψ_m is often included in Ψ_π.

The gravitational component (Ψ_g) results from differences in potential energy due to a difference in height from the reference level, being positive if above the reference, and negative below:

$$\Psi_g = \rho_W g h \qquad (4.9)$$

where ρ_W is the density of water and h is the height above the reference plane. Although often neglected in plant systems, Ψ_g increases by 0.01 MPa m^{-1} above the ground, so it must be included when considering movement in tall trees.

Cell water relations

Plant cells behave as osmometers with an inner compartment, the protoplast, bounded by the semi-permeable plasma membrane which is relatively permeable to water and impermeable to solutes. The degree of semi-permeability of a membrane to any solute is described by the **reflection coefficient** (σ) which varies between 0 for a completely permeable membrane, to 1 for a perfectly semi-permeable membrane. Since water permeates the plasma membrane relatively easily, the water potential within cells equilibrates with the immediate environment within seconds, though it takes longer for all cells in a tissue to equilibrate when the bathing solution is changed.

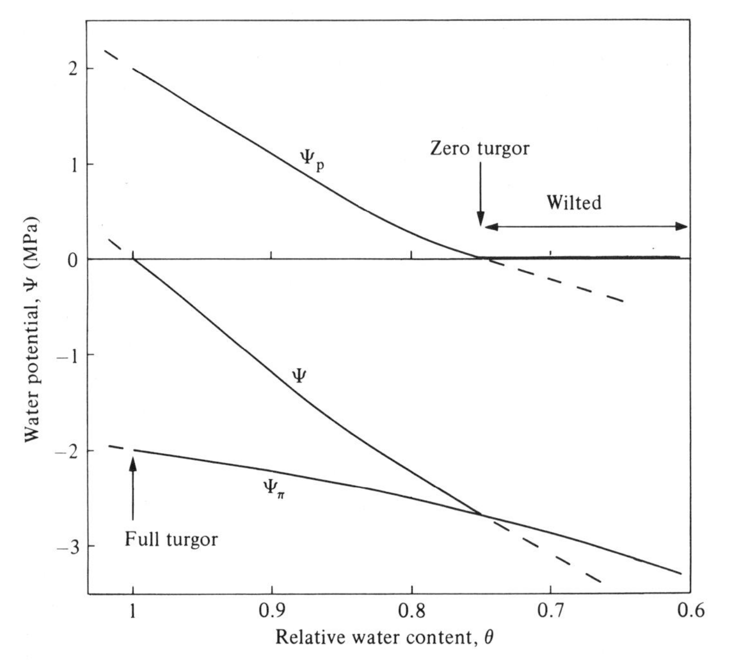

Fig. 4.3. Höfler–Thoday diagram illustrating the relationships between total water potential (Ψ), turgor potential (Ψ_p), osmotic potential (Ψ_π) and relative water content (θ) as a cell or tissue loses water from a fully turgid state. The dotted line below zero turgor represents possible negative turgor in rigid cells.

Another important characteristic of plant cells is that they are encased in a relatively rigid cell wall that resists expansion, thus enabling the generation of an internal hydrostatic pressure. The main components of water potential that are relevant in plant cells are the osmotic and pressure components so that

$$\Psi = \Psi_p + \Psi_\pi \tag{4.10}$$

The pressure difference between that inside and that outside the cell wall is usually positive and is commonly called the turgor pressure (P). For a given cell solute content the turgor pressure decreases as the total water potential falls. Water potentials in transpiring leaves are usually between -0.5 and -3.0 MPa.

The water relations of plant cells (and tissues) are conveniently described by the Höfler–Thoday diagram (Fig. 4.3) which shows the interdependence of cell volume, Ψ, Ψ_π and Ψ_p as a cell loses water. In a fully turgid cell $\Psi = 0$, so that $\Psi_\pi = -\Psi_p$. At this point the water content expressed as a fraction of the water content at full turgidity (the **relative water content**, θ)

is 1 (by definition). As water is lost, the cell volume decreases so that the turgor pressure generated as a result of elastic extension of the cell wall decreases approximately linearly with cell volume until the point of zero turgor (when Ψ_p equals zero). In most plant cells, with any further decreases in water content turgor pressure remains close to zero (though see discussion of evidence for negative turgor by Acock & Grange 1981). It is probable that negative pressures develop in certain rigid cells such as the ascospores of *Sordaria* (Milburn 1979). As the cell volume decreases, the osmotic potential declines curvilinearly as expected from equation 4.8, which indicates that $-\Psi_\pi$ is approximately inversely related to cell volume.

Visible wilting of leaves is usually observed when the point of zero turgor is reached. This is sometimes also called the point of limiting plasmolysis, from the observation that when immersed in solutions of lower water potential the cells plasmolyse (i.e. the cell membrane separates from the cell wall), possibly doing irreversible damage. In normal aerial tissues, however, plasmolysis does not occur because the capillary forces at the air–water interface in the cell wall microcapillaries prevent them draining (see above), so that all the tension is supported by the wall rather than the membrane. Plasmolysis also does not occur when cells are immersed in solutions in which the solutes are too large to penetrate the cell walls because, like in normal aerial tissues, all the tension generated is supported by the cell wall. As solute concentrations increase, the tension eventually becomes insupportable so that the cells collapse – a condition known as cytorrhysis. In such a system, small negative turgor pressures of much less than 0.1 MPa cause collapse of mesophytic cells, but the small thickened cells of some desert plants can withstand in excess of 1.6 MPa (Oertli *et al.* 1990).

An important character determining the shape of the curves in Fig. 4.3 is the elasticity of the cell walls. If the cell wall is very rigid, the water potential and its components change relatively rapidly for a given water loss. The wall rigidity as measured in a tissue is described by the **bulk modulus of elasticity of the cell** (ε_B), which can be defined by

$$\varepsilon_B = \mathrm{d}P/(\mathrm{d}V/V) \tag{4.11}$$

though some authors normalise it to V_o, the turgid volume of the cell, rather than to V. It should be noted that this bulk modulus of elasticity is distinct from the modulus of elasticity of the cell wall material itself, and in addition that it depends to some extent on the tissue structure and the nature of the interactions between the cells. Values of ε_B for plant cells are normally in the range 1–50 MPa, where the larger values indicate relatively inelastic cells, or tissues with small cells. By analogy with Young's modulus for linear expansion (see Chapter 12), a bulk modulus of elasticity for a homogeneous solid is constant with volume. Plant tissues are neither solid nor homogeneous and when they are compressed they lose water (i.e. matter is not conserved as it is in a solid), so it is not surprising that ε_B shows a marked non-linear behaviour, often increasing with turgor pressure in an approximately hyperbolic fashion from near zero at zero turgor.

Although the Höfler–Thoday diagram is appropriate for single cells, the various cells in any tissue are of different sizes and may have different wall elasticities and solute contents. In addition, in a tissue there is a pressure component caused by neighbouring cells pressing on each other. Therefore the properties of a tissue, although they can be represented by this type of diagram, can be very different from those of the component cells. For example, the individual cells in the mesocarp of a cherry fruit are thin-walled and elastic (low ε_B), but normally in the fruit these cells are constrained by a relatively rigid skin, so that the ε_B measured in intact tissue may be much larger than that for the component cells.

The solute concentrations in the cell walls and in the major long distance transport pathway, the xylem, are usually very low, giving rise to less than 0.1 MPa of water potential. In the xylem conduits, the dominant component of Ψ is the pressure, which can reach very large negative values (below -6.0 MPa in some severely stressed desert plants). The xylem vessels are rigid enough to withstand these tensions without undergoing serious deformation.

Further details of cell and tissue water relations may be found in texts such as Slatyer (1967) (still the best), Kramer (1983) and Nobel (1991) and in papers by Cheung *et al.* (1975), Richter (1978) and Tomos (1987).

Growth and cell water relations

Volumetric growth depends mainly on cell expansion. A theoretical framework was developed by Lockhart (1965) and has been reviewed in more detail by Cosgrove (1986). The water flux into any cell and consequent growth depends on the driving force for water uptake, the hydraulic conductivity of the cell membrane and also on the rheological properties of the cell wall (which themselves depend on cell wall biochemistry).

In order to determine the driving force for water uptake by a cell it is necessary to take account of the degree of semi-permeability of the membrane. Where the membrane is perfectly semi-permeable, the driving force is the total water potential difference across the membrane ($\Delta\Psi$, equal to the sum of the differences in pressure and osmotic potentials)

$$\Delta\Psi = \Delta\Psi_p + \Delta\Psi_\pi = (\Psi_{p(o)} - \Psi_{p(i)}) + (\Psi_{\pi(o)} - \Psi_{\pi(i)}) \tag{4.12}$$

where the $\Delta\Psi$ terms refer to differences between water potentials outside (o) and inside (i) the plasmalemma. Where the membrane is not perfectly semi-permeable and lets through some of the osmotica, then that component of $\Delta\Psi$ which depends on solute gradients (the second term of equation 4.12) becomes less effective at driving water flow. Remembering that the cell turgor pressure, P, equals $\Psi_{p(i)} - \Psi_{p(o)}$, the overall driving force for water flow can be written as

$$\sigma\Delta\Psi_{\pi} + \Delta\Psi_{p} = \sigma\Delta\Psi_{\pi} - P \qquad (4.13)$$

When σ is 0 the driving force for water flux reduces to the hydrostatic pressure difference, and when σ equals 1, the driving force is given by the difference in total water potential ($\Delta\Psi$).

From the general transport equation (equation 3.19; Table 3.1) the volume flux density of water ($\mathbf{J}_v$, $m^3\ m^{-2}\ s^{-1}$) is given by the product of the effective driving force and a proportionality constant

$$\mathbf{J}_v = L_p(\sigma\Delta\Psi_{\pi} - P) \qquad (4.14)$$

In this case the proportionality constant L_p ($m\ s^{-1}\ Pa^{-1}$) is termed a **hydraulic conductance**. For plant cells L_p ranges from about 10^{-13} to $2 \times 10^{-12}\ m\ s^{-1}\ Pa^{-1}$ (Tomos 1988; Nobel 1991).

The rate of volume increase of a cell is therefore the product of $\mathbf{J}_v$ and the cell surface area (A). Dividing through by the cell volume (V) gives the relative rate of increase of cell volume as

$$(1/V)(dV/dt) = (A/V)\,L_p(\sigma\Delta\Psi_{\pi} - P) = L_v(\sigma\Delta\Psi_{\pi} - P) \qquad (4.15)$$

where L_v ($s^{-1}\ Pa^{-1}$) is a volumetric hydraulic conductance obtained by multiplying L_p by the ratio of cell surface to volume (A/V).

As well as being dependent on cell water relations, the rate of volume increase is also dependent on wall rheological properties

$$(1/V)(dV/dt) = \phi(P - Y) \qquad (4.16)$$

where Y is the yield threshold (Pa) or turgor that must be exceeded before any extension occurs, and ϕ is the cell wall **extensibility** ($s^{-1}\ Pa^{-1}$), which describes the rate at which cells undergo irreversible expansion whenever Y is exceeded. (Extensibility contrasts with elasticity (ε) which refers to reversible changes in cell dimensions.) Equations 4.15 and 4.16 can be equated to eliminate P, giving, on rearrangement

$$(1/V)(dV/dt) = (\phi L_v/(\phi + L_v))\,(\sigma\Delta\Psi_{\pi} - Y) \qquad (4.17)$$

When water transport is not limiting (i.e. $L_v \gg \phi$) turgor pressure approaches the maximum value and equation 4.17 reduces to 4.16 and growth is determined by the rheological properties Y and ϕ which depend on metabolic processes such as the synthesis of wall materials. Conversely, when water supply is limiting, turgor pressure approaches the yield threshold, and growth depends on the rate of water supply.

Measurement of plant water status

Techniques for the estimation of plant water status are described in detail elsewhere (Barrs 1968; Brown & van Haveren 1972; Slavik 1974; Ritchie & Hinckley 1975; Turner 1988) and will be only outlined briefly.

Water potential is the most frequently used measure of plant water status and is particularly relevant in studies of water movement, though in certain situations, such as where there is no semi-permeable membrane in the flow path, a component (in this case pressure potential) may be more useful. Although Ψ has the advantage of being a rigorous measure, there is strong evidence that it is often not directly involved in the control of physiological processes such as growth or photosynthesis (Hsiao 1973; Jones, 1990). More usually such processes are related to turgor pressure, rather than to water activity (or Ψ); this is not altogether surprising in view of the small changes in a_w that correspond to physiologically important water deficits (equation 4.7). Although it is difficult to distinguish between turgor and cell volume (or relative water content) as major controlling factors, experiments using isolated protoplasts (Kaiser 1982) have demonstrated that cell volume may be the important variable in some cases. How this might be sensed is not certain, though there is evidence for the existence of membrane 'stretch' sensors (see Tomos 1987; Schroeder & Hedrich 1989).

Although leaf water potential (Ψ_ℓ) is probably not as useful a measure of water status as had once been thought, it is still of value, especially in studies of water flow, or as an indicator of a component such as turgor pressure. The two main instruments used for its measurement are the thermocouple psychrometer and the pressure chamber, though liquid-phase equilibrium (Shardakov method) and various indirect techniques, such as the β-gauge, can also be used.

Psychrometer

The principle of the psychrometer is that a sample of tissue is allowed to come to water vapour equilibrium with a small volume of air in a chamber and the humidity of this air is measured using thermocouples set up to measure the wet bulb depression or the dewpoint of the air (see Chapter 5). These results can then be used to give the tissue water potential by calibration against solutions of known water potentials. Brown & van Haveren (1972) give details of the differences in procedure when using different instruments and include many tables with useful conversions. With growing tissues psychrometry may lead to underestimates of Ψ because growth requires extra water if Ψ is to be maintained at the original value. Osmotic potentials of plant tissues may also be obtained using the psychrometer after first disrupting the cell membranes (e.g. by rapid freezing and thawing) so that there is no longer a turgor component of cell water potential and $\Psi = \Psi_\pi$. Osmotic potentials of extracted sap may also be determined with a psychrometer or else with an osmometer that measures the freezing point depression. It is, however, difficult to extract a representative sample of the cell sap since it is liable to be diluted by the extracellular water during collection and an appropriate correction must be made (see e.g. Markhart *et al.* 1981). A further problem with psychrometers is that they tend to be laboratory instruments requiring long periods (often more than 8 h) of equilibration and carefully controlled temperature facilities.

Pressure chamber

A much more adaptable technique is the pressures chamber (see reviews by Ritchie & Hinckley 1975; Turner 1988) which is rapid and can be used for estimating leaf water potential in the field, though it is necessarily destructive. If a leaf is cut from a plant, the tension in the xylem (which is particularly large if the plant is transpiring rapidly or is under stress) causes the xylem sap to be withdrawn from the cut surface. This leaf may then be sealed into a pressure chamber (Fig. 4.4) with the cut end exposed; the chamber is then pressurised until the sap just wets the cut surface thus restoring the sap to its position *in vivo*. At the time of collection the original average leaf water potential would be given by

$$\Psi_\ell(\text{original}) = \Psi_{\mathrm{p}} + \Psi_\pi \qquad (4.18)$$

where Ψ_{p} and Ψ_π are leaf averages of turgor and intracellular osmotic potentials. As the chamber is pressurised the water potential is raised by the amount of pressure applied so that, at the balance pressure (P^*), the water potential is zero (there is free water at the cut surface) so

$$\Psi_\ell(\text{original}) + P^* = 0 \qquad (4.19)$$

The negative of the balance pressure therefore equals the original Ψ_ℓ. The method is rapid and accurate, generally giving results comparable to those given by the psychrometer.

For precise work it is necessary to correct equation 4.19 for the osmotic potential of the apoplast (the water-filled space outside the cell membranes and including the xylem sap). Because the apoplast contains some solutes, at the balance pressure the water potential is still below zero by an amount equal to the osmotic potential in the apoplast. Since $|\Psi_\pi|$ of the apoplast is usually less than 0.1 MPa, the consequent overestimation of Ψ_ℓ is usually negligible.

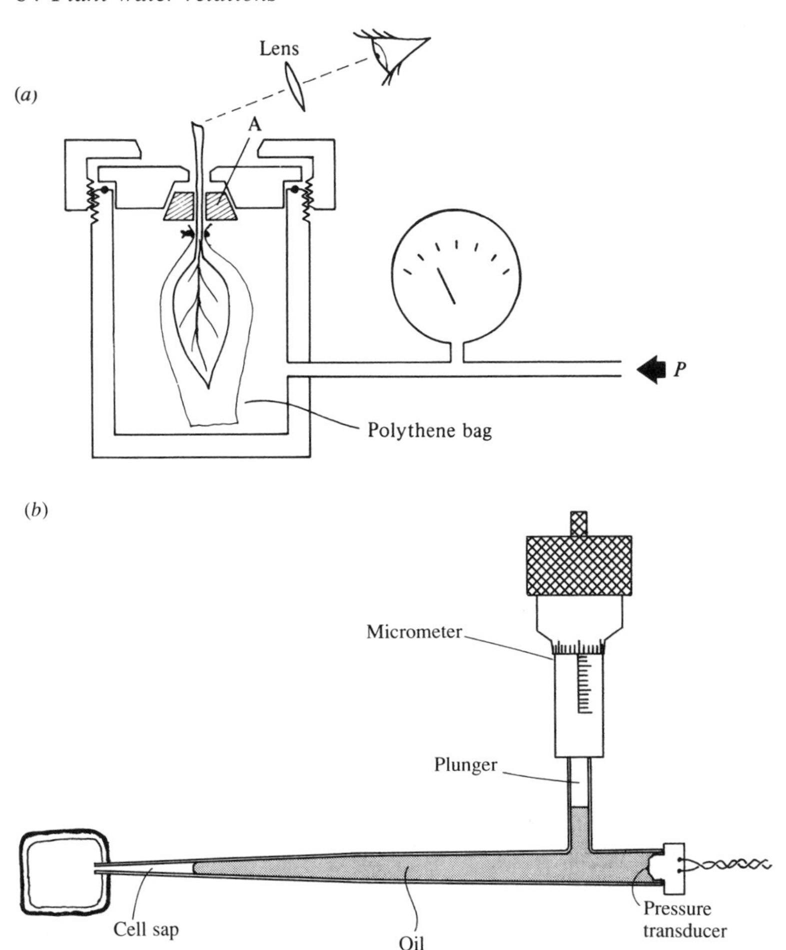

Fig. 4.4. (*a*) Pressure chamber showing one possible method of sealing with a rubber bung (A). The leaf is sealed in a polyethylene bag to minimise evaporation. (*b*) Schematic diagram of a pressure probe. The intracellular hydrostatic pressure is transmitted to the pressure transducer via an oil-filled microcapillary inserted into the cell. Volume is adjusted using the micrometer and observing the interface between the cell sap and the oil under a microscope. The elastic modulus is obtained from the initial pressure change on applying a step change in volume.

It is also possible to dehydrate tissues by raising the pressure above P^* and, from measurements of the volume of sap expressed (V_e) for any pressure increment, it is possible to construct a Höfler–Thoday diagram as in Fig. 4.3. This depends on the assumption that turgor pressure falls to and

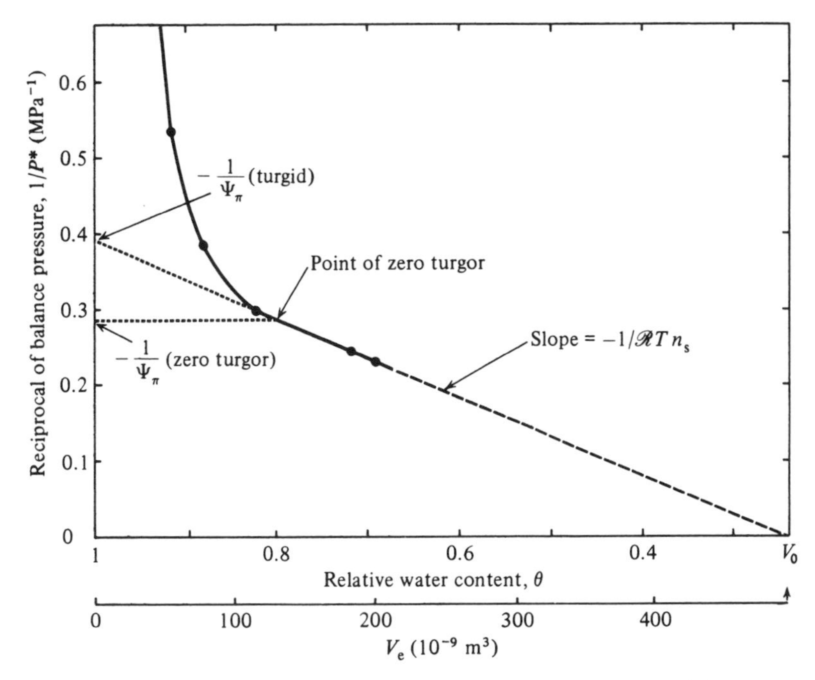

Fig. 4.5. Pressure–volume curve for an apple leaf (cv. Golden Delicious) on 5 October 1978. The volume of sap expressed (V_e) may be obtained from the relative water content (θ) by multiplying by the turgid water content (W = 652 mg for the given leaf).

remains at zero for all water potentials below that giving zero turgor. In this region, from equation 4.8

$$P^* = -\Psi_\pi = \mathcal{R}Tc_s = \mathcal{R}Tn_s/V \tag{4.20}$$

where n_s is the total number of moles of solute present and V is the volume of water within the symplast of the cells in the tissue. Rearranging and substituting $V_o - V_e$ for V (where V_o is the original volume of the symplast when the tissue was cut) gives

$$1/P^* = -1/\Psi_\pi = (V_o/\mathcal{R}Tn_s) - (V_e/\mathcal{R}Tn_s) \tag{4.21}$$

This equation shows that a plot of $1/P^*$ against V_e gives the curve shown in Fig. 4.5. The intercept of the straight line portion ($V_o/\mathcal{R}Tn_s$) is equal to the inverse of the original osmotic potential, while the intercept on the V-axis gives the total osmotic volume of the system. The curved portion of the 'pressure–volume' curve occurs where turgor is positive.

A particularly interesting version of the pressure chamber has been developed by Passioura and his colleagues (e.g. Gollan *et al.* 1986), where by use of a split lid it is possible to seal the roots of a growing plant into the chamber with the shoot emergent from the seal. By applying pressure to the roots it is possible to keep the shoot turgid irrespective of the water potential of the soil, and thus to investigate the response of plants to soil and leaf water potentials independently. An example of its use is presented in Chapter 6 (Figure 6.11).

Pressure probe

The introduction of the pressure-probe for direct measurement of the turgor of individual cells (Hüsken *et al.* 1978; Fig. 4.4*b*) has, for the first time, permitted direct measurements of individual cell turgor in higher plants, and in combination with analysis of the extracted cell contents (Malone *et al.* 1989) provides a powerful tool for studies of plant water relations, though it is not yet applicable to plants in the field. In essence the instrument consists of a micro-capillary that can be inserted into individual cells; the pressure in the liquid in the capillary (in contact with the cell sap) can be monitored with a pressure transducer and the pressure changed by means of a motorised plunger. The pressure probe can be used to estimate a number of water relations quantities: the water potential of individual cells can be estimated (using equation 4.10) from direct measurements of cell turgor pressure and the osmotic potential of extracted sap; the elastic modulus, ε, can be determined from the magnitude of the initial (before significant water flow through the cell membrane) pressure change on rapidly changing the volume of the cell by a small known amount (using equation 4.11); while L_p can be obtained from the rate of relaxation of volume after a pressure change (from equation 4.14). Recent evidence suggests that it might even be possible to measure xylem tensions using a pressure probe if the oil is replaced by water (Balling & Zimmermann 1990).

Other measures

The water content of plant tissues expressed as a fraction of the fully turgid water content of that tissue (i.e. when $\Psi = 0$), the relative water content (θ), is a particularly useful measure of plant water status since it is closely related to cell volume (Fig. 4.3) and is likely to be relevant in many studies of the metabolic effects of plant water deficits (e.g. Hsiao 1973) being often more important than water potential. The measurement of relative water content is discussed in detail by Barrs (1968).

Where should water status be measured?

Plant physiologists have concentrated on the measurement of plant water status, especially Ψ_ℓ, on the grounds that it is the water status of the plant or, where one is concerned with a leaf process such as photosynthesis, the leaf, that should determine any response. Unfortunately this simplistic view is complicated by the signalling that can be involved in growth coordination, with, for example, functioning of the leaves being partly dependent on transport of growth regulators from the roots. There are a number of lines of evidence that suggest that the use of leaf water status may often be inappropriate as a measure of plant water status (see Jones 1990 for a review):

(a) Leaf water status shows much short-term variability as a result of its dependence on environmental conditions with Ψ changing by as much as two-fold within minutes in response to the passage of clouds. In contrast Ψ_ℓ at any one time is often maintained relatively constant over a wide range of soil water potential (Ψ_s) (e.g. Bates & Hall 1981). It is hard to envisage how the small differences in mean Ψ_ℓ between soil water treatments can be distinguished from such large background fluctuations.

(b) On occasion, Ψ_ℓ may even be *higher* in droughted plants with more closed stomata than in well-watered plants (Jones 1983); this stomatal closure cannot be explained by a feedback control of stomatal aperture acting through leaf water status, though it could be explained by a response to Ψ_s.

(c) In a number of split-root experiments it has been possible to maintain leaf water status at control levels by keeping a proportion of the root system well watered while drying only part of the root zone; in such cases stomata may still close (Blackman & Davies 1985) or leaf growth may be inhibited (Gowing *et al.* 1990) in response to soil drying even with no depression of Ψ_ℓ. Analysis of xylem sap obtained from such split-root experiments has provided evidence that abscisic acid (ABA) transported from the droughted roots may provide the root-sourced signal causing stomatal closure in such situations (Zhang *et al.* 1990), though the identity of the signal is still controversial.

(d) Some other types of experiment where stomatal behaviour has been shown to be more closely related to soil water content than to Ψ_ℓ are presented in Chapter 6 (e.g. Figs 6.10 and 6.11).

For these reasons measures of water status that are closely related to soil water status are often more relevant to plant functioning than is midday Ψ_ℓ.

Unfortunately soil water content is very heterogeneous, so it is difficult to measure an effective mean Ψ_s within the rooting zone weighted for the root distribution. The measurement of Ψ_ℓ predawn (when Ψ_ℓ approaches equilibrium with the effective Ψ_s) therefore provides a particularly valuable measure of water status in drought studies. An alternative method for calculating an effective mean Ψ at the root surface of transpiring plants has been proposed by Jones (1983). This is based on the flow models described below (equation 4.25) and uses measurements of Ψ_ℓ and transpiration at any time of day.

Integration of water status over time to obtain a measure of the degree of stress to which plants have been subjected can be particularly useful for longer-term studies. Perhaps the most convenient method is to sum predawn Ψ_ℓ (Schulze & Hall 1981). The use of 'crop water stress index' as a measure of water stress that can readily be integrated over time is discussed in Chapter 10.

Hydraulic flow

Mass flow of water can be described by the familiar transport equation. For flow in porous media and in capillaries, the appropriate driving force is the hydrostatic pressure gradient ($\partial P/\partial x$), so

$$\mathbf{J}_v = -L\partial P/\partial x \qquad (4.22)$$

where $\mathbf{J}_v$ is the volume flux density ($m^3\ m^{-2}\ s^{-1}$) which is equal to an average velocity ($m\ s^{-1}$), and L is a **hydraulic conductivity coefficient** ($m^2\ s^{-1}\ Pa^{-1}$) corresponding to the diffusion coefficient in Fick's First Law (see equation 3.10, p. 48). The choice of hydrostatic pressure rather than total water potential as the driving force in equation 4.22 can be understood if one realises that a gradient of Ψ_π can only affect volume flow when there is a semi-permeable membrane present to generate a pressure. Equation 4.22 is known as *Darcy's Law* when it is applied to flow in soils and can be used for any porous medium.

As with Fick's First Law, it is often more convenient to apply equation 4.22 in the integrated form, and to include the pathlength in the coefficient to give

$$\mathbf{J}_v = L\Delta P/\ell = L_p\Delta P \qquad (4.23)$$

where L_p is a hydraulic conductance ($m\ s^{-1}\ Pa^{-1}$) which, for a uniform path is given by L/ℓ (where ℓ is the pathlength) so is analogous to diffusive conductance (g). L_p therefore depends on pathlength while L is a property of the material through which flow occurs and of fluid viscosity. Equation 4.23 is a special case of equation 4.14 for a system with $\sigma = 0$.

For hydraulic flow through cylindrical tubes, the hydraulic conductance can be expressed in terms of tube radius and fluid viscosity to give

$$\mathbf{J}_v = (r^2/8\eta\ell)\,\Delta P \tag{4.24}$$

where r is the radius of the tube (m) and η is the dynamic viscosity of the fluid (kg m^{-1} s^{-1} or Pa s and $\simeq 1 \times 10^{-3}$ for water at 20 °C). This equation or the corresponding version of equation 4.22 is referred to as *Poiseuille's Law*, and shows that the average flow rate through the unit cross-section of a tube increases as the square of the radius. The total flow *per conduit* will therefore increase by the fourth power of the radius. Although this relationship is only valid for laminar flow conditions, that is when the Reynolds number (see Chapter 3 – in this case given by $\mathbf{J}_v \rho r/\eta$) is less than about 2000, this is generally appropriate for liquid flow in plant systems. The frictional drag at the walls of any capillary leads to most rapid flow at the centre. This maximum flow velocity is twice the average velocity given by equation 4.24. The hydraulic conductance, L_p, for any capillary can be easily calculated as $r^2/8\eta\ell$. As with other transport processes electrical analogues can be used to analyse flow in complex systems, again using resistances ($R = 1/L_p$) in preference to conductance where the pathway consists primarily of several components in series. (A capital italic R is used to indicate hydraulic resistances.)

It is worth noting that application of equation 4.23 to flow in tubes and in porous media usually involves a different areal basis for the expression of $\mathbf{J}_v$. In the former it is the actual cross-section of the conducting tube, while in the latter it is the total cross-section including any non-conducting matrix.

Water flow in plants

The pathway for the transpirational flux through plants is illustrated in Fig. 4.6 (see Slatyer 1967; Milburn 1979). The main pathway for longitudinal flow is the xylem, in which the conducting elements are primarily the non-living and heavily thickened and lignified tracheids and xylem vessels. Vessels, which are found only in angiosperms, consist of files of cells whose end walls have broken down to form continuous tubes that may vary from a few centimetres to many metres in length, and from about 20 μm to as much as 500 μm in diameter. The tracheids, on the other hand, originate from single cells and are found in all vascular plants so they constitute the main conduits in gymnosperms such as coniferous trees. Tracheids are typically only a few millimetres long and between about 15 μm and 80 μm in diameter. In woody plants the largest conduits tend to occur in the early

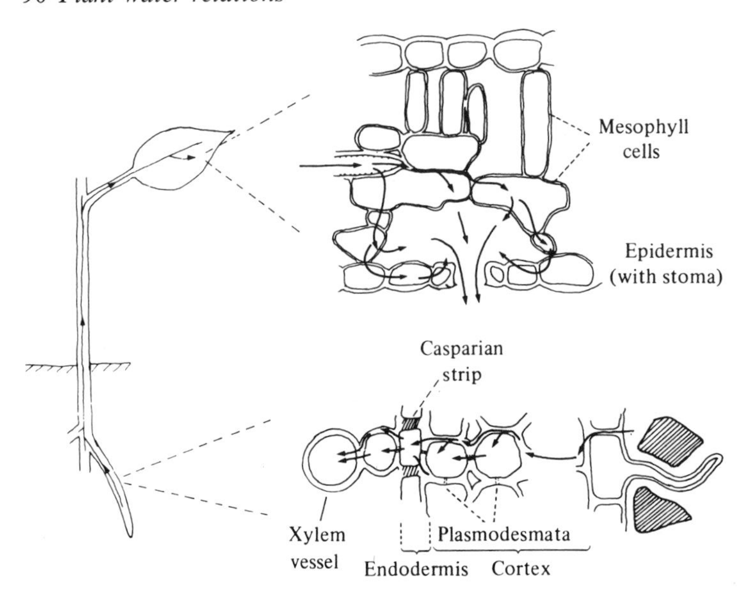

Fig. 4.6. The pathway of water flow from the soil to the leaves, showing the parallel paths for flow in the cell walls and in the symplast within the roots and leaves.

formed wood of an individual growth ring but are much smaller in the 'late' wood. Typical examples of early and late wood diameters are 35 μm and 14 μm for tracheids of Scots pine, while the range of vessel size in an English oak was 268 μm down to 34 μm (Jane 1970). Adjacent conduits are connected by pits through which sap can pass, though the pore diameter in the pit membranes that separate conduits is probably less than 0.2 μm.

The radial movement of water in the roots from the soil to the xylem occurs through the cortical tissue, partly in the water-filled free space of the cell wall, and partly within the symplasm (the connected protoplasm within the cell membrane). The separate cells forming the symplasm are interconnected by narrow cytoplasmic connections, the plasmodesmata. The symplasm is separated from the apoplast by the plasma membrane. Between the cortex and the vascular tissue in the root is a specialised cell layer, the endodermis where the cell wall pathway for water movement is blocked by a band of suberised tissue, the Casparian strip (Fig. 4.6). At this point all the water must pass through a membrane into the cytoplasm. Similarly, the path from the vascular tissue to the evaporating sites in the leaves may involve some movement in the symplast and some in the cell wall. The relative contribution of the two pathways is difficult to quantify because of the anatomical complexity, as well as because of lack of information on permeability of cell membranes and other components (see Canny 1990).

It is easy to see why the xylem provides such an efficient long-distance transport pathway. If one assumes a typical pressure gradient in a transpiring plant of -0.1 MPa m^{-1}, it can be calculated from Poiseuille's Law that the flow velocity in 100 μm radius conduits would be 125 mm s^{-1} (i.e. $(100 \times 10^{-6})^2 \times 10^5/(8 \times 10^{-3} \times 1)$ m s^{-1}); similarly, in 20 μm conduits it would be 5 mm s^{-1}, while, if the same law holds in the cell wall interstices which are only about 5 nm, flow velocity would be down to about 3.1×10^{-7} mm s^{-1}, or about seven orders of magnitude less than in an equal area of xylem vessel. Observed maximum flow velocities in stems of different trees actually range from about 0.3–0.8 mm s^{-1} in conifers and 0.2–1.7 mm s^{-1} in so-called diffuse-porous hardwoods (e.g. *Populus*, *Acer*) to 1.1–12.1 mm s^{-1} in ring-porous hardwoods (e.g. *Fraxinus*, *Ulmus*) (see Zimmermann & Brown 1971). In herbaceous species the xylem water stream may even reach velocities of 28 mm s^{-1} (100 m h^{-1}). These values are rather below those expected, partly as a result of the flow resistance in the pit membranes. In ring-porous trees especially, the majority of flow occurs in the most recent wood. For example in *Ulmus* 90% of the flux has been observed in the most recent year's wood, even though some conduction occurred in wood up to four years old (Ellmore & Ewers 1986).

Sap flow measurement. A number of methods have been developed for measuring in intact plants the velocity of xylem flow or the mass flow rate of water through the stem (see Pearcy *et al.* 1989 for a simple summary). The most widely used techniques involve either the heat pulse technique (reviewed by Jones *et al.* 1988) pioneered by Huber in the 1930s, which is based on the measurement of the time between a pulse of heat being applied to a section of the stem and the detection of the temperature rise a certain distance downstream, or continuous heating (e.g. Granier 1987) or heat-balance (e.g. Čermák & Kucera 1981) approaches.

Cavitation. In spite of its enormous tensile strength, the water in xylem can rupture (cavitate) under the extreme tensions that occur naturally, if suitable nucleation sites, either on the vessel wall or as a result of air entry through pit membranes, are available. Once initiated, the bubble then rapidly expands forming an embolism within the vessel or tracheid until it is stopped at the pit membranes (Fig. 4.7). Further movement is usually prevented by the capillary effects in the narrow pores – for example 0.1 μm pores in the membrane will prevent passage of the vapour–water interface under as much as 1.5 MPa tension (see equation 4.3).

Soon after formation the embolisms that block the vessels contain only

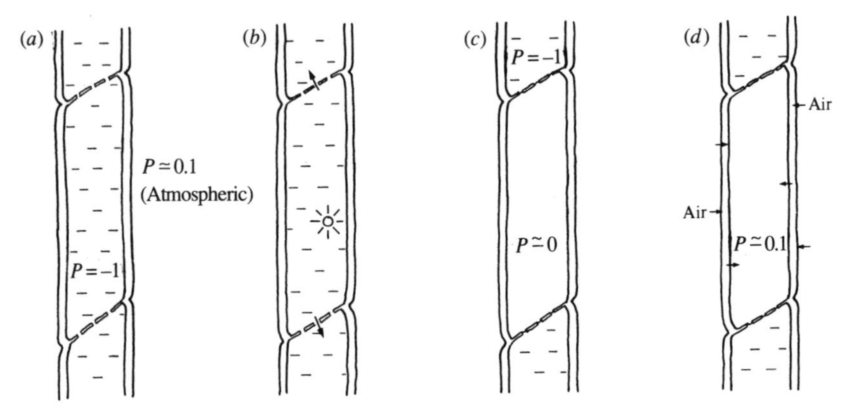

Fig. 4.7. Schematic illustration of events during and after cavitation in a xylem vessel. The initial situation is shown in (*a*) where the absolute pressure (P) in the xylem sap is -1 MPa (the equivalent Ψ_p is -1.1 MPa because it is referred to atmospheric pressure) and the air pressure outside is *c*. 0.1 MPa. On the occurrence of a cavitation (*b*) within the vessel, the sap is rapidly withdrawn until the meniscus is held at the pit membranes (*c*). At this early stage the space is probably filled with water vapour at a few kilopascal absolute pressure ($P \simeq 0$ MPa); subsequently slow diffusion of air into the conduit probably raises the internal pressure to atmospheric ($P \simeq 0.1$ MPa). Refilling of a cavitated vessel is likely to be much easier in the early stages, since it can occur without the requirement for positive pressures. (Modified after Milburn 1979.)

water vapour, and can presumably refill readily as soon as xylem sap pressure recovers to atmospheric or above, for example as a response to root pressure at night. Once air diffuses into the vessel replacing the water vapour, however, refilling will be more difficult so that embolisms, at least in woody plants, are probably irreversible at this stage. Air entry is likely to be complete within about 1000 s (Tyree & Sperry 1989). Air entry can act as a stimulus for the formation of tyloses (ingrowths from the xylem parenchyma cells that completely block xylem vessels that have cavitated or been damaged by infection). These cause the irreversible loss of xylem function in older wood.

The occurrence of cavitations can be detected with a microphone as audible 'clicks' resulting from the shock waves created as the vessel walls relax after a cavitation (Milburn 1979). The use of ultrasonic detectors that are sensitive in the region 0.1–1.0 MHz, appears to provide a better detection method because the lack of environmental noise in this frequency range means that such sensors can be used in the field (Tyree & Dixon 1983). Unfortunately the profiles of audible and ultrasonic emissions produced as stems dry do not always coincide, indicating that they may

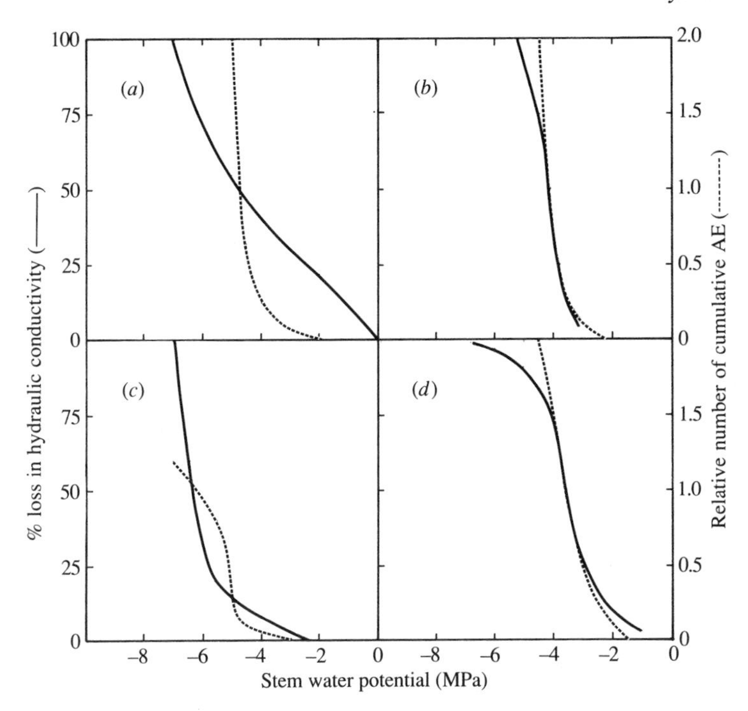

Fig. 4.8. Some xylem vulnerability profiles showing the loss in hydraulic conductivity (———) and acoustic emissions (AE,·····················) with falling water potential for (*a*) *Cassipourea elliptica*, (*b*) *Acer saccharum*, (*c*) *Rhizophora mangle*, and (*d*) *Thuja occidentalis*. Acoustic emissions are expressed relative to the number observed when Ψ had fallen to a value giving a 50% reduction in hydraulic conductance. (Data recalculated from eye-fitted curves in Sperry *et al.* 1988b, Tyree & Dixon 1986, Tyree & Sperry 1989.)

represent cavitations in different tissue elements. If the acoustic events are to have any value as indicators of xylem dysfunction, it is necessary to show that the events being detected occur within the main functional xylem elements, rather than predominantly in fibres or small tracheids. This has not yet been adequately demonstrated and there is good evidence that many acoustic events often originate in non-conducting elements. For example, when 5 cm lengths of apple wood were allowed to desiccate, high acoustic emission rates were observed (Sandford & Grace 1985), even though the number of complete vessel elements in such short pieces of wood would have been very small. In spite of such uncertainties the number of cavitations detected as a plant dehydrates is often related to the loss in stem hydraulic conductivity as measured on excised stems, though as indicated

Table 4.1. *Variation in susceptibility of different plants to xylem cavitation*

Approximate values of Ψ giving different amounts of cavitation (the threshold, 50 % cavitation and 90 % cavitation) were read from published graphs. Data were obtained using either loss of hydraulic conductivity or acoustic emissions, with values in parentheses representing acoustic data.

	Ψ (MPa)			
	Threshold	50 %	90 %	Ref.
Acer saccharum	−3.0 (−2.5)	−4.1	−4.7	5
Acer pseudoplatanus	(−1.5)	(−1.8)	—	1
Cassipourea elliptica	−0.0 (−2.0)	−4.1	−6.6	3
Juniperus virginiana	−3.5 (−4.2)	−6.4	−8.8	5
Lycopersicon esculentum (tomato)	(−0.2)	(−0.4)	—	1
Malus × domestica (apple)				
on M25 rootstock (unstressed)	(−2.0)	—	—	4
on M9 rootstock (unstressed)	(−0.9)	—	—	2, 4
on M9 rootstock (prestressed)	(−2.5)	—	—	2
Ricinus communis	(−0.5)	(−0.8)	—	1
Rhododendron ponticum				
mature leaf	(−1.7)	(−2.1)	—	1
immature leaf	(−0.8)	(−1.0)	—	1

References: 1. Crombie *et al.* 1985; 2. Jones & Peña 1987; 3. Tyree & Sperry 1989; 4. Jones *et al.* 1989*b*; 5. Tyree & Dixon 1986.

in Fig. 4.8 the relationship is often not close. Because of the difficulties with acoustic detection as an indicator of xylem embolism it is safest to use loss of hydraulic conductivity as the basic measure of embolism (using methods described by Sperry *et al.* 1988*a*).

There is often a threshold water potential for cavitation; this varies both with the stress prehistory of the plant and with the species, and even the cultivar (Table 4.1). Vulnerability to cavitation is closely related to the size of pores in the pit membranes, with those vessels having the largest pores being most vulnerable. This suggests that air entry provides the seeding mechanism that initiates cavitations (Tyree & Sperry 1989). Although it has been suggested that the presence of small vessels favours avoidance of cavitation, thus acting as a safety mechanism, and smaller latewood vessels within a species do tend to be least vulnerable, the correlation of vulnerability with vessel size does not appear to hold across species.

Cavitations in conducting vessels decrease the hydraulic conductivity of the stem; this in turn tends to a decline in Ψ_ℓ, which would itself favour further embolism. In the absence of stomatal closure this cycle can continue until all the conducting tissue is lost, though stomatal closure normally prevents such catastrophic xylem failure. There is a tendency for cavitations to occur most in young twigs.

Steady-state flow. The transpirational flow of water through the soil-plant system is driven by evaporation which lowers the leaf water potential. This in turn sets up a gradient of water potential between the soil and the leaf so that water flows. [Note that the actual driving force for hydraulic flow in the apoplast, $\partial\Psi_p/\partial x$ ($= \partial P/\partial x$), is similar to $\partial\Psi/\partial x$, because matric and osmotic components are small in the xylem sap where Ψ_π is generally above -0.1 MPa. Note also that in a vertical vessel at equilibrium (hence Ψ is constant and there is no flow) there is a gradient of P of -0.01 MPa m^{-1} to counteract the gravitational potential $\rho_w gh$.]

Because of the complexity of the flow pathway (Figs. 4.6 and 4.9), steady-state flow is usually analysed in terms of rather simplified 'black-box' resistance models (Fig. 4.9), rather than by equation 4.24. Most workers since van den Honert (1948) have based their analyses on the simple catenary series model in Fig. 4.9*c*. Using this simplification the following equation can be used to describe the relationships between steady-state flow and water potential within the system:

$$\mathbf{E} = \frac{\Psi_s - \Psi_\ell}{R_s + R_r + R_{st} + R_\ell} = \frac{\Psi_s - \Psi_r}{R_s} = \frac{\Psi_r - \Psi_{st}}{R_r} = \frac{\Psi_{st} - \Psi_x}{R_{st}} = \frac{\Psi_x - \Psi_\ell}{R_\ell} \qquad (4.25)$$

where **E** is the water flux (evaporation rate) through the system, Ψ_s, Ψ_r, Ψ_{st}, Ψ_x and Ψ_ℓ refer, respectively, to the water potentials in the bulk soil, at the surface of the roots, at the base of the stem, at the top of the stem and at the evaporating sites within the leaves. The hydraulic resistances in the soil (R_s), root (R_r), stem (R_{st}) and leaf (R_ℓ) refer to the flow resistances shown in Fig. 4.9*c*. Since **E** is usually expressed as a volume flux density (m^3 m^{-2} s^{-1}), R has units of MPa s m^{-1}, though the area basis used can be leaf area, stem cross-sectional area or ground area. If **E** is expressed as a flux per plant, then R has units MPa s m^{-3}. All these are valid alternative methods of expression.

The flux through the system is controlled by the rate of evaporation, so leaf water potential can only be regarded as indirectly *controlling* the flow through the plant by any effect it may have on the gas-phase (mainly stomatal) resistance. Using another analogy with electrical circuits, evaporation can be represented by a constant current generator (Fig. 4.9*c*)

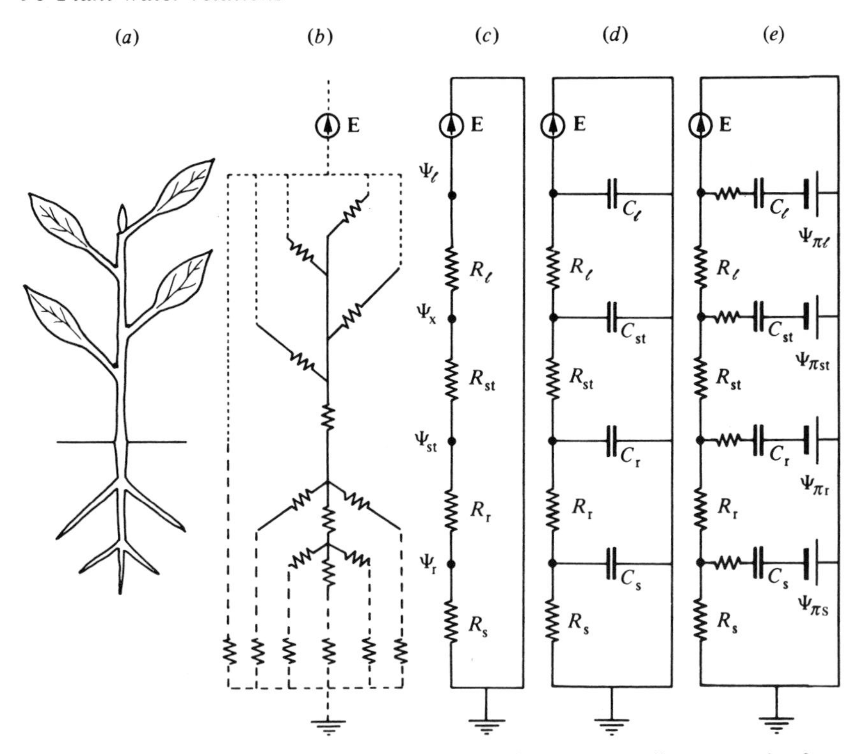

4.9. (*a*) Simplified representation of a plant; (*b*) the corresponding network of flow resistances, including resistances in the soil, the roots, the stem and the leaves, with evaporation being driven by the constant current generator E; (*c*) simplified catenary model with the complex branched pathway of (*b*) represented as a linear series with the hydraulic resistances in the soil (R_s), roots (R_r), stem (R_{st}) and leaves (R_ℓ) each being represented by a single resistor; (*d*) same as (*c*) but including capacitances (C) of the appropriate tissues; (*e*) as (*d*) but including resistances to transfer to or from storage and voltage sources (Ψ_π) that represent the osmotic potentials of each component. In this case the voltage drop across each capacitor represents the turgor pressure.

because the potential drop (and the resistance) across the vapour phase is typically more than an order of magnitude greater than that in the liquid phase. Therefore typical changes in the liquid-phase resistance have a negligible effect on the total resistance (and hence flow). Typical values of Ψ are between -1 and -2.5 MPa, giving $\Delta\Psi$ in the liquid phase of 1 to 2 MPa, while water potential in the atmosphere is commonly lower than -50 MPa ($\simeq 69\%$ relative humidity – see Chapter 5), giving a $\Delta\Psi$ in the vapour phase more than an order of magnitude greater. This comparison between gas- and liquid-phase resistances is only approximate because, as

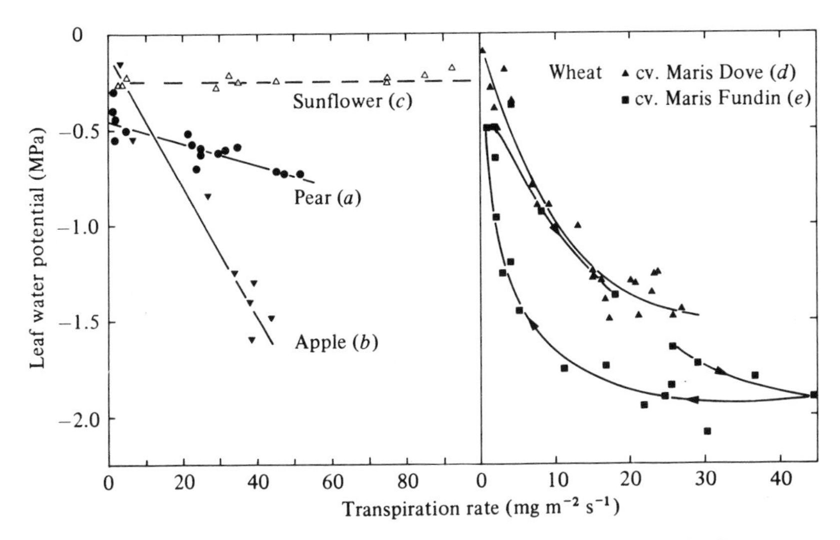

Fig. 4.10. Examples of some of the types of relationship between leaf water potential and transpiration rate that have been observed. The sunflower and pear data are from Camacho-B *et al.* (1974), the apple data from Landsberg *et al.* (1975) and the wheat data from Jones (1978).

seen in Chapter 3, the driving force for gas-phase diffusion is partial vapour pressure which is not linearly related to Ψ (Chapter 5).

Experimental results relating the water potential drop in the liquid phase in the soil–plant system to evaporation rates are illustrated in Fig. 4.10. The water potential drop in the soil–plant system is sometimes linearly related to the evaporation rate (Fig. 4.10*a*,*b*). Such results indicate that R_{sp} (the soil–plant resistance) is constant and independent of flow rate. In many cases, however, the relationship shows a marked curvature (Fig. 4.10*d*) while, in extreme cases, water potential is nearly constant over a wide range of evaporation rates (Fig. 4.10*c*). These results can be interpreted as R_{sp} decreasing at high flow rates. The reason for this flow-rate dependence of R_{sp} is not known, though several hypotheses have been suggested (e.g. Fiscus 1975; Passioura 1984; Boyer 1985). Both types of response have been reported in many species. The tendency for resistance to decrease with increasing flow rate is mechanistically slightly surprising, since it might be expected that the soil component of the resistance would increase, but it probably has an adaptive advantage in preventing the occurrence of severe plant water stress at high fluxes.

Measurements of water potential gradients up the stem of a transpiring plant are usually obtained with a pressure chamber using excised leaves. If attached leaves are sealed in plastic bags for approximately half an hour before the measurement, Ψ_ℓ equilibrates with the stem xylem (Ψ_x) and gives a true measure of the Ψ gradient up the stem. If the leaf is transpiring at the time of sampling, Ψ_ℓ also depends on the resistance within the leaf petiole and can be used to estimate R_ℓ. The total plant resistance R_p ($= R_r + R_{st} + R_\ell$) is often obtained from measurements of Ψ_ℓ in plants grown in solution culture as this eliminates any R_s. The gradient of Ψ_ℓ for leaves at different insertion levels up a transpiring plant ranges from less than 0.03 MPa m^{-1} for branches of trees to about 0.1 MPa per node in a wheat plant (about 1 MPa m^{-1}) (Jones 1977*a* and unpublished). Table 4.2 gives some estimated values for flow resistances in different parts of the soil–plant system in apple trees, illustrating that the major proportion of the resistance is in the roots, though there is also a large component in the petiole or leaf.

Dynamic responses. Fig. 4.10*e* shows an example where the relationship between the water potential drop and **E** is not unique; that is there is hysteresis. This indicates a failure of the steady-state model to simulate the true dynamic behaviour when **E** is varying. The hysteresis where liquid flow lags behind evaporative demand may be modelled by incorporating capacitors into the circuit analogue (Fig. 4.9*d*).

The **capacitance** (C) of any part of the system may be defined as the ratio of the change in tissue water content (W) to the change in water potential, i.e.

$$C = \frac{\mathrm{d}W}{\mathrm{d}\Psi} = W_{max}\frac{\mathrm{d}\theta}{\mathrm{d}\Psi} = W_{max}\, C_r \qquad (4.26)$$

where W_{max} is the maximum (turgid) tissue water content and θ is the relative water content. The term $\mathrm{d}\theta/\mathrm{d}\Psi$ can be called a **relative capacitance** (C_r), which is an intrinsic property of the tissue so is useful when comparing tissues of very different shape or size. When the water content is expressed per unit area (leaf, stem cross-section or ground) the units for C would be m^3 m^{-2} MPa^{-1} (= m MPa^{-1}) but when expressed in absolute terms C would be in m^3 MPa^{-1}. For inclusion in flow models, the basis for C must be the same as that used for **E**.

Some examples of relationships between Ψ and tissue water content are shown in Fig. 4.11. Although these curves are in general non-linear, it is often possible to approximate C_r by a constant value (i.e. a straight line) over much of the relevant physiological range. Values of C_r taken from this figure range from about 5% MPa^{-1} for apple leaves to 33% MPa^{-1} for wheat and tomato leaves. Corresponding values for C_r for conifers have been estimated at 4.7% for *Larix* and 6.3% for *Picea* (Schulze *et al.* 1985).

Inclusion of the capacitances of various tissues in the model of Fig. 4.9*c*

Table 4.2. *Flow resistances in apple trees calculated from data in Landsberg* et al. (1976)
All resistances were converted to a single plant basis and expressed as a percentage of $R_{st}+R_r$ (which included a soil component). Values in brackets represent the range of observed values.

	$R_{st}+R_r$ (MPa s m^{-3} × 10^{-7})	R_r (%)	R_{st} (%)	R_ℓ (%)
Potted trees (2 year old)				
Experiment 1	10.7	66	34	41
	(8.2–12.9)	(48–74)	(26–52)	
Experiment 2	30.5	52	48	21
	(26–34)	(46–59)	(41–54)	
Orchard tree (9 year old)	0.8	60	50[a]	35

[a] Rounding error.

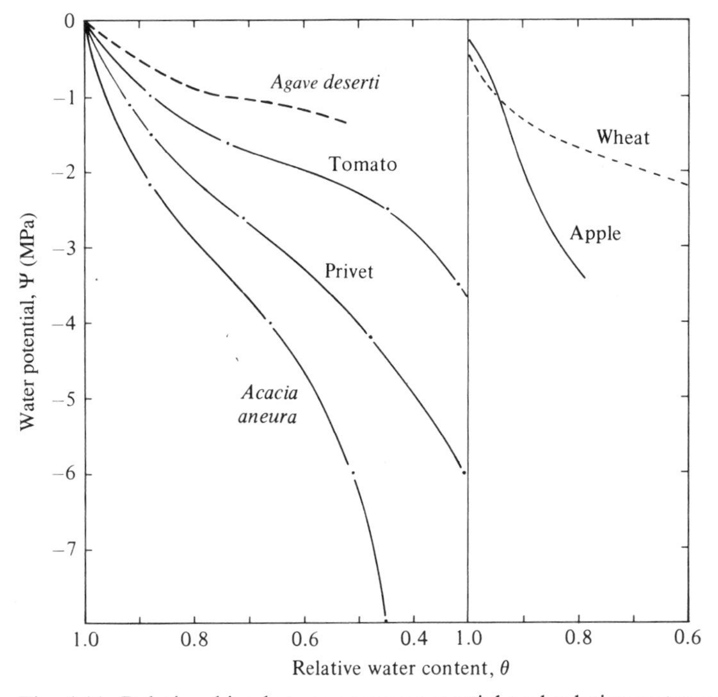

Fig. 4.11. Relationships between water potential and relative water content for different species. (Data from Slatyer 1960, Jones 1978, Jones & Higgs 1980, Nobel & Jordan 1983.)

gives that shown in Fig. 4.9*d*. Although even this model is rather complex, a number of authors have extended the approach even further to include large networks of resistors and capacitors representing detailed anatomical and morphological measurements of hydraulic architecture. Another extension to this approach (Smith *et al.* 1987) is illustrated in Fig. 4.9*e*. These authors showed that by including voltage sources it is possible to incorporate cell osmotic potential as an explicit variable. A consequence of this is that, since the voltage drop across the voltage source is Ψ_π, and the voltage drop between the nodes (—•—) and ground is Ψ, then the voltage drop across the capacitor represents the turgor pressure (Ψ_p). These complex models can be used to simulate the dynamics of water relations of different tissues; examples where such an approach has been used include the dynamic simulation of water relations of plants as diverse as *Agave* (Smith *et al.* 1987) and trees (e.g. *Thuja*: see Tyree 1988). Although numerical solution of these models can be readily achieved by modern computer routines for electrical network analysis, their complexity often limits their value in understanding the underlying control processes. The general behaviour of such resistance–capacitance models is best illustrated using a simplified lumped parameter model (Fig. 4.12) where the complex network in Fig. 4.9*d* is approximated by a circuit with only one resistance and one capacitor. To analyse this (from equation 4.25), flow through the plant ($\mathbf{J}_p$) can be written

$$\mathbf{J}_p = (\Psi_s - \Psi_\ell)/R_p \tag{4.27}$$

The rate of change of leaf water content ($\mathrm{d}W_\ell/\mathrm{d}t$) is given by the difference between the flow of water into the leaf and that lost by evaporation, so

$$\frac{\mathrm{d}W_\ell}{\mathrm{d}t} = \mathbf{J}_p - \mathbf{E} \tag{4.28}$$

Substituting from equation 4.27 gives

$$\frac{\mathrm{d}W_\ell}{\mathrm{d}t} = \frac{\Psi_s - \Psi_\ell}{R_p} - \mathbf{E} \tag{4.29}$$

If one now assumes a constant capacitance one can write (from equation 4.26)

$$\frac{\mathrm{d}\Psi_\ell}{\mathrm{d}t} + \frac{\Psi_\ell}{R_p C} = \frac{\Psi_s - \mathbf{E}R_p}{R_p C} \tag{4.30}$$

This is a first-order differential equation of the type commonly encountered in the analysis of electrical circuits. It can be solved using standard

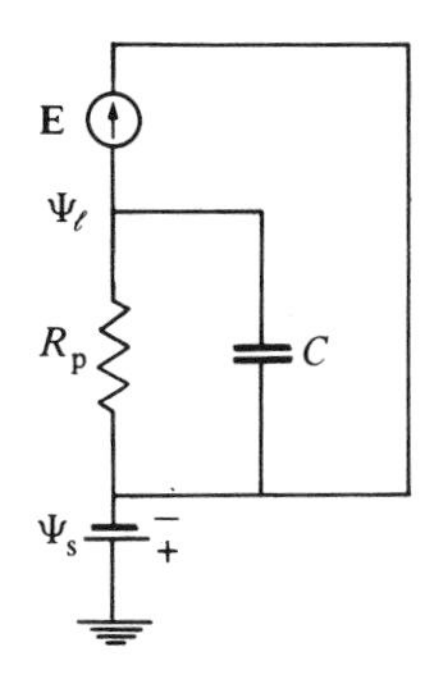

Fig. 4.12. A simple lumped parameter model of the plant hydraulic system, where flow is driven by the constant current generator E, and the soil water potential Ψ_s is lowered below the reference value (ground) by a battery.

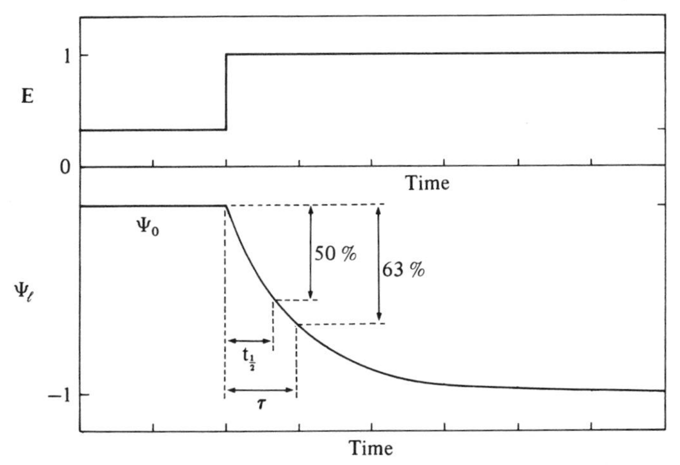

Fig. 4.13. The time course of changes in Ψ_ℓ for an instantaneous change in E, as predicted by the model in Fig. 4.12, illustrating the half-time ($t_{\frac{1}{2}}$ – the time required for 50 % of the total change) and the time constant (τ).

mathematical techniques to give the time dependence of Ψ_ℓ after a step change in **E** (Fig. 4.13):

$$\Psi_\ell = A + Be^{-t/\tau} \tag{4.31}$$

where $A = (\Psi_s - \mathbf{E}R_p)/R_pC$, $B = \Psi_o + R_p\mathbf{E} - (\Psi_s/R_pC)$ and $\tau = R_pC$. In this, Ψ_o is the value of Ψ_ℓ before the step change in **E**. The value τ ($= RC$) is called the **time constant** and is the time for 63 % of the total change (see Fig. 4.13). It can be easily shown that the half-time, or time for 50 % of the total change to occur, is equal to $\tau \times -\ln 0.5 = 0.693\tau$.

In practice, the assumptions of constant capacitance and resistance and the use of the simplified 'lumped parameter' model are adequate for many purposes even though realistic models may be much more complex (see e.g. Denmead & Millar 1976; Jones 1978). In particular, the relatively small number of parameters required in fitting this model to experimental data is a major advantage. Methods for solving equation 4.30 in the much more realistic cases where **E** is continuously varying are described by Powell & Thorpe (1977) and Jones (1978).

The dynamics of water exchange by single cells can also be treated in the same way as tissues. As for tissues (equation 4.31), the time constant for equilibration of water potential (τ) is given by the product of the resistance to water uptake ($R = 1/(A\,L_p)$,where A is the cell surface area) and the cell capacitance ($C = V/(\varepsilon - \Psi_\pi)$, where V is the cell volume and ε is the cell volumetric elastic modulus), so that

$$\tau = V/(A\,L_p(\varepsilon - \Psi_\pi)) \qquad (4.32)$$

It follows from this equation that the rate of hydraulic equilibration of a cell increases with both L_p and ε.

Liquid phase transport processes

The movement of solutes from tissue to tissue or over shorter distances from cell to cell within plants is essential for normal growth and development. For example, minerals from the soil must reach the leaves and other aerial tissues while carbohydrates from the leaves must be transported downwards to build new roots. Details of the various transport processes can be found in Nobel (1991), Zimmermann & Milburn (1975), Lüttge & Pitman (1976) or Moorby (1981), as well as other texts on transport mechanisms. Here, only a brief outline will be presented to place the different processes in perspective.

In Chapter 3 (p. 57) it was stated that, when a finite quantity of material is released in a plane, the concentration profile is the shape of the Gaussian curve, and the distance from the origin (x) at which the concentration falls to 37% of that at the origin is given by

$$x = \sqrt{(4\boldsymbol{D}t)} \qquad (4.33)$$

Because diffusion coefficients in solution are approximately four orders of magnitude smaller than in air (see Appendix 2), diffusion is much slower in solution and effective only over short distances. For example, solving equation 4.33 for a typical solute $\boldsymbol{D}$ of 1×10^{-9} m^2 s^{-1}, gives the time to diffuse 100 μm as 2.5 s (i.e. $(100 \times 10^{-6})^2/(4 \times 10^{-9})$ s), 2.5×10^4 s ($\simeq$ 6.9 h) to diffuse 1 cm and 2.5×10^8 s ($\simeq$ 8 years) to diffuse 1 m. Clearly, therefore, simple diffusion cannot be an important mechanism for long-distance transport in plants, though it is important for cell-to-cell or within-cell transport. Within individual cells the cytoplasm is continually moving and this cytoplasmic streaming can act to speed transfer over short distances in a way analogous to turbulent transfer in a crop boundary layer. As long as there is no *net* cytoplasmic movement, the only effect is to lower the apparent value of ***D***.

Although the transport of solutes across plant cell membranes, whether by active or passive mechanisms, is important for plant functioning, a discussion of this topic is outside the scope of this book and details may be found in the texts referred to above.

The long-distance movement of solutes is primarily in the bulk flow of water in the xylem or in the other specialised conducting tissue, the phloem. The bulk flow of water in the xylem is important for the rapid movement of not only mineral nutrients from the roots to the shoots, but also metabolites and plant growth regulators such as cytokinins. However, because transpiration is unidirectional, it cannot be involved in basipetal transport so that another pathway is required for redistribution of solutes, particularly in a downward direction.

This alternative long-distance transport pathway, which is particularly important for carbohydrate movement to meristematic or storage tissues, is in the specialised (living) sieve tubes of the phloem. The mechanism of phloem transport is still incompletely understood but it is thought that this also involves mass flow, though in this case the pressure gradient that drives the flow may be generated by active (energy-requiring) membrane transport pumps that concentrate the solutes in the phloem sieve tubes of the source tissues. The high solute concentrations lower the water potential so that water follows through the semi-permeable sieve-tube membrane, thus raising the hydrostatic pressure and causing flow out of the source tissue. There may also be active solute unloading at the sink tissues, further aiding the development of a pressure gradient.

For both xylem and phloem, therefore, the flux of the *i*th solute ($\mathbf{J}_i$, $kg\,m^{-2}\,s^{-1}$) across a plane is given by the product of concentration and velocity

$$\mathbf{J}_i = c_i \mathbf{J}_v \tag{4.34}$$

Typical concentrations of different solutes in phloem and xylem sap are presented in Table 4.3. This table shows that by far the major component of the phloem is sugar while there is none in the xylem. Although this is true for many plants, some trees such as sugar maple do have significant amounts of sugar in the xylem, particularly in spring. The concentration of almost all substances, with the notable exception of calcium, are markedly higher in the phloem. Therefore, a given $\mathbf{J}_v$ in the phloem will transport more solute than an equivalent velocity in the xylem, though transport in the phloem does require expenditure of energy by the plant while the much greater bulk flow in the xylem is driven by energy from elsewhere.

Table 4.3. *Comparison of concentrations of some solutes in xylem and phloem sap from* Nicotiana glauca (*data from Hocking* 1980)

	Xylem sap ($g\ m^{-3}$)	Phloem sap ($g\ m^{-3}$)
Chloride	64	486
Sulphur	43	139
Phosphorus	68	435
Ammonium	10	45
Calcium	189	83
Magnesium	34	104
Potassium	204	3673
Sodium	46	116
Amino compounds	283	11×10^3
Sucrose	0	$155–168 \times 10^3$
Total dry matter	$1.1–1.2 \times 10^3$	$170–196 \times 10^3$

Calcium is somewhat exceptional in that most workers have found fairly similar concentrations of calcium in the two types of sap, or else calcium is more concentrated in the xylem (e.g. Table 4.3). A consequence of the rather low calcium concentration in the phloem appears to be that low transpiring tissues, such as apple fruits or the enclosed leaves of lettuce, can suffer severe calcium deficiency, which gives rise to disorders such as bitterpit and breakdown in apple and tipburn in lettuce (see e.g. Bangerth 1979).

Sample problems

4.1 (i) To what height would water rise in (*a*) a wettable vertical capillary 1 mm diameter, (*b*) a similar capillary tilted at 45°, (*c*) a vertical capillary where the wall material has a contact angle of 50°, (*d*) a wettable capillary 1 μm in diameter? (ii) What pressure would need to be applied to the column in (i, *d*) to prevent any capillary rise?

4.2 A single cell as a Ψ_π of -1.5 MPa when $\Psi = -1$ MPa. The volume increases by 25% as Ψ increases to -0.5 MPa. What are (i) the original value of Ψ_p; (ii) the new value of Ψ_π; (iii) the new value of Ψ_p; (iv) θ, the initial relative water content of the cell (assume Ψ is linearly related to volume); (v) ε_B, the volumetric elastic modulus at full turgor?

4.3 (i) What would be the volume flow rate of water through a smooth cylindrical pipe 1 m long, 0.2 mm in diameter, with an applied pressure differential between the ends of 5 kPa? (ii) What are the values of L, L_p and R? (iii) What diameter tube would be required to maintain the same flow if the pressure differential decreased to 1 kPa?

4.4 If $\Psi_{soil} = -0.1$ MPa, $\Psi_{\ell} = -1.2$ MPa and transpiration rate $= 0.1 \times 10^{-6}$ m^3 H_2O (m^2 leaf)$^{-1}$ s^{-1}, (i) calculate total hydraulic resistance (*a*) on a leaf area basis, (*b*) on a plant basis (assuming a leaf area per plant of 0.1 m^2), (*c*) on a ground area basis (assuming 30 plants m^{-2}). (ii) Estimate Ψ_{ℓ} if half the plants are removed without affecting total crop evaporation. (iii) Estimate Ψ_{ℓ} if, instead, half the shoots on each plant are removed (assuming that half the hydraulic resistance is normally in the soil–root part of the flow path).

5

Energy balance and evaporation

The preceding chapters outlined the basic principles of mass transfer and described the application of electrical analogues. The application of these principles to evaporation requires the extension of the simple analogues to include the flow of heat as well as water vapour. This is because energy is required to supply the latent heat of evaporation. For this reason, as a preliminary to the treatment of evaporation, I first outline the energy balance equation and introduce the concepts of isothermal net radiation and the radiative heat transfer resistances, as well as define the various quantities that are used to specify the amount of water in air. The energy balance equation is then used to derive a general equation for the description of evaporation from single leaves and from plant communities, and this is used to investigate particular aspects of the environmental and physiological control of evaporation and of dewfall.

Energy balance

The principle of the conservation of energy (the *First Law of Thermodynamics*) states that energy cannot be created or destroyed, but only changed from one form to another. Applying this to a plant leaf (or canopy) it can be seen that the difference between all the energy fluxes into and out of the system must equal the rate of storage, thus

$$\mathbf{\Phi}_n - \mathbf{C} - \lambda\mathbf{E} = \mathbf{M} + \mathbf{S} \tag{5.1}$$

where $\mathbf{\Phi}_n$ is the net heat gain from radiation (shortwave plus longwave), $\mathbf{C}$ is the net 'sensible' heat *loss*, $\lambda\mathbf{E}$ is the net latent heat *loss*, $\mathbf{M}$ is the net heat stored in biochemical reactions, $\mathbf{S}$ is the net physical storage. It is convenient to express all these fluxes per unit area (of leaf or ground) to give units of flux density (W m^{-2}).

Net radiation. $\mathbf{\Phi}_n$ is the dominant term in equation 5.1, not only because it is often the largest, but it drives many of the other energy fluxes. Typical diurnal changes in $\mathbf{\Phi}_n$ for different surfaces have been presented in Chapter 2 (Table 2.3).

Sensible heat flux. The sensible heat loss, **C**, is the sum of all heat loss to the surroundings by conduction or convection. For example, whenever a leaf is warmer than the surrounding air, heat is lost and **C** is positive. The total sensible heat flux may be partitioned into that lost to the air (by conduction and convection), for which the symbol **C** is retained, and that lost by conduction to other surroundings, particularly the soil, which is given the symbol **G**. **G** is negligible for individual leaves. For plant canopies, however, **G** refers to the soil heat flux which is positive during much of the day (representing a loss from the canopy) and may range from 2% of $\mathbf{\Phi}_n$ for a dense canopy to more than 30% of $\mathbf{\Phi}_n$ in sparse canopies with little shading of the soil. At night **G** is negative and of similar absolute magnitude to daytime values.

Latent heat flux. The rate of heat loss by evaporation ($\lambda\mathbf{E}$) is that required to convert all water evaporated from the liquid to the vapour state and is given by the product of the evaporation rate and the latent heat of vaporisation of water ($\lambda = 2.454$ MJ kg^{-1} at 20 °C).

Storage. The rate of metabolic storage (**M**) represents the storage of heat energy as chemical bond energy and is dominated by photosynthesis and respiration. Typical maximum rates of net photosynthesis of 0.5–2.0 mg CO_2 m^{-2} s^{-1} correspond to **M** between 8 and 32 W m^{-2} (see Chapter 7), values which are usually less than 5% $\mathbf{\Phi}_n$. At night, **M** takes smaller negative values associated with dark respiration, except in certain species of the Araceae where extremely high respiration rates can occur during the spring (see Chapter 9). The physical storage (**S**) includes energy used in heating the plant material as well as (for a canopy) heat used to raise the temperature of the air. In general this physical storage is small except for massive leaves or stems (e.g. cacti) and forests. For example, a very dense cereal crop might contain 3 kg water m^{-2}, so that an estimate of **S** when canopy temperature is changing at 5 °C h^{-1} would be only 17.5 W m^{-2} ($3 \times 4200 \times 5/3600$ = mass × specific heat × $\mathrm{d}T/\mathrm{d}t$).

Isothermal net radiation

The value of the net radiation, $\mathbf{\Phi}_n$, for a given incoming radiation varies as a function of factors such as windspeed, humidity and leaf resistance that affect surface temperature (see Chapter 9), because (from equation 2.12, p. 26)

$$\mathbf{\Phi}_n = \mathbf{\Phi}_{absorbed} - \varepsilon\sigma(T_s)^4 \qquad (5.2)$$

Where T_s is the Kelvin temperature of the surface, ε is the emissivity and σ

is the Stefan–Boltzmann constant. For predictive studies it is useful to define an 'environmental' net radiation that is independent of surface temperature. The isothermal net radiation ($\mathbf{\Phi}_{ni}$) can be defined as the net radiation that would be received by an identical surface in an identical environment *if it were at air temperature*. Therefore

$$\mathbf{\Phi}_{ni} = \mathbf{\Phi}_{absorbed} - \varepsilon\sigma(T_a)^4 \tag{5.3}$$

Substituting from equation 5.2 gives the relationship between $\mathbf{\Phi}_{ni}$ and $\mathbf{\Phi}_n$ as

$$\mathbf{\Phi}_{ni} = \mathbf{\Phi}_n + \varepsilon\sigma(T_s^4 - T_a^4) \tag{5.4}$$

If one now makes the substitution $T_s = T_a + \Delta T$, and multiplies out, one gets

$$\mathbf{\Phi}_{ni} = \mathbf{\Phi}_n + \varepsilon\sigma(T_a^4 + 4T_a^3(\Delta T) + 6T_a^2(\Delta T)^2 + 4T_a(\Delta T)^3 + (\Delta T)^4 - T_a^4) \tag{5.5}$$

Terms in T_a^4 cancel out and because $\Delta T \ll T_a$, all terms in $(\Delta T)^2$ and higher powers can be neglected, so that

$$\mathbf{\Phi}_{ni} \simeq \mathbf{\Phi}_n + 4\varepsilon\sigma T_a^3(\Delta T) \tag{5.6}$$

The second term in equation 5.6 represents a longwave radiative heat loss and can be put into a form analogous to the usual equation for sensible heat loss:

$$\text{radiative heat loss} = \left(\frac{4\varepsilon\sigma T_a^3}{\rho c_p}\right)(\rho c_p)(T_s - T_a) \tag{5.7}$$

where ρ is the density of air, c_p is the specific heat of air and $(4\varepsilon\sigma T_a^3/\rho c_p)$ is a 'conductance' to radiative heat transfer (g_R), so that

$$\mathbf{\Phi}_n = \mathbf{\Phi}_{ni} - g_R \rho c_p(T_s - T_a) \tag{5.8}$$

In addition to this radiative heat loss, an isolated leaf can also lose sensible heat by convection, a pathway that is in parallel with that for radiant heat loss (see Fig. 5.1). Because radiant and sensible heat transfer are each proportional to the leaf-to-air temperature difference, we can now define a total thermal conductance (g_{HR}), as the parallel sum of g_H and g_R ($= g_H + g_R$). The effect of temperature on g_R is shown in Table 5.1, illustrating that it has a small effect on the total thermal conductance, except when the boundary layer conductance for sensible heat transfer is rather small. As will be seen later, isothermal net radiation is particularly useful in modelling studies designed to determine the consequences for leaf temperature or evaporation rate of changing environmental or physiological factors.

Table 5.1. *Temperature dependence of the 'radiative' conductance* g_R, *and some typical values of the total thermal conductance* (g_{HR}) *for a range of values of* g_H. *The value in brackets is* g_H *as a percentage of* g_{HR}

		g_{HR} (mm s^{-1})		
Temperature (°C)	g_R (mm s^{-1})	$g_H = 2$	20	200 (mm s^{-1})
0	3.54	5.5 (36)	23.5 (85)	204 (98)
10	4.10	6.1 (33)	24.1 (83)	204 (98)
20	4.69	6.7 (30)	24.7 (81)	205 (98)
30	5.37	7.4 (27)	25.4 (79)	205 (97)
40	6.10	8.1 (25)	26.1 (77)	206 (97)

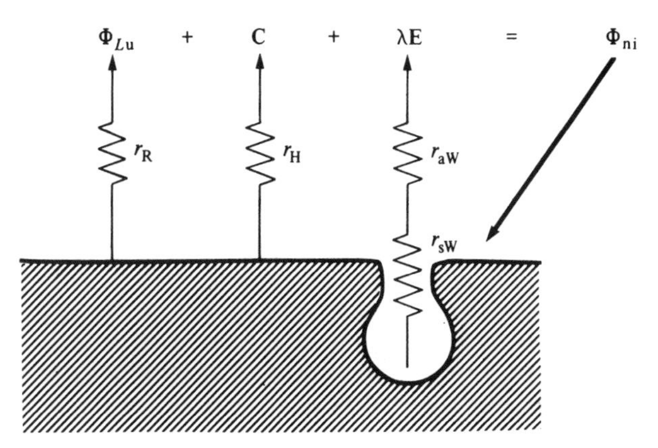

Fig. 5.1. Energy exchanges for a leaf, where the radiative heat loss $\mathbf{\Phi}_{Lu}$ is the difference between actual net radiation and isothermal net radiation.

Measures of water vapour concentration

In addition to the measures of gas composition described in Chapter 3, there are several that are specifically used for describing the water vapour content of air. The mass concentration of water vapour (c_W) is called the **absolute humidity**. This is related to the water vapour partial pressure (usually given the symbol e) by equation 3.6 (p. 47) which simplifies to

$$c_W = eM_W/\mathscr{R}T = (2.17/T)e \tag{5.9}$$

Where c_W is in g m^{-3}, e is in Pa. An alternative expression relating c_W and

e, that will be used in several subsequent derivations, may be obtained using equation 3.6, remembering that the partial pressure of dry air $(p_a) = P - e$:

$$c_W = \rho_a(M_W/M_A)[e/(P-e)] \simeq \rho_a(M_W/M_A)(e/P) \tag{5.10}$$

where ρ_a is the density of dry air and M_A is the effective molecular weight of dry air ($\simeq 29$), so $M_W/M_A = 0.622$. The approximation in equation 5.10 usually introduces negligible error since $e \ll P$.

The moisture status of air may also be described in terms of water potential (Ψ). When air is allowed to equilibrate with liquid water at the same temperature, the equilibrium state is

$$\Psi_{\text{liquid}} = \Psi_{\text{vapour}}$$

so that vapour in equilibrium with free water has a water potential of zero. In this case the air is *saturated* with water vapour. When the air contains less than the saturation amount, water potential is given by

$$\Psi = (\mathscr{R}T/\bar{V}_W)\ln(e/e_s) \tag{5.11}$$

where e_s is the saturation partial pressure of water vapour (or saturation vapour pressure). The value of e_s is a function of temperature ($e_{s(T)}$), closely approximated by the following empirical relationship (see Appendix 4 for further details and for tabulated values for e_s and for the corresponding concentration, c_{sW}):

$$e_{s(T)} = \mathrm{a}\exp\left\{\frac{\mathrm{b}T}{\mathrm{c}+T}\right\} \tag{5.12}$$

where T is in °C, $e_{s(T)}$ is in Pa, and the empirical coefficients are

a = 613.75, b = 17.502, c = 240.97.

Where air is not saturated with water vapour, the degree of saturation is often expressed as the **relative humidity** (h), which is the vapour pressure expressed as a fraction (or often percentage) of the saturation vapour pressure at that temperature:

$$h = e/e_{s(T)} \tag{5.13}$$

Another commonly used term is the **vapour pressure deficit** (δe) given by

$$\delta e = e_{s(T)} - e \tag{5.14}$$

The dewpoint temperature (T_d) is the temperature at which the water vapour pressure equals the saturation vapour pressure. As the air is cooled

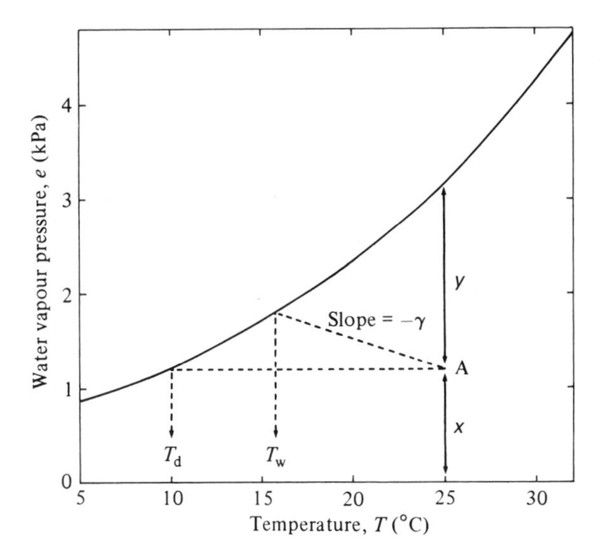

Fig. 5.2. Curve of saturation water vapour pressure against temperature (solid curve) illustrating the relationship between various measures of humidity. For the point marked A – vapour pressure (e) = x; vapour pressure deficit (δe) = y; relative humidity (h) = $x/(x+y)$; dew point (T_d) = temperature where saturation vapour pressure equals x; wet bulb temperature (T_w) is also shown.

below its dewpoint, condensation occurs. The relationships between vapour pressure, vapour pressure deficit, dewpoint and relative humidity are illustrated in Fig. 5.2.

Another useful term is the wet bulb temperature (T_w), which is the temperature that a moist surface reaches when it evaporates into an unsaturated atmosphere (see Fig. 5.2). The value of the wet bulb temperature depends on vapour pressure, air temperature and on the rate of air movement over the evaporating surface. The wet bulb temperature is an experimentally measurable quantity that is commonly used for estimating atmospheric humidity using the relation

$$e = e_{s(T_w)} - \gamma(T_a - T_w) \tag{5.15}$$

where γ is the psychrometer constant ($Pc_p/0.622\,\lambda = 66.1$ Pa K^{-1} for a well-ventilated surface at 100 kPa pressure and 20 °C; other values are given in Appendix 3). Psychrometric tables are widely available for determining e directly from dry bulb (T_a) and wet bulb (T_w) temperatures measured either in normal meteorological screens or in aspirated enclosures. The theoretical derivation of γ is discussed by Monteith & Unsworth (1990). There are many other methods available for measuring air humidity, including those that depend on the absorption of IR or UV radiation by water vapour, or the equilibrium adsorption of water onto solids with consequent changes in mechanical properties (e.g. length, as in the hair hygrometer) or electrical properties. Details of water vapour measurement may be found in Fritschen & Gay (1979).

Evaporation

Penman–Monteith equation

In Chapter 3 it was established that mass transfer is proportional to a concentration difference, so that for evaporation from a moist surface

$$\mathbf{E} = g_W \, \Delta c_W \tag{5.16}$$

where g_W is the total conductance of the pathway between the evaporating sites and the bulk air (for plants it includes stomatal and boundary layer components) and Δc_W is the water vapour concentration difference between the surface and the bulk air. Since vapour pressure is a more appropriate driving force where air temperature does not equal surface temperature, equation 5.10 can be used to replace Δc_W by $(\rho_a M_W / M_A P)\,\Delta e$, giving

$$\mathbf{E} = g_W (0.622\, \rho_a / P)\,(e_{s(T_s)} - e_a) \tag{5.17}$$

where $e_{s(T_s)}$ is the saturation vapour pressure at surface temperature.

This equation provides the basis for measurements of leaf or boundary layer conductance in cuvettes (see Chapter 6) or, where independent estimates of g_W are available, it can be used to estimate **E**. However, it does require a knowledge of surface temperature (for determining $e_{s(T_s)}$). If T_s is not measured, it can be determined from energy balance considerations (see Chapter 9), or else it is possible to eliminate the need for a knowledge of surface conditions by combining equation 5.17 with the energy balance equation (5.1) and using an approximation originally suggested by Penman (1948). In this, the surface–air vapour pressure difference $(e_{s(T_s)} - e_a)$ is replaced by the vapour pressure deficit of the ambient air $(e_{s(T_a)} - e_a)$ plus a term that depends on the temperature difference between the surface and the air (see Fig. 5.3):

$$\begin{aligned} e_{s(T_s)} - e_a &= (e_{s(T_a)} - e_a) - s(T_a - T_s) \\ &= \delta e + s(T_s - T_a) \end{aligned} \tag{5.18}$$

where s is the slope of the curve relating saturation vapour pressure to temperature (which is assumed to be approximately constant over the range T_a to T_s), and δe is the vapour pressure deficit of the ambient air. For values of s see Appendix 4.

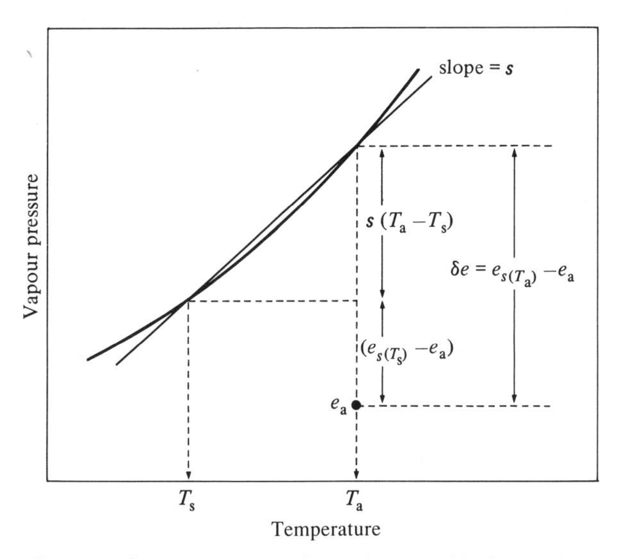

Fig. 5.3. The Penman transformation. In this figure the solid curve represents the relationship between saturation vapour pressure and temperature. The slope of this curve over the range T_s to T_a is approximated by a straight line of slope s. The value of the surface to air vapour pressure difference $(e_{s(T_s)} - e_a)$ is given by the vapour pressure deficit of the ambient air (δe) minus the difference between the saturation vapour pressures at T_a and T_s (approximated by $s\,(T_s - T_a)$).

Using this substitution in equation 5.17 gives

$$\mathbf{E} = g_W(0.622\,\rho_a/P)\,[\delta e + s(T_s - T_a)] \tag{5.19}$$

It is possible to write an equivalent expression for the sensible heat flux between the surface and the air (equation 3.29, with the volumetric heat capacity of dry air ($\rho_a c_p$) used to approximate the heat capacity of air):

$$\mathbf{C} = g_H\,\rho_a c_p (T_s - T_a) \tag{5.20}$$

Eliminating $(T_s - T_a)$ from these two equations gives

$$\mathbf{E} = g_W(0.622\,\rho_a/P)\left[\delta e + \left(\frac{s\,\mathbf{C}}{g_H\,\rho_a c_p}\right)\right] \tag{5.21}$$

In the steady state (and neglecting **M**) the energy balance (equation 5.1) reduces to

$$\mathbf{C} = (\mathbf{\Phi}_n - \mathbf{G}) - \lambda \mathbf{E} \tag{5.22}$$

substituting this for **C** in equation 5.22 and rearranging gives the following equation for **E**:

$$\mathbf{E} = \frac{\{s(\mathbf{\Phi}_n - \mathbf{G}) + \rho_a c_p g_H \delta e\}}{\lambda[s + (\gamma g_H / g_W)]} \tag{5.23a}$$

This equation is often termed the Penman–Monteith equation and it can be expressed in a number of alternative forms. For example, by using the approximate substitution in equation 5.10, it can be expressed in terms of the absolute humidity deficit of the ambient air (δc_W) as

$$\mathbf{E} = \frac{\{[\varepsilon(\mathbf{\Phi}_n - \mathbf{G})/\lambda] + g_H \delta c_W\}}{\varepsilon + g_H / g_W} \tag{5.23b}$$

where $\varepsilon = s/\gamma$. Where the boundary layer conductances for heat and water vapour are similar, as in a fully turbulent boundary layer, one can make use of the rules for adding conductances in series (Fig. 3.2) to replace g_H/g_W by $(1 + g_a/g_\ell)$ where g_ℓ is the leaf conductance (otherwise referred to as a physiological or surface conductance) largely determined by stomatal aperture.

The original derivation (Penman 1948) was in a form to describe evaporation from a free water surface ($\mathbf{E}_o$). For that surface there is no surface resistance to water loss, so that the denominator can be simplified using the approximation $g_H \simeq g_W$ (see Chapter 3). Practical application of this simplified equation (which will be called the standard Penman equation) involves the use of a series of empirically determined conversions to estimate the available energy ($\mathbf{\Phi}_n - \mathbf{G}$) and g_H from readily available climatic data (mean air temperature and humidity, duration of bright sunshine and windrun) together with information on daylength and latitude (for the estimation of $\mathbf{\Phi}_n$). The free water evaporation calculated in this way can be used as a basis for estimating evaporation from well-watered crops that completely cover the ground (often called the potential evapotranspiration). Many studies have shown that the actual evaporation from a wide variety of vegetation types is a relatively constant function of $\mathbf{E}_o$ (normally between 0.6 and 0.8 in temperate climates). The coefficient may be much smaller at times when there is incomplete ground cover such as after planting or near harvest. Values for crop coefficients for a wide range of crops have been tabulated by Doorenbos & Pruitt (1984).

The advantage of equation 5.23 (which was first applied to leaves by Penman (1953) and to plant canopies by Monteith (1965*b*)), as compared with the free water version as a basis for predicting water loss by leaves or vegetation, is that it removes the need for the empirical conversion factor by introducing the concept of a surface conductance to water vapour that

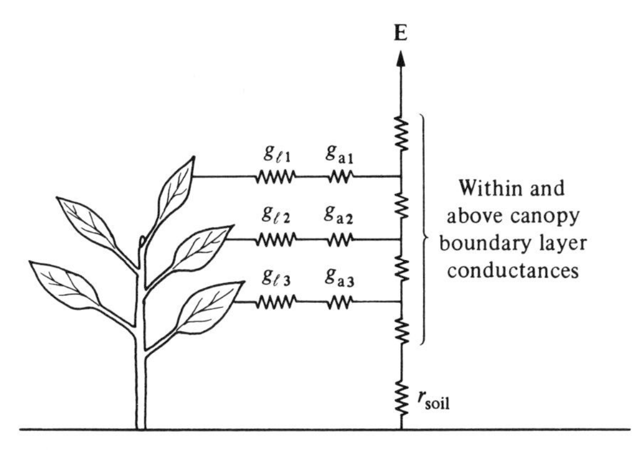

Fig. 5.4. The pathways of water loss and the associated resistances for evaporation (**E**) from a plant canopy. It is not possible to replace this network by a single physiological conductance g_L in series with a single canopy boundary layer component g_A.

depends, among other things, on stomatal aperture. Recent work has emphasised the importance of the physiological, as opposed to the purely environmental, control of evaporation. Some historical perspective is given by Monteith (1981*a*).

When applied to single leaves, g_H is simply the boundary layer conductance for heat and g_W is the series sum of the leaf (mainly stomatal) and boundary layer conductances (i.e. $g_W = g_{\ell W}\, g_{aW}/(g_{\ell W} + g_{aW})$). In canopies where the pathway of water loss is more complex (Fig. 5.4) interpretation of g_W is more difficult. This is because it is not possible to separate completely the boundary layer from the 'physiological' or 'leaf' component (which may even include a soil component). In addition the method assumes that the source and sink distributions within the canopy are identical for each entity used in evaluating resistances (e.g. water vapour, heat, momentum: see Thom 1975). In practice, however, this physiological conductance (g_{LW} – note that we distinguish canopy conductances or resistances per unit ground area by capital italic subscripts) is reasonably well approximated by the parallel sum of the individual leaf conductances:

$$g_{LW} = \Sigma(\overline{g_{\ell i}}\, L_i) \tag{5.24}$$

where $\overline{g_{\ell i}}$ is the mean leaf conductance per unit projected leaf area in a given stratum and L_i is the leaf area per unit ground area, or leaf area index, in that stratum of canopy. The leaf conductance may be measured using diffusion porometers (see Chapter 6) and the boundary layer resistance by

Table 5.2. *Some published values for canopy conductance* (g_{LW}) *for different crops and natural communities* (*see Jarvis* et al. 1976; *Wallace* et al. 1981; *and Miranda* et al. 1984)

Crop	Location	Canopy conductance g_{LW} (mm s^{-1})
Alfalfa	Munich, Germany	14–40 (seasonal range)
Alfalfa	Arizona, USA	0.4–50 (seasonal range)
Barley	Nottingham, England	1–8 (diurnal range)
Barley	Nottingham, England	5–67 (seasonal range)
Grassland	Matador, Canada	1.7–10 (seasonal range)
Heather moorland	S. Scotland	4–20 (diurnal range)
Lentil	Madhya Pradesh, India	1–17 (seasonal range)
Maple/beech	Montreal, Canada	0.8–10 (seasonal range)
Norway spruce	Munich, Germany	6.7–10 (seasonal range)
Rice	Kaudulla, Sri Lanka	16–117 (diurnal range)
Wheat	Madhya Pradesh, India	1–10 (seasonal range)
Wheat	Rothamsted, England	2.5–40 (diurnal range)

any of the methods outlined in Appendix 8. Alternatively g_{LW} may be estimated from equation 5.23 if **E** is known. Some typical values for g_{LW} are presented in Table 5.2. Low values can result from either stomatal closure or poor ground cover.

Although the behaviour of equation 5.23 is complex, it is readily apparent that **E** increases with increasing radiant energy and with increasing vapour pressure deficit of the ambient air. Similarly **E** increases with increasing total water vapour conductance, but changes in the boundary layer conductance can increase or decrease **E** as g_a occurs in both the numerator and denominator. The same equation predicts dewfall when **E** is negative (see below).

Equation 5.23 is extremely flexible and of wide application, particularly in modelling studies where it is necessary to predict the water balance of leaves or plant communities in different conditions. In many cases it is advantageous to eliminate the effect of surface temperature from the radiation term in the same way that equation 5.23 eliminates the surface temperature effect from the humidity term. This can be done by making use of the concept of isothermal net radiation and substituting $\mathbf{\Phi}_{ni}$ for $\mathbf{\Phi}_n$ and g_{HR} for g_H in equation 5.23 (see Jones 1976).

Coupling to the environment

The fact that the total conductance to water vapour loss occurs as a proportionality constant in equation 5.16 has led to the belief that stomata play a dominant role in controlling transpiration. Inspection of equation 5.23, however, shows that other factors (e.g. boundary layer conductance, radiation, and humidity) are also important. The simple approach of equation 5.16 is misleading because the driving force for evaporation (Δc_{W}) is itself dependent on the evaporation rate (and hence on a wide range of these other factors), and this is taken into account in the Penman–Monteith equation. A major step forward in quantifying the degree of stomatal control of evaporation from different sized leaves and from crops was made by McNaughton & Jarvis (1983), who rewrote equation 5.23 in a form that partitioned crop evaporation into two components: a so-called **equilibrium evaporation rate** ($\mathbf{E}_{\mathrm{eq}}$) that depends only on the energy supply (radiation), and an **imposed evaporation rate** ($\mathbf{E}_{\mathrm{imp}}$). The relative importance of these two depends on the degree of coupling of the evaporating surface (leaf or crop) to the environment.

(*a*) *Imposed evaporation.* Where the boundary layer conductance is large, heat and mass transfer are very efficient so that leaf temperature approaches air temperature whatever the input radiation, and the surface is said to be well coupled to the environment. In this case, as the boundary layer conductance tends to infinity, leaf temperature tends to air temperature, and equation 5.23 reduces to a form similar to that of equation 5.16:

$$\mathbf{E}_{\mathrm{imp}} = (\rho_{\mathrm{a}} c_p / \lambda \gamma) g_{\ell} \delta e \qquad (5.25)$$

where g_{ℓ} (or $g_{L\mathrm{W}}$ for a canopy) is the physiological conductance. The efficient transfer between the surface and the atmosphere means that the conditions of the bulk atmosphere are 'imposed' at the leaf surface, so that evaporation is proportional to the leaf conductance as expected from equation 5.16.

(*b*) *Equilibrium evaporation.* At the other extreme, when the boundary layer conductance is very small, heat and mass transfer between the surface and the atmosphere is extremely poor and the surface is said to be poorly coupled to the environment. In this case evaporation tends to the equilibrium rate. In the extreme of complete isolation, as the boundary layer conductance tends to zero, $\mathbf{E}_{\mathrm{imp}}$ decreases to zero and equation (5.23) reduces to

$$\mathbf{E}_{\mathrm{eq}} = \varepsilon \mathbf{\Phi}_{\mathrm{n}} / \lambda(\varepsilon + 1) = s \mathbf{\Phi}_{\mathrm{n}} / \lambda(s + \gamma) \qquad (5.26)$$

Although the lack of any effect of stomatal aperture or atmospheric humidity on evaporation in this case is somewhat surprising, it can be

Table 5.3. *Likely values for the decoupling coefficient* (Ω) *and for the sensitivity of evaporation to leaf or canopy physiological conductance* $(dE/E)/(dg_\ell/g_\ell)$ *for different leaves and crops* (*see Jarvis* 1985; *Jarvis & McNaughton* 1986; *Jones* 1990)

Single leaves	Leaf width (mm)	Ω	$(dE/E)/(dg_\ell/g_\ell)$
Rhubarb	500	0.8	0.2
Cucumber	250	0.7	0.3
Bean	60	0.5	0.5
Onion	8	0.3	0.7
Asparagus	1	0.1	0.9
Open field crops	**Crop height (m)**	Ω	$(dE/E)/(dg_\ell/g_\ell)$
Grass	0.1	0.9	0.1
Strawberry	0.2	0.85	0.15
Tomato	0.4	0.7	0.3
Wheat	1.0	0.5	0.5
Raspberry	1.5	0.4	0.6
Citrus orchard	5.0	0.3	0.7
Forest	30	0.1	0.9
Other situations		Ω	$(dE/E)/(dg_\ell/g_\ell)$
Uncontrolled glasshouse		0.9–1.0	0–0.1
Lysimeters or 1 m² plots		< 0.1	> 0.9
Controlled environment chamber		< 0.1	> 0.9

explained by considering what happens to evaporation from a crop growing in a sealed glasshouse. In this case the crop and its adjacent air is completely decoupled from the outside air, with no transfer of water vapour through the glass and little or no heat loss. When there is no radiation input into the glasshouse, evaporation from the crop raises the humidity until eventually, when the air is saturated, the leaf-to-air vapour difference falls to zero and evaporation rate eventually falls to zero, as one would expect. In contrast, when there is an input of radiation, this raises the temperature inside the glasshouse at a rate that is proportional to the irradiance. As the temperature increases, the water vapour holding capacity of the air also increases (Fig. 5.2), so that it is possible to maintain a nearly constant leaf-to-air vapour pressure gradient even though the vapour pressure of the air in the glasshouse may be continually increasing. This steady-state rate of

Table 5.4. *Typical evaporation rates for reasonably dense plant canopies well supplied with water. Values in brackets represent maximum rates*
Data from Denmead 1969; Denmead & McIlroy 1970; Grant 1970; Kanemasu *et al.* 1976; Körner & Mayr 1981; Kowal & Kassam 1973; Lang *et al.* 1974; Lewis & Callaghan 1976; Monteith 1965*b*; Rauner 1976; Ritchie & Jordan 1972; Rosenberg 1974.

	Daily **E** ($mm\ day^{-1}$)		Hourly **E** ($mm\ h^{-1}$)
Central England	January	0.2	—
	April	1.3 (3)	—
	July	3 (6)	0.25
Continental Europe	Summer	5	—
European Alps – 500 m elevation	Summer	4.5 (9)	—
(bright days only) – 2500 m elevation	Summer	3.8 (8.2)	—
Arctic tundra – meadow	July	2 (3)	—
USA great plains	July	6 (10–12)	— (1.0)
North Nigeria	July	5	—
South East Australia	Summer	5.6 (9–10)	0.4 (0.9)

evaporation depends on the rate of energy input into the system according to equation 5.26.

In practice, evaporation from a leaf, a crop or any area of vegetation operates between these limits and can be expressed as the sum of an imposed component and an equilibrium component by rewriting equation 5.23 as

$$\mathbf{E} = \Omega \mathbf{E}_{eq} + (1-\Omega)\,\mathbf{E}_{imp} \tag{5.27}$$

where

$$\begin{aligned}\Omega &= (\varepsilon+1)/(\varepsilon+g_H/g_W)\\ &\simeq (\varepsilon+1)/(\varepsilon+1+g_a/g_\ell)\end{aligned} \tag{5.28}$$

(if one assumes that $g_{aH} \simeq g_{aW}$). The Ω, or **decoupling coefficient**, is a measure of the coupling between conditions at the surface and in the free airstream and can vary between 0 (for perfect coupling) and 1 (for complete isolation). It is apparent from equation 5.28 that Ω depends on the ratio between the surface conductance and that of the boundary layer, rather than on their absolute values.

Sensitivity of evaporation to stomata. A particular value of this approach is that it enables one to estimate the degree to which stomata control evaporation. Because g_ℓ is largely determined by stomatal conductance, a

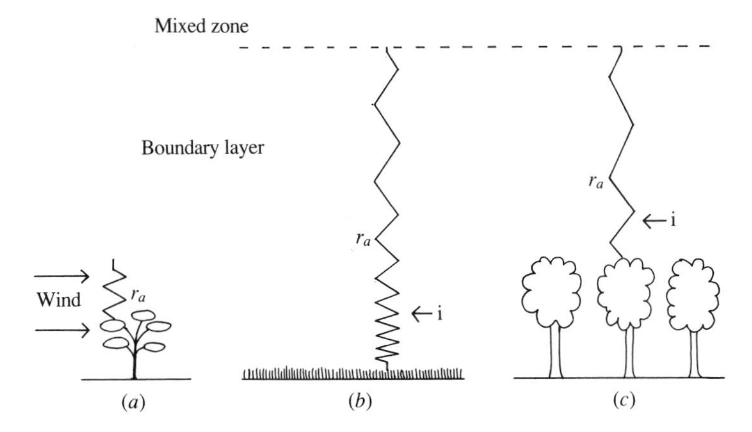

Fig. 5.5. An illustration of the relative magnitudes of the boundary layer resistances appropriate for (*a*) an isolated plant; (*b*) an extensive short, smooth canopy such as mown grass; and (*c*) an extensive tall, rough canopy. Also shown is the normal height for micrometeorological measurements (i), indicating that there is a significant resistance between this height and the mixed layer that is not normally included in micrometeorological estimates of the boundary layer resistance or conductance.

measure of the stomatal control of evaporation is the relative sensitivity of **E** to a small change in g_ℓ (i.e. $(d\mathbf{E}/\mathbf{E})/(dg_\ell/g_\ell)$). This can be obtained by differentiating equation 5.23, which, after some rearrangement (Jarvis & McNaughton 1986) gives

$$(d\mathbf{E}/\mathbf{E})/(dg_\ell/g_\ell) = (1-\Omega) \tag{5.29}$$

The sensitivity of evaporation to stomatal or surface conductance for different situations is summarised in Table 5.3.

Evaporation from plant communities

Some typical rates of evaporation from different plant canopies in a range of climates are presented in Table 5.4. In arid environments rates of evaporation as high as 10 mm day^{-1} or 1 mm h^{-1} (which is equivalent to an energy requirement of 680 W m^{-2}) are reasonably common for well-irrigated crops, while in temperate climates in winter, rates fall to less than 0.3 mm day^{-1}. Factors affecting crop evaporation are discussed in relation to the principles outlined above.

Canopy type and area. Considering first a tall isolated plant in an otherwise non-vegetated area (Fig. 5.5), it is apparent that the appropriate air vapour pressure deficit (δe) in equation 5.23 will be that of the airstream in the immediate vicinity of the plant, and the appropriate value for g_a is that for transfer from the individual leaves to the bulk airstream (perhaps 100 mm s^{-1} when not sheltered within a canopy: see Chapter 3). By substituting in equation 5.28, this gives an Ω of between 0.14 and 0.24 for typical unsheltered leaves of well-watered plants (assuming a typical range of g_ℓ of 5–10 mm s^{-1} and a temperature of 20 °C – i.e. $(2.2+1)/(2.2+1+100/5)$).

As the airstream moves over an extensive area of homogeneous vegetation, transpiration will raise the humidity of the air near the crop surface, with the extent of this effect increasing with the area of vegetation. This has a feedback effect on **E**, tending to reduce it. It follows that the relevant g_A, which here needs to describe transfer from the leaves right up to the mixed layer where δe is unaffected by the crop, is correspondingly smaller than g_a of the individual leaves (Fig. 5.5) and decreases with increasing area of uniform vegetation. The value of Ω is therefore much larger than for an isolated plant. Figure 5.5 also shows that Ω is strongly dependent on the aerodynamic characteristics of the vegetation, with the improved transfer (higher g_A) above a rough canopy such as a forest (typical g_A 100–300 mm s^{-1}: see Chapter 3) as compared with a smooth canopy such as grassland (typical g_A perhaps 10 mm s^{-1}), tending to decrease Ω.

Some estimated values of Ω for a range of leaves and crops are shown in Table 5.3. It is clear from this that the degree of coupling increases (Ω decreases and the sensitivity of **E** to g_ℓ increases) as leaf size decreases, as canopy height increases, and as the area of vegetation decreases. The dependence of transpiration on canopy type and on canopy conductance is illustrated in Fig. 5.6. Transpiration from forests is often rather less than from field crops because of the relatively low values of canopy conductances (especially in coniferous forest). Figure 5.6 confirms the result in Table 5.3 that evaporation from short crops would be expected to be particularly insensitive to g_L as long as g_L remains above about 10 mm s^{-1}.

Another effect of canopy boundary layer conductance is that the vapour pressure term ($\rho_a c_p g_H \delta e$) in the numerator of equation 5.23 is greater than the available energy term ($s(\mathbf{\Phi}_n - \mathbf{G})$) in forests but smaller in short grass. Figure 5.7 shows that **E** for a forest would be much more sensitive to δe than to $\mathbf{\Phi}_n$, while the converse is true for grass.

Another important factor in the water loss from plant canopies is the evaporation of intercepted rainfall. The difference between tall and short crops is particularly apparent when one considers the *total* evaporation from the canopy including that occurring directly from the surface of leaves wetted by rainfall. This loss of intercepted rainfall can be greater than total transpiration for forests in humid areas (Calder 1976), and may lead to total forest evaporation being as much as twice that from grass. The reason

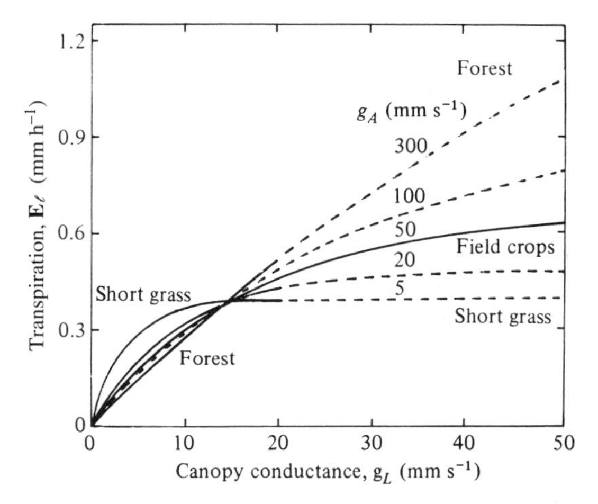

Fig. 5.6. Calculated relationships between transpiration rate and canopy conductance at different boundary layer conductances, for 400 W m^{-2} available energy, 1 kPa vapour pressure deficit and 15 °C. The solid lines represent the probable range of values of canopy conductance for different crops, being up to 50 mm s^{-1} for some field crops. (After Jarvis 1981.)

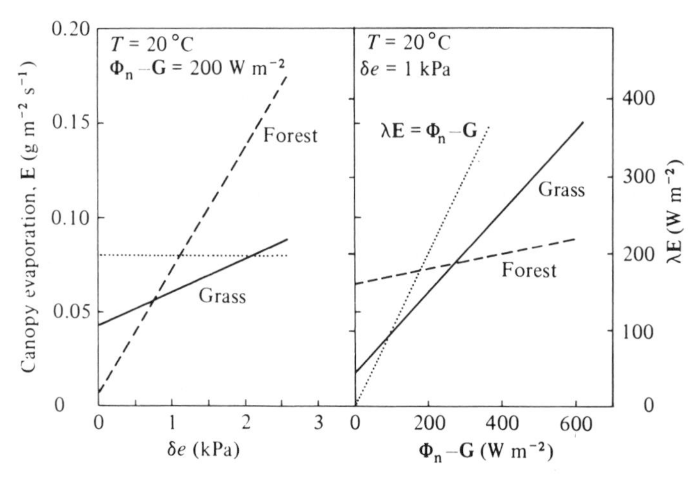

Fig. 5.7. Dependence of canopy evaporation on humidity deficit (δe) and available energy ($\mathbf{\Phi}_n - \mathbf{G}$) for short grass ($g_{AH}$ = 10 mm s^{-1}) for forest (g_{AH} = 200 mm s^{-1}), with a constant physiological conductance (g_{LW}) of 10 mm s^{-1}. The line where evaporation consumes all available radiant energy is shown dotted.

for the different behaviour can be seen in Fig. 5.6 where, as g_L increases (as happens when the surface gets wet and stomatal control is eliminated) forest evaporation may increase three-fold while grass evaporation changes relatively little.

Advection. In arid regions, where crops may be surrounded by relatively unvegetated areas, it is possible for the air to be significantly hotter and drier than in the crop, thus providing an extra source of energy so that the latent heat loss can significantly exceed any net radiative gain. This advection, or transfer of heat from the surroundings, is often called the 'oasis' effect and is adequately treated by the Penman equation, as long as measurements are restricted to the crop boundary layer. The magnitude of this advection effect is illustrated by results for rice growing at Griffith, New South Wales, Australia (Lang *et al.* 1974), where $\lambda \mathbf{E}$ could be as much as 170% of $\mathbf{\Phi}_n$ near the edge of the crop.

Soil evaporation. The evaporation from the soil is often an important component of canopy water loss, but is strongly dependent on soil wetness and on plant cover. Several reports indicate that, even with wet soil, soil evaporation is only about 5% of the total when the leaf area index (L) reaches 4 (for crops as diverse as wheat and *Pinus radiata*: Denmead 1969; Brun *et al.* 1972). When L is 2 or less, however, wet soil evaporation can be as much as half the total. Ritchie (1972) has suggested that after rain about 10 mm of soil evaporation can occur at a high rate (i.e. the soil surface resistance is small or zero). As the soil dries, the surface resistance increases and soil evaporation decreases significantly. With a dry soil surface, soil evaporation is relatively unimportant, and evaporation is solely transpiration from plant leaves; this is approximately proportional to leaf area up to $L \simeq 3$. Soil evaporation is discussed further by van Bavel & Hillel (1976) and others.

Soil water availability. Soil drying also eventually leads to physiological stresses that decrease transpiration from plant canopies by effects such as stomatal closure. As has been pointed out above, the sensitivity of canopy evaporation to stomatal closure as the soil dries is very dependent on canopy and environmental characteristics that affect the degree of coupling (Ω), while the stomatal response to soil drying is itself very species dependent (see Chapter 6).

Implications for water relations experimentation

It is important to recognise the difficulties involved in extrapolating results from physiological studies on single leaves, potted plants or even single plants or small plots in the field to the prediction of the behaviour of large areas of vegetation in the field. This is especially true for studies of evaporation. A couple of examples will be presented to illustrate the pitfalls.

(*a*) *Plant breeding*. A plant breeder may wish to select plants that use water sparingly. Two lines may be found that differ in stomatal conductance ($\simeq g_\ell$) by, say, 20%, but it is not appropriate to infer that one would transpire at a rate 20% lower than the other. If they are compared in a controlled environment, or even in typical single plant field plots ($\Omega \simeq 0.05$), the difference in transpiration would be about 19% (from equation 5.29, $d\mathbf{E}/\mathbf{E}$ would be equal to $(1-\Omega)\, dg_\ell/g_\ell$ or 19%). If instead large field plantings of several hectares ($\Omega \simeq 0.7$) of each line are compared, $d\mathbf{E}/\mathbf{E}$ would be $0.3\, dg_\ell/g_\ell$ or only 6%. The advantage of the selection would decrease with increasing area of planting.

(*b*) *Sensitivity to soil water*. An understanding of coupling has also helped to resolve the controversy that has surrounded the sensitivity of crop evaporation to soil water. For example, Denmead & Shaw (1962) reported that crop evaporation started to fall below the potential rate when only 10% of soil water was removed, while others (see Ritchie 1973) have found that as much as 80% of 'available' soil water may be removed before crop evaporation falls significantly. It is worth noting that Denmead & Shaw's experiment was conducted on maize plants grown in large pots in the field amongst a continuous canopy of unstressed plants. This experimental design resulted in the vapour pressure deficit around each individual plant being determined by the weather and by the evaporation from the *unstressed* plants, with there being no opportunity for feedback from stomatal conductances of the stressed plants on the canopy humidity. Although grown in the field the droughted plants would have behaved as isolated plants (small Ω) with a much greater sensitivity of $\mathbf{E}$ to stomatal closure than would have been the case for a canopy of identical stressed plants.

Estimating evaporation rates

Calculations based on the Penman–Montieth equation (equation 5.23) probably provide the most powerful general technique for estimating evaporation rates from vegetation, though estimation of the physiological conductance can be a major limitation. The use of the standard Penman equation for estimating free water evaporation ($\mathbf{E}_0$), together with empirical crop coefficients (largely required to correct for differences in ground cover), is adequate in many cases. Various refinements are available. For example a widely used system in the UK (MORECS: Thompson *et al.* 1982) incorporates a number of corrections to the Penman–Monteith equation: these include both the isothermal radiation correction (equation 5.8) and a water balance calculation (to modify evaporation via changes in the physiological conductance that result from stomatal closure as soil water is depleted).

Evaporation from large areas of many types of vegetation, at least when well supplied with water, is approximated by the Priestley–Taylor relationship:

$$\lambda \mathbf{E} \simeq 1.26 s \mathbf{\Phi}_n/(s+\gamma) = 1.26 \varepsilon \mathbf{\Phi}_n/(\varepsilon+1) \qquad (5.30)$$

This form of equation is already familiar as the equilibrium evaporation rate (equation 5.26) though the theoretical basis for the factor of 1.26 is unclear (see Monteith 1981*a*), and there can be large deviations from this value, particularly in dry conditions and for very rough canopies.

An alternative method that is still widely used for estimating a potential evaporation rate is the use of evaporimeters or evaporation pans. These consist of pans of open water whose evaporation rate can be monitored. Unfortunately there are several types of pan in use with differences in size, shape and recommended exposure. Evaporation rates from such instruments tend to be 10–45 % higher than even the Penman $\mathbf{E}_0$, because they are too small to achieve an equilibrium evaporation estimate. For most purposes it is more convenient and reliable to estimate $\mathbf{E}_0$ directly by means of the standard Penman equation or its modifications.

There is also a wide range of empirical formulae for estimating evaporation rates; these usually have a less rigorous physical basis than the Penman equation (see Doorenbos & Pruitt 1984 for a summary), but have the advantage of correspondingly smaller data requirements. For example, readily available climatic data such as temperature, and sometimes also humidity, duration of bright sunshine or solar radiation, may be all that are required. It is also possible to estimate total evaporation from vegetation by means of soil water budget approaches based on actual measurements of soil water using neutron scattering or weighing lysimeters. Unfortunately plants in lysimeters, for example, often do not respond identically to those grown in a normal soil profile. Information on evaporation rates and even on sources of water (for example the depth in the soil profile from which it is extracted), can be obtained from studies of the natural deuterium/hydrogen (D/H) ratios in water (see Ehleringer & Osmond 1989). Other techniques available include the use of crop enclosures (see Chapter

7 – though the unnatural environmental coupling in such devices means that very great care needs to be taken in interpreting any data), sap flow meters (see Jones *et al.* 1988) or even silicon accumulation in leaves (Hutton & Norrish 1974).

Three micrometeorological approaches are worth discussing in a little more detail, as the principles involved are instructive.

Estimation from leaf temperature. In principle energy balance considerations enable one to estimate evaporation rates from leaf or canopy temperature (as measured by IR thermometry or other remote sensing techniques), though in practice the requirement for estimates of variables in addition to leaf temperature (e.g. some of net radiation, air temperature, boundary layer conductance, or canopy conductance) often limits the precision that can be obtained. One approach that can be useful for studies of leaf transpiration is to compare the temperature of a non-transpiring leaf (or leaf model) with that of a transpiring leaf (or model). The use of completely wet leaves or models enables one to estimate the potential evaporation rate.

The energy balance (equation 5.1) for a dry surface at the steady state reduces to

$$\mathbf{\Phi}_{n(d)} = \mathbf{C}_{(d)} \tag{5.31}$$

and for an evaporating surface it is

$$\mathbf{\Phi}_{n(w)} = \mathbf{C}_{(w)} + \lambda\mathbf{E} \tag{5.32}$$

using subscripts (d) and (w) to refer to dry and evaporating surfaces, respectively. Substituting for $\mathbf{\Phi}_n$ using equation 5.8 and for $\mathbf{C}$ using equation 5.20 gives, for the dry surface

$$\mathbf{\Phi}_{ni} - \rho_a c_p g_R (T_{s(d)} - T_a) = \rho_a c_p g_{aH} (T_{s(d)} - T_a) \tag{5.33}$$

and for the evaporating surface

$$\mathbf{\Phi}_{ni} - \rho_a c_p g_R (T_{s(w)} - T_a) = \rho_a c_p g_{aH} (T_{s(w)} - T_a) + \lambda\mathbf{E} \tag{5.34}$$

Subtracting 5.33 from 5.34 and rearranging gives the following expression for evaporation

$$\mathbf{E} = \rho_a c_p (g_{aH} + g_R)(T_{s(d)} - T_{s(w)})/\lambda \tag{5.35}$$

If an independent estimate of g_a is available (see Appendix 8) it is possible to estimate the evaporation rate from a surface from the temperature difference between it and a similar non-evaporating surface.

In many field situations, especially where remote sensing is being used, no equivalent dry surface is available so it may be more convenient to make use of equation 5.34, using independent estimates of the net radiation and of air temperature. These latter estimates are normally ground based, though methods are being developed for estimating them remotely (Curran 1985).

Bowen ratio. A widely used technique, based on the energy balance equation, depends on measurement of the **Bowen ratio** ($\boldsymbol{\beta}$) – the ratio of sensible to latent heat losses ($\mathbf{C}/\lambda\mathbf{E}$). Substituting this in equation 5.22 gives

$$\lambda\mathbf{E} = (\boldsymbol{\Phi}_{\mathrm{n}} - \mathbf{G})/(1+\boldsymbol{\beta}) \tag{5.36}$$

Assuming that the transfer coefficients for heat and water vapour in the turbulent boundary layer above a crop are equal (Chapter 3) and substituting the appropriate driving forces for **C** and **E** gives

$$\boldsymbol{\beta} = \frac{\mathbf{C}}{\lambda\mathbf{E}} = \left(\frac{P\rho c_p}{0.622\rho}\right)\frac{\Delta T}{\lambda\Delta e} = \gamma\frac{\Delta T}{\Delta e} \tag{5.37}$$

where γ (= $Pc_p/0.622\lambda$) is the psychrometer constant. Therefore all that is needed to estimate **E** is a knowledge of net radiation and soil heat flux together with the gradients of T and e in the boundary layer. When β is small this method is particularly insensitive to errors in measurement of T or e, because all energy is then being used for evaporation and the error is just that in determining ($\boldsymbol{\Phi}_{\mathrm{n}}$–**G**) (see equation 5.36). As well as not requiring information on surface conditions, this technique does not even require a knowledge of the actual physiological or boundary layer conductances.

Eddy correlation. Perhaps the most direct method available for estimating fluxes above a canopy is the eddy correlation method proposed by Swinbank (1951). If one considers a point within the crop boundary layer, at any instant the air may be moving in any direction as a consequence of the random eddies. The net flux of an entity such as water vapour across a horizontal plane is given by the integral over time of the product of the instantaneous vertical velocity (u_z) and the instantaneous concentration at that level. The integral of u_z will, over a long period, be zero as there is no net upward or downward movement of air. If the fluctuations in concentration are not correlated with u_z, then the net flux will also be zero, but where vertical velocity and concentration are correlated there will be a net flux (Fig. 5.8). The eddy correlation technique therefore requires that u_z and c_{W} are measured accurately over very short periods. The necessary sensor response times depend on the size of the eddies that carry the flux and may range from about 1 Hz several metres above rough forest canopies down to about 0.001 Hz close to relatively smooth surfaces. In principle the technique may be applied to any transport process though, in practice, the availability of sensors with a rapid enough response limits the precision

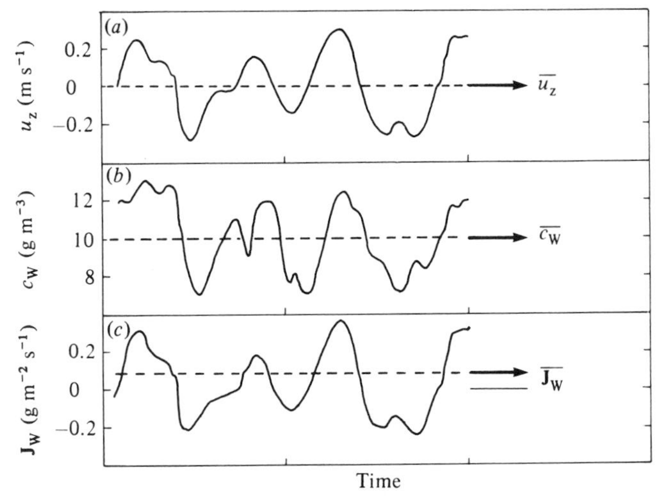

Fig. 5.8. Eddy correlation estimation of water vapour flux $\mathbf{J}_W$. (*a*) time course of u_z, (*b*) time course of c_W at height *z*, and (*c*) time course of $\mathbf{J}_W = u_z \times c_W$. Bars over symbols indicate long-term mean values.

attainable. With recent developments in sensor technology and in electronics and computing there is now increasing emphasis on this technique.

Dew

The condensation of water vapour on plant leaves or other surfaces is governed by the same physical principles as evaporation, so that equation 5.23 is applicable, with dew forming whenever **E** is negative. However, as there is no equivalent to the physiological resistance (i.e. $g_{LW} = \infty$), equation 5.23 (using net isothermal radiation) simplifies to

$$\mathbf{E} = \frac{s(\mathbf{\Phi}_{ni} - \mathbf{G}) + \rho_a c_p g_{HR}\,\delta e}{\lambda(s+\gamma)} \tag{5.38}$$

This predicts that dew occurs when $-s(\mathbf{\Phi}_{ni} - \mathbf{G}) > \rho_a c_p g_{HR}\,\delta e$.

If one assumes that $(\mathbf{\Phi}_{ni} - \mathbf{G})$ falls to a minimum of about -100 W m^{-2} on a clear night, substitution of this in equation 5.38 predicts maximum rates of dewfall of almost 0.1 mm h^{-1} when the air is saturated. This compares with maximum observed rates of dewfall of up to about 0.4 mm per night (Monteith 1963; Burrage 1972), while more typical amounts on clear nights may be 0.1 to 0.2 mm per night (Tuller & Chilton 1973). There are several reasons why actual rates of dewfall tend to be well below the theoretical maximum. Most important is the fact that air is often not completely saturated, so that some of the heat required to satisfy the net radiation loss is obtained by cooling the air to the dewpoint temperature before any condensation can occur. In addition, as the humidity deficit increases, dewfall becomes more sensitive to increasing windspeed (equation 5.38 and Fig. 5.9). This is because a high windspeed increases sensible heat exchange, thus reducing the amount that the temperature of a leaf can fall below air temperature for a given radiation loss. In the extreme case where g_A tends to infinity, $T_\ell \simeq T_a$, so that condensation can only occur when the air is saturated.

There is, however, a conflict between the requirement for calm conditions for maximum condensation according to equation 5.38, and the requirement for adequate transfer of water vapour down through the atmosphere to the surface. The magnitude of this downward flux can be appreciated from the fact that 0.4 mm of liquid water represents the *entire* water vapour content of the bottom 30 m of a saturated atmosphere at 15 °C. Therefore significant wind is required to maintain the humidity near the surface high enough for condensation to continue at high rates. At very low windspeeds most of the condensation probably originates by distillation from the soil, rather than as dewfall from the atmosphere, though dewfall usually predominates in England (Monteith 1957). Both types of condensation must be distinguished from guttation – the droplets of water exuded from hydathodes on plant leaves.

Quantitatively, dewfall is usually at least an order of magnitude less than potential rates of evaporation and so rarely contributes significantly to the water balance of plant communities. However, there are areas such as the dew deserts of Chile where condensation can equal or exceed rainfall. Taken together with evidence that dew can be absorbed by leaf tissues (Stone 1957) and raise leaf water potentials (Kerr & Beardsell 1975), it is probable that dew is a factor contributing to plant survival in such climates. A more important ecological effect results from the favourable micro-environment that wet leaf surfaces provide for fungal infection. In this case, the duration of surface wetness is likely to be more significant than the total amount. At least during summer in England, dew typically persists for periods of 6–12 h.

There is evidence that some plants can condense water from a non-saturated atmosphere by the excretion of hygroscopic salts onto the leaf surface. Even when the ambient relative humidity was no higher than 82%, significant water was found to accumulate on the leaves of *Nolana mollis* in the Atacama Desert in Chile, though the quantities were only of the order of 0.1 mm per night (Mooney *et al.* 1980).

The most direct methods for dew estimation include the use of weighing lysimeters or actual collection using absorbent tissue. Alternatively, dew

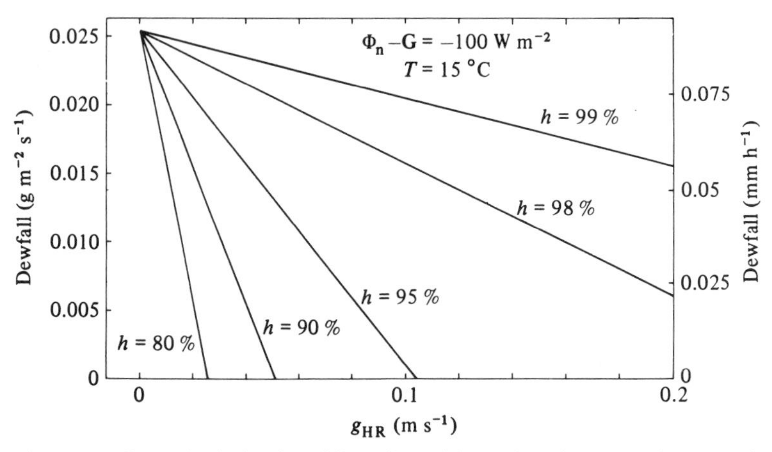

Fig. 5.9. Effect of relative humidity (h) and boundary layer conductance (g_{HR}) on rate of dewfall as predicted by equation 5.38 for a temperature of 15 °C and an available energy of -100 W m^{-2}.

may be estimated using an artificial surface positioned on a recording balance, but it is necessary to ensure that the radiative and aerodynamic properties of the collector resemble those of real leaves. Another very useful technique for estimation of dew duration is to measure the electrical resistance between two electrodes (e.g. strips of copper wire) attached to the surface of a leaf. When the leaf is wet the resistance decreases dramatically. These and other techniques are outlined by Rosenberg *et al.* (1983).

Sample problems

5.1 Assuming an air temperature (T_a) of 30 °C and a relative humidity, h, of 40 %, estimate (i) e_s, (ii) e, (iii) c_W, (iv) δe, (v) T_w, (vi) T_d, (vii) m_W, (viii) x_W, (ix) Ψ.

5.2 The net radiation absorbed by a given leaf at 22 °C is 430 W m^{-2}. What is (i) the corresponding isothermal net radiation ($\mathbf{\Phi}_{ni}$) if $T_a = 19$ °C, (ii) the radiative conductance (g_R)?

5.3 (i) Using the Penman–Monteith equation, calculate the evaporation rate from a forest ($g_A = 200$ mm s^{-1}) or from short grass ($g_A = 10$ mm s^{-1}) for ($\mathbf{\Phi}_n - \mathbf{G}) = 400$ W m^{-2}, $T_a = 20$ °C, $\delta e = 1$ kPa; (*a*) when the surface is wet, (*b*) when $g_L = 30$ mm s^{-1}. (ii) What are the corresponding values of the Bowen ratio?

5.4 Assuming an area of homogeneous crop where the total boundary layer conductance is 15 mm s^{-1}, the crop physiological conductance is 5 mm s^{-1} and the air temperature is 20 °C, calculate (i) the value of the decoupling coefficient (Ω), and (ii) the relative reduction in crop transpiration rate if the stomata close to such an extent that the physiological conductance decreases by 50 %. (iii) What particular assumption have you had to make in the calculation of (ii)?

6

Stomata

The evolution of the stomatal apparatus was one of the most important steps in the early colonisation of the terrestrial environment. Even though the stomatal pores when fully open occupy between about 0.5 and 5% of the leaf surface, almost all the water transpired by plants, as well as the CO_2 absorbed in photosynthesis, passes through these pores. It is only in rare cases such as in the fern ally, *Stylites* from the Peruvian Andes, that CO_2 may be absorbed through the roots (Keeley *et al.* 1984). The central role of the stomata in regulating water vapour and CO_2 exchange by plant leaves is illustrated in Fig. 6.1. This figure also shows some of the complex feedback and feedforward control loops that are involved in the control of stomatal aperture and hence of diffusive conductance. It is the extreme sensitivity of the stomata to both environmental and internal physiological factors that enables them to operate in a manner that optimises the balance between water loss and CO_2 uptake.

This chapter outlines the fundamental aspects of stomatal physiology, their occurrence in plants, morphology, response to environmental factors and mechanics of operation, including a description of the various control loops illustrated in Fig. 6.1. The role of the stomata in the control of photosynthesis and of water loss is discussed in more detail in Chapters 7 and 10.

Further information on stomata, their responses and mechanism of operation, may be found in texts by Jarvis & Mansfield (1981), Willmer (1983), Zeiger *et al.* (1987) and Weyers & Meidner (1990), and in the review by Cowan (1977).

Distribution of stomata

True stomata are distinguished from other epidermal pores such as hydathodes, lenticels and the pores in the thalli of some liverworts by their marked capacity for opening and closing movements. These changes in aperture depend on alterations in the size and shape of specialised

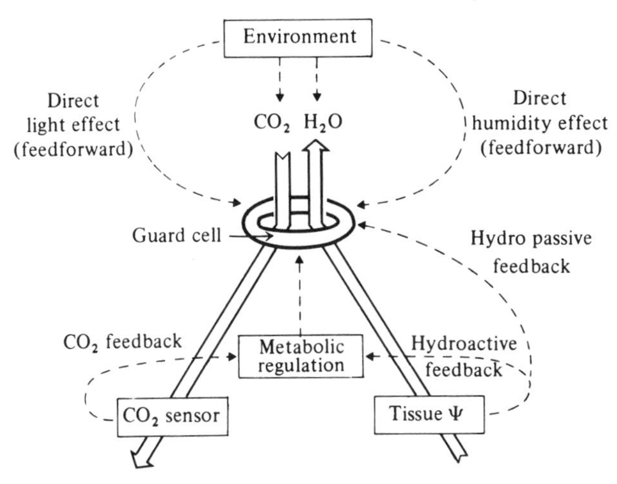

Fig. 6.1. Simplified diagram illustrating the role of stomata in regulating CO_2 and H_2O fluxes, showing the feedback and feedforward control pathways (dashed lines). For details see text. (Modified after Raschke 1975).

epidermal cells, the guard cells (Fig. 6.2). Stomata are present in the aerial parts of practically all the land flora, being found in the sporophytes of mosses, in ferns and in both gymnosperms and angiosperms. Although they are most frequent on leaves, they also occur in other green tissues such as stems, fruits and parts of inflorescences (e.g. awns of grasses and sepals of angiosperms). They tend to be most frequent on the lower surface of plant leaves, while in some species, especially trees, they occur only on the lower epidermis. Leaves with stomata on both sides are called amphistomatous, and those with stomata restricted to the lower epidermis are hypostomatous.

The two main types of stomata found in higher plants (Fig. 6.2) are (*a*) the elliptical type and (*b*) the graminaceous type, which is found in the Gramineae and Cyperaceae and has distinctive dumbbell-shaped guard cells arranged in rows. In many species, the stomata have ante-chambers outside the pore or special protective structures such as outer lips or even membrane 'chimneys' (see Fig. 6.2*d*–*h*), or else they are partially occluded by wax. All these features increase the effective diffusive resistance of the pore.

Representative examples of dimensions and frequencies of stomata in different species and on different leaves are presented in Table 6.1. It is clear that frequency and size vary as a function of leaf position and growth conditions. Even within one species there may be a large genetic component of the variation between different cultivars or ecotypes.

Table 6.1. *Examples showing the range of values for stomatal frequency* (ν, mm^{-2}) *and pore length* (ℓ, μm) *for different species, leaf position and growing conditions* (– *no data*)

	Adaxial surface		Abaxial surface		
	ν	ℓ	ν	ℓ	Ref.
Trees:					
Carpinus betulus	0	—	170	13	1
Malus pumila (cv. Cox)	0	—	390	21	2
Malus pumila – variation during season	0	—	230–430	—	2
Malus pumila – variation within leaf	0	—	170–360	—	2
Malus pumila – different cultivars	0	—	350–600	—	3
Pinus sylvestris	120	20	120	20	1
Picea pungens	39	12	39	12	1
Other dicots:					
Beta vulgaris	111	14.6	131	15.3	4
Tomato – low–high light	2–28	—	83–105	—	5
Soybean – range for 43 cultivars	81–174	21–23	242–385	19.5–21.7	6
Soybean – well-watered–stressed	149–158	—	357–418	—	6
Ricinus communis	182	12	270	24	1
Tradescantia virginiana	7	49	23	52	1
Grasses:					
Sorghum bicolor – mean of 6 cultivars	—	22.6	135	23	7
Hordeum vulgare – flag leaf	54–98	17–24	60–89	17	1, 8, 9
Hordeum vulgare – 5th leaf below flag	—	—	27–42	—	9

References: 1. Meidner & Mansfield 1968; 2. Slack 1974; 3. Beakbane & Mujamder 1975; 4. Brown & Rosenberg 1970; 5. Gay & Hurd 1975; 6. Ciha & Brun 1975; 7. Liang *et al.* 1975; 8. Miskin & Rasmusson 1970; 9. Jones 1977*b*.

Stomatal mechanics

Stomatal movements depend on changes in turgor pressure inside the guard cells and in the adjacent epidermal cells (which are sometimes modified to form distinct subsidiary cells). The changes in turgor can result either from

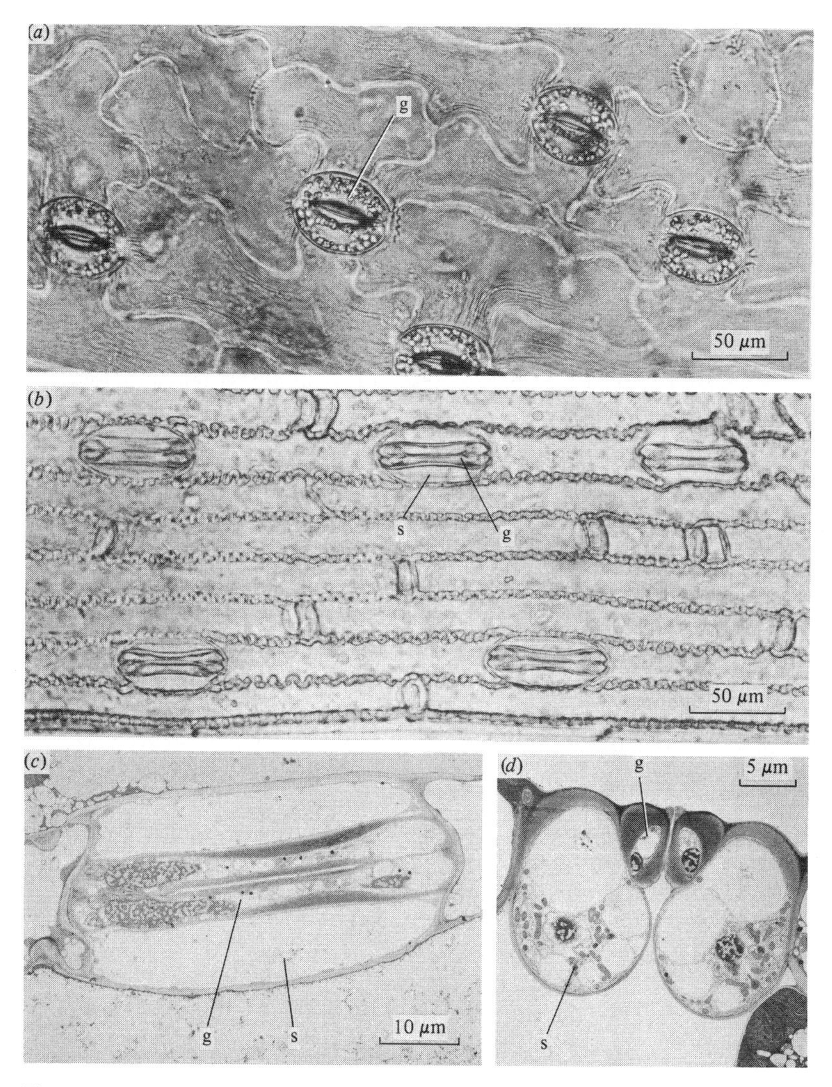

Fig. 6.2(*a*)–(*d*). For legend see opposite.

a change in total water potential (Ψ) of the guard cells as the supply or loss of water changes, or from active changes in osmotic potential (Ψ_π). The former mechanism, relying on changes *outside* the guard cells has been termed 'hydropassive' (Stålfelt 1955), while the latter has been termed 'hydroactive'. Both involve movement of water into or out of the guard cells.

Changes in guard cell turgor cause alteration in pore aperture as a

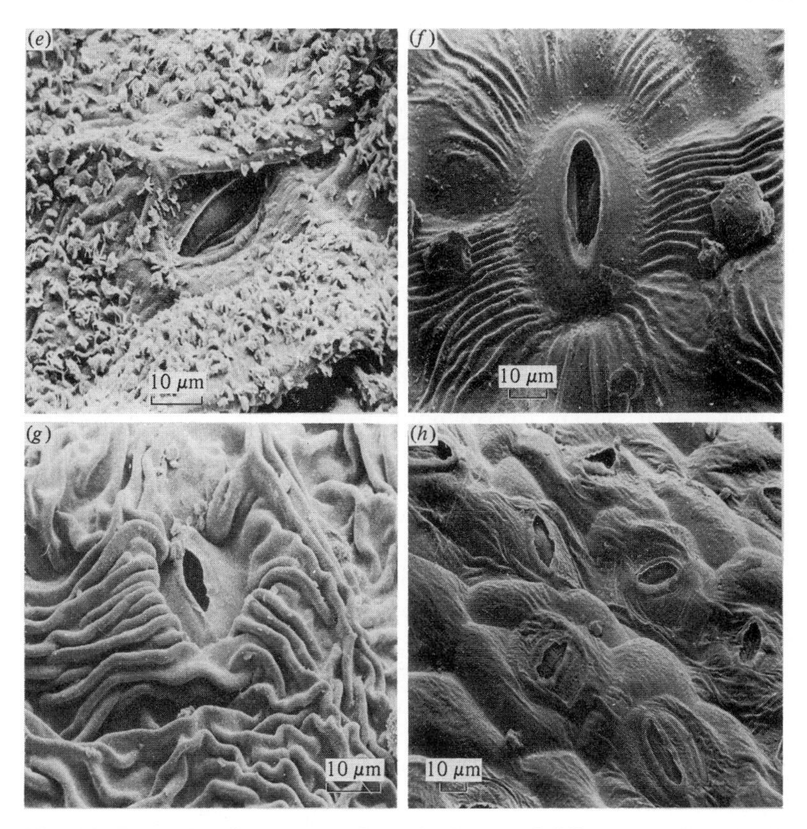

Fig. 6.2. Representative stomata from the leaves of different species: (*a*) elliptical type (abaxial surface of *Gloriosa superba*); (*b*) graminaceous type (abaxial surface of wheat, *Triticum aestivum*) (light micrographs of stripped epidermis by M. Brookfield); (*c*) and (*d*) paradermal and transverse sections, respectively, of wheat stomata showing the thickened guard cells (g) and the subsidiary cells (s) (courtesy of Dr M. L. Parker, Plant Breeding Institute, Cambridge). (*e*) to (*h*) Scanning electron micrographs of lower epidermes of some arid zone species (courtesy of Dr B. M. Joshi, University of Jodhpur) showing surface structure: (*e*) *Acacia senegal* (note the wax platelets); (*f*) *Echinops echinatus*; (*g*) *Tribulus terrestris*; (*h*) *Chorchorus tridens*.

consequence of the specialised structure and geometry of the stomatal complex. Two significant features of the commonest types of stomata are the presence of inelastic radially oriented micellae in the cell walls and a markedly thickened ventral wall (adjacent to the pore). Although it appears that neither of these characters is essential to the operation of elliptical stomata, they do influence the details of their movement (see Sharpe *et al.* 1987 for a detailed analysis of stomatal mechanics). At least for elliptical stomata, movement probably involves deformation of the guard cells out of the plane of the epidermis, though some bulging into the subsidiary cells may also occur. The changes in guard cell volume that occur during opening are not well documented but, at least in *Vicia faba*, anatomical studies indicate that the lumen volume may double when opening from closed to an aperture of 18 μm (Raschke 1975). Stomatal apertures are approximately linearly related both to guard cell volume and, more approximately, to guard cell turgor pressure.

An important feature of stomatal operation is the role of the subsidiary cells. Analysis of stomatal mechanics has shown that in many cases the subsidiary cells have a mechanical advantage over the guard cells, such that equal increases in pressure in guard and subsidiary cells cause some closure. This implies that closure cannot normally occur as a simple hydraulic response to declining bulk leaf water status, and that all stomatal movements normally result from an active process. This conclusion is supported by the well-known observation of transient stomatal opening on excising leaves (Iwanoff 1928), which could partly be explained in terms of this mechanical advantage as bulk leaf turgor falls. The antagonism between guard and subsidiary cells has been recognised for over a century (e.g. von Mohl 1856).

Stomatal guard cells generally contain chloroplasts, though they are commonly less frequent, smaller and of different morphology to those in mesophyll cells. In spite of this widespread occurrence of chloroplasts in guard cells, however, there is little evidence that photosynthetic carbon reduction occurs or is involved in stomatal opening. Nevertheless it is likely that photophosphorylation and $NADP^+$ reduction provide energy for stomatal opening. Certain orchids (*Paphiopedilum* spp.) are exceptional in that their guard cells lack chlorophyll yet their stomata are functional (at least in intact leaves).

As we have seen, most stomatal movements, including those in response to changes in water status, involve active changes in guard cell osmotic potential. A general feature of active stomatal opening movements is that a large proportion of the osmotic material consists of potassium ions. In *Vicia faba*, for example, the K content increases from about 0.3 to 2.4 pmol per guard cell (90×10^{-6} to 680×10^{-6} mol m^{-3}) as stomata open (MacRobbie 1987). It is now generally accepted that this K^+ uptake is driven by ATP-powered primary proton extrusion at the plasmalemma. This sets up an electrical driving force for cation entry and a pH gradient (cytoplasm alkaline). The pH gradient may be dissipated by synthesis of malic acid in the cytoplasm or by chloride uptake by co-transport with protons. The malate is generated within the guard cells from storage carbohydrates such as starch, though in *Allium* species (which lack starch in their guard cells), Cl^- provides the counterion for K^+.

It has been estimated for *Vicia faba* (Allaway 1973; Allaway & Hsiao 1973) that an increase in pore width of 10 μm requires a 1.25 MPa decrease in osmotic potential, which can be largely accounted for by the increase in potassium malate. The well-known decrease in starch content that occurs in the guard cells of many species as stomata open may provide both organic anions and energy for the ion pumps. Stomatal closure is also usually an active metabolic process and does not just rely on ion leakage.

There is now strong circumstantial evidence that the plant growth regulator abscisic acid (ABA) is involved in regulating stomatal responses, especially those involving water stress. For example, externally applied ABA closes stomata, levels of endogenous ABA increase rapidly in stress (often in parallel to stomatal closure), and mutants that are deficient in the capacity for ABA synthesis (for example *sitiens* and *flacca* in tomato, and *droopy* in potato) are not able to close their stomata, though the lesion can be reverted by the supply of exogenous ABA (see e.g. Addicott 1983). Although there are several reports (e.g. Henson 1981) that endogenous ABA concentrations do not always correlate well with stomatal aperture, particularly during recovery from stress, it is possible that these observations may be explained in terms of compartmentation into physiologically active and inactive pools, or that other compounds are involved. The inhibition of stomatal opening by ABA requires the presence of calcium ions; this observation, taken together with evidence that two groups of compounds (calcium channel blockers and calmodulin antagonists) that are known to interfere with the action of calcium as a 'second messenger', also reduce the ability of stomata response to ABA, suggests that calcium may be involved in stomatal responses to ABA (de Silva *et al.* 1985; Davies & Jones 1991).

Methods of study

The stomatal pathway and the corresponding resistance (r_s) or conductance (g_s) to mass transfer is only one component of the total leaf resistance (r_ℓ) (Fig. 6.3). The cuticular transfer pathway (r_c) is in parallel with the stomata, while there is also a transfer resistance within the intercellular spaces (r_i) and, in series with that, a possible wall resistance (r_w) at the surface of the mesophyll cells. Many methods for determining r_s include these other resistances so more correctly provide a measure of r_ℓ; furthermore they often include a boundary layer component. Throughout this book r_ℓ will be used as a measure of r_s, even though these terms are not exactly equivalent. The relative values of stomatal and other components of r_ℓ are discussed in detail below. It follows from discussion in Chapter 3 that stomatal resistances for different gases are inversely proportional to their diffusion coefficients.

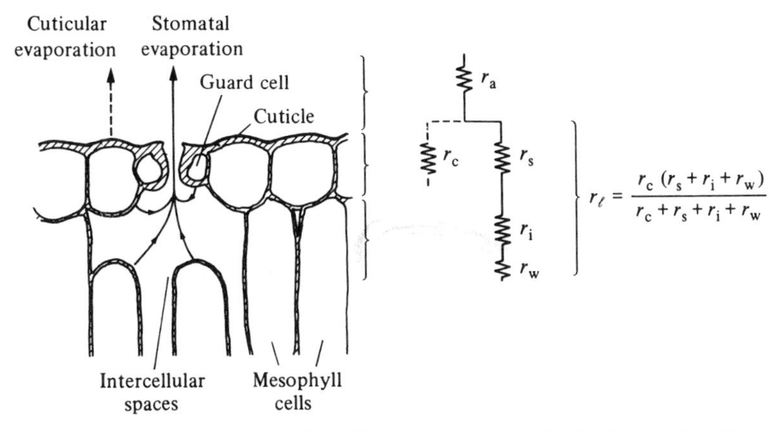

Fig. 6.3. Pathways for water loss from one surface of a leaf, showing the boundary layer (r_a), cuticular (r_c), stomatal (r_s), intercellular space (r_i), wall (r_w), and leaf (r_ℓ) resistances. The total leaf resistance is the parallel sum of r_ℓ for upper and lower surfaces.

Although the most physiologically meaningful measure of stomatal functioning is the diffusion resistance (or conductance), techniques that measure other parameters are widely used (Weyers & Meidner 1990). Important techniques include the following.

Microscopic measurement

Anatomical measurements *in vivo*, as well as on fixed and cleared leaves or epidermal imprints, allow one to quantify the size of the stomatal complexes and their frequency on leaf surfaces. Much of our current knowledge of stomatal physiology derives from studies using isolated epidermal strips where changes in pore size can be followed microscopically. A convenient method for obtaining epidermal imprints is to spread a solution such as nail varnish over the epidermis, allow it to dry, peel it off and store in small envelopes for subsequent microscopic examination. Although useful for counting stomata and for measuring their dimensions, it is difficult to determine apertures from imprints as the impression material may not rupture at the narrowest part of the pore. It is always necessary to demonstrate that microscopic measurements, whether on living or prepared material, isolated epidermes or replicas, are representative of the situation *in vivo*, though it has been suggested that accurate measurements can be obtained using low-temperature scanning electron microscopy where ultra-rapid cryofixation can avoid aperture changes (van Gardingen *et al.* 1989).

Anatomical measurements may be used to derive estimates of the stomatal diffusion resistance (r_s) using diffusion theory (see e.g. Penman & Schofield 1951). For the simplest case of a cylinder or other shape of constant cross-section, the resistance of the pore ($s\,m^{-1}$) is given by equation 3.21 as

$$r_s = \ell / \boldsymbol{D} \quad (6.1)$$

where ℓ is the length of the tube and $\boldsymbol{D}$ is the diffusion coefficient.

This resistance per unit pore area can be converted to a resistance per unit area of leaf by dividing by the ratio of average pore cross-sectional area to leaf surface area. For circular pores this ratio in $v\pi r^2$, where v is the frequency of stomata per unit leaf area and r is the pore radius. This gives the stomatal resistance on a leaf area basis as

$$r_s = \ell / v\pi r^2 \boldsymbol{D} \quad (6.2)$$

Because the pore area is much less than the leaf area there is a zone close to the pore where the lines of flux converge on the pore. Therefore this zone of the boundary layer is not effectively utilised for diffusion and there is an extra resistance or 'end-effect' associated with each pore. The magnitude of this extra resistance can be derived from three-dimensional diffusion theory as being proportional to pore radius and equal to $\pi r / 4\boldsymbol{D}$. Therefore converting to a leaf area basis and adding this extra resistance to the pore resistance gives

$$r_s = [\ell + (\pi r/4)] / v\pi r^2 \boldsymbol{D} \quad (6.3)$$

This approach can be extended to other shaped pores and to the estimation of the small intercellular space resistance (Fig. 6.3). It is also possible to correct for alterations in $\boldsymbol{D}$ that occur when the pore size is small, and for interactions with other diffusing species (see Jarman 1974), but these effects are small and usually neglected.

Equations 6.1–6.3 can readily be converted to obtain molar resistances if we remember that the molar resistance ($r^m = r$) can be obtained from r by multiplying by ($\mathcal{R}T/P$) (see equation 3.23*b*).

Infiltration

Graded solutions of differing viscosity (e.g. various mixtures of liquid paraffin and kerosene) have been widely used to estimate relative stomatal opening (see Hack 1974). The viscosity of the solution that just infiltrates the pores provides a measure of aperture. Because of differences in anatomy of pores in different species and differences in cuticular composition, this approach is only of use for studying qualitative differences within one species.

Viscous-flow porometers

The mass flow of air through stomata under a pressure gradient has been widely used as a measure of stomatal aperture since the first viscous-flow 'porometer' was constructed by Darwin & Pertz in 1911. Air may be forced through the stomata in either of the ways illustrated in Fig. 6.4 and the viscous-flow resistance derived from the measured flow rate or the rate of pressure change (see Meidner & Mansfield 1968). The mass flow rate is inversely proportional to the viscous-flow resistance. This resistance in turn depends on the apertures of the stomata, which constitute the major resistance, and to a lesser extent on resistance to mass flow through the intercellular spaces in the leaf mesophyll (see Fig. 6.4).

Unfortunately there are difficulties in absolute calibration because of the complexity of the flow paths. Furthermore the resistance obtained with trans-leaf flow porometers (Fig. 6.4*a*) is the *sum* of the resistances of the upper and lower surfaces, since they are in series, and is dominated by the larger (usually that of the upper epidermis). However, for diffusive exchange of CO_2 or water vapour, the upper and lower pathways are in parallel, so that the diffusive resistance is largely dependent on the smaller of the two resistances (usually the lower epidermis). Results obtained with viscous-flow porometers must, therefore, be interpreted with care. Nevertheless viscous-flow porometers are particularly useful for continuous recording (a typical example is illustrated in Fig. 6.4*c*). Theoretical and experimental results indicate that diffusive resistances are proportional to the square root of the viscous-flow resistance.

Diffusion porometers

These instruments measure diffusive transfer and are therefore the most relevant and useful for studies of leaf gas exchange. The several trans-leaf diffusion porometers that measure diffusion of gases such as hydrogen, argon or nitrous oxide (see Meidner & Mansfield 1968) are, however, subject to many of the criticisms of the trans-leaf viscous-flow porometers and are little used nowadays. Currently the designs of most of the widely used instruments are based on the measurement of the rate of diffusive water loss from plant leaves. As such they all measure the leaf resistance to water vapour (including any cuticular component) together with any boundary layer resistance in the porometer chamber.

The leaf resistance may be obtained for any sample by subtracting the

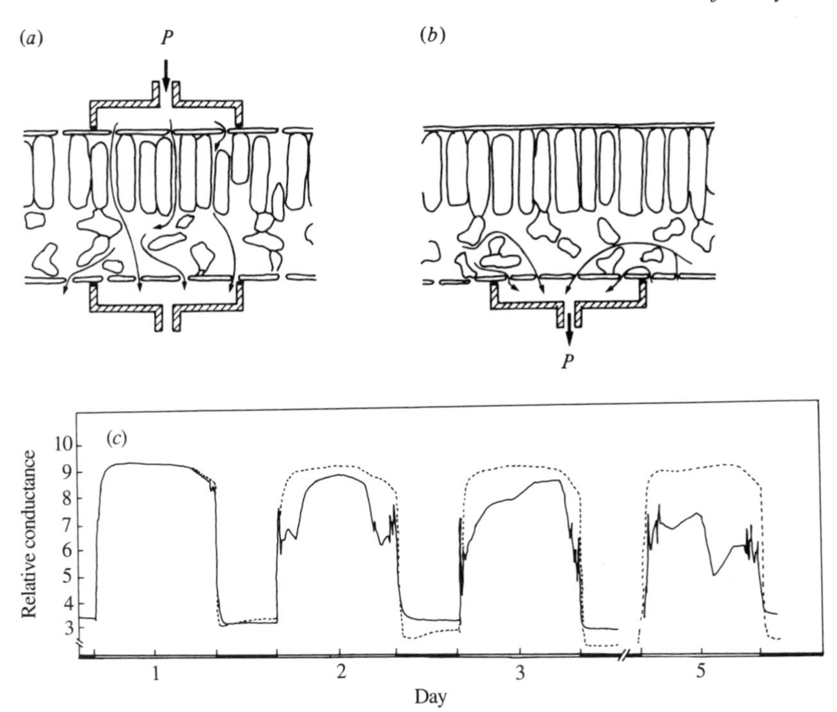

Fig. 6.4. Mass-flow (or viscous-flow) porometers: (*a*) trans-leaf type, (*b*) one-cup type. The applied pressure (*P*) may be above or below atmospheric, and either the flow rate or the rate of change of *P* measured. (*c*) Typical viscous-flow porometer traces for a control leaf of *Helianthus* (dashed line) and for a corresponding leaf on a plant whose roots had been placed in moist air near the start of day 1 (solid line) (data from Neales *et al.* 1989).

chamber boundary layer resistance (a chamber constant usually determined by using water-saturated blotting paper in place of the leaf as described in Appendix 8) from the total resistance observed. The value of this chamber boundary layer resistance may be minimised by stirring the air in the chamber with a small fan.

(i) *Transit-time instruments*. The principle of this type of instrument is that when a leaf is enclosed in a sealed chamber, evaporation will tend to increase the humidity in the chamber at a rate dependent, among other things, on the stomatal diffusion resistance. The time taken for the humidity to increase over a fixed interval can be converted to resistance by the use of a previously obtained calibration curve. Calibration involves replacing the leaf by a wet surface (e.g. wet blotting paper) covered with a calibration plate or microporous membrane (e.g. Celgard 2400) that has a known

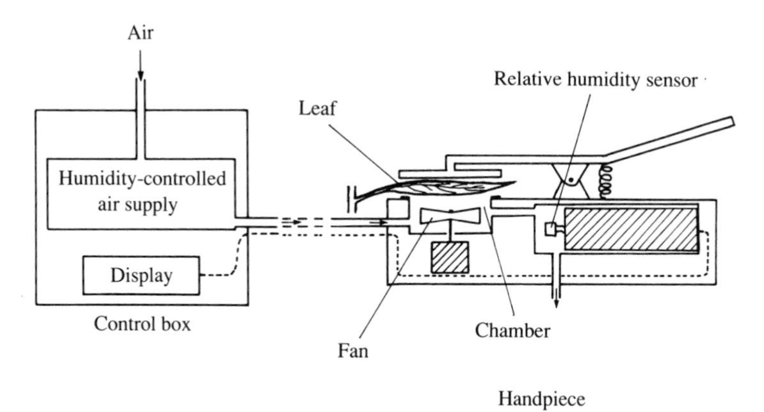

Fig. 6.5. Flow diagram of a typical continuous-flow porometer.

diffusion resistance. Calibration plates with a range of resistances are obtained by varying the number and size of precision drilled holes, with the resistance for any plate being obtained from theory (equation 6.3).

There are several designs of transit-time porometer, many of which are available commercially. Different instruments may include stirring in the chamber to minimise the chamber boundary layer resistance and may be used on irregular leaves or even conifer needles, while others include automatic timing, automatic purging of the chamber with dry air between measurements, or the facility for changing the humidity range over which they operate so as to mimic natural conditions as closely as possible, together with microprocessors to calculate and store data as necessary. In practice the main source of error occurs when leaf temperature (T_ℓ) is not the same as the cup temperature. In such cases porometer estimates of g_ℓ may underestimate the real differences between droughted and well-watered plants (Meyer *et al.* 1985).

(ii) *Continuous-flow* (= *steady-state*) *porometers.* The principle of this equipment has long been used in laboratory gas exchange equipment and in recent years has been adapted to field instruments (Fig. 6.5). When a leaf is enclosed in a chamber through which air is flowing, the evaporation rate into the cuvette (in molar units) can be calculated approximately from the product of the flow entering the chamber and the concentration differential across the chamber as

$$\mathbf{E}^{\mathrm{m}} = u_{\mathrm{e}}(e_{\mathrm{o}} - e_{\mathrm{e}})/(PA) = u_{\mathrm{e}}(x_{\mathrm{Wo}} - x_{\mathrm{We}})/A \tag{6.4}$$

where $\mathbf{E}^{\mathrm{m}}$ is the evaporation rate (mol m^{-2} s^{-1}), u is the molar flow rate (mol s^{-1}), x_{W} is the mole fraction of water vapour (mol mol^{-1}), A is the leaf surface area (m^2) exposed in the chamber, and the subscripts 'o' and 'e' refer to the outlet and inlet flows respectively. This equation is only approximate because the water transpired by the leaf changes the flow rate across the cuvette, so it is better to use the full mass balance

$$\mathbf{E}^{\mathrm{m}} = (u_o x_{\mathrm{Wo}} - u_e x_{\mathrm{We}})/A \tag{6.5}$$

The difference between u_o and u_e equals the evaporation rate (the CO_2 exchange by the leaf may be ignored because it is comparatively small and is largely balanced by O_2 exchange) so that one can write

$$u_o = u_e + (u_o x_{\mathrm{Wo}} - u_e x_{\mathrm{We}}) \tag{6.6a}$$

which rearranges to

$$u_o = u_e \frac{(1 - x_{\mathrm{We}})}{(1 - x_{\mathrm{Wo}})} \tag{6.6b}$$

which on substituting back into equation 6.5 gives

$$\mathbf{E}^{\mathrm{m}} = \frac{u_e(x_{\mathrm{Wo}} - x_{\mathrm{We}})}{A(1 - x_{\mathrm{Wo}})} \tag{6.7}$$

The term $(1 - x_{\mathrm{Wo}})$ in the denominator of equation 6.7 adjusts the flow rate for the amount added by the transpiration from the leaf, though for typical conditions this leads to a correction of only 2–4%. Equivalent equations can be written for **E** (expressed as a mass flux density) using volume flow rates and concentrations (see Chapter 3).

The evaporation rate is related to the total resistance to water vapour loss $(r_{\ell\mathrm{W}} + r_{\mathrm{aW}})$, where r_{aW} is the chamber boundary layer resistance, by

$$\mathbf{E}^{\mathrm{m}} = \frac{(x_{\mathrm{W}s} - x_{\mathrm{Wo}})}{(r_{\ell\mathrm{W}} + r_{\mathrm{aW}})} \tag{6.8}$$

where the water vapour pressure at the evaporating sites within the leaf is assumed equal to the saturation vapour pressure at leaf temperature $(e_{s(T\ell)})$ and $x_{\mathrm{W}s}$ is the corresponding mole fraction, and x_{Wo} is assumed to be representative of the mole fraction of water vapour in the chamber air.

Eliminating $\mathbf{E}^{\mathrm{m}}$ from equations 6.7 and 6.8 leads to

$$(r_{\ell\mathrm{W}} + r_{\mathrm{aW}}) = g_{\mathrm{W}}^{-1} = \frac{A(x_{\mathrm{W}s} - x_{\mathrm{Wo}})(1 - x_{\mathrm{Wo}})}{u_e(x_{\mathrm{Wo}} - x_{\mathrm{We}})} \tag{6.9}$$

Application of this equation can be greatly simplified if air temperature equals leaf temperature (i.e. the system is isothermal) and if dry air is input to the chamber (i.e. $x_{\mathrm{We}} = 0$). Making the approximation that the term $(1 - x_{\mathrm{Wo}}) \simeq 1$, this equation simplifies to

$$r_{\ell\mathrm{W}} = 1/g_{\ell\mathrm{W}} = \{[(1/h) - 1]\, A/u_e\} - r_{\mathrm{aW}} \tag{6.10}$$

where h is the relative humidity of the outlet air $(= x_{\mathrm{Wo}}/x_{\mathrm{W}s})$.

Unfortunately equations 6.9 and 6.10, though adequate for many purposes, do not take account of the fact that the total transpiration from the leaf is made up of a diffusive component given by equation 6.8 and an additional mass flow equal to the mean water vapour mole fraction along the diffusion pathway from the intercellular spaces times the evaporation rate. Adding this correction to equation 6.8 gives (von Caemmerer & Farquhar 1981)

$$\mathbf{E}^{\mathrm{m}} = \left\{\frac{x_{\mathrm{W}s} - x_{\mathrm{Wo}}}{r_{\ell\mathrm{W}} + r_{\mathrm{aW}}}\right\} + \mathbf{E}^{\mathrm{m}}\left\{\frac{x_{\mathrm{W}s} + x_{\mathrm{Wo}}}{2}\right\} \tag{6.11}$$

Rearranging equation 6.11, and substituting from equation 6.7 gives the following complete expression for the total resistance or conductance to water vapour

$$g_{\mathrm{W}} = (r_{\ell\mathrm{W}} + r_{\mathrm{aW}})^{-1} = \frac{u_{\mathrm{e}}(x_{\mathrm{Wo}} - x_{\mathrm{We}})\{1 - (x_{\mathrm{W}s} + x_{\mathrm{Wo}})/2\}}{A(1 - x_{\mathrm{Wo}})(x_{\mathrm{W}s} - x_{\mathrm{Wo}})} \tag{6.12}$$

The correction involved is normally of the order of 2–4 %, so is only important for accurate work, especially where one is attempting to calculate an intercellular space CO_2 concentration (see Chapter 7).

Estimation of leaf temperature in porometers. Accurate estimation of g_ℓ is dependent on precise estimation of T_ℓ. Although it is common practice to estimate T_ℓ in leaf chambers and porometers by means of thermocouples appressed to the leaf surface, these tend to lead to an estimate of T_ℓ intermediate between the true value and T_a, with consequent errors in g_ℓ. An alternative approach is to estimate the leaf–air temperature difference by means of the leaf energy balance (Parkinson & Day 1980).

Figure 6.5 shows a typical flow diagram of a continuous-flow porometer. These instruments may be used with a constant flow rate (e.g. Parkinson & Legg 1972; Day 1977), in which case the relative humidity of the outlet air is uniquely related to r_ℓ (irrespective of temperature). Alternatively the instruments may be operated in a null-balance mode where the flow rate is adjusted to give a particular relative humidity (see Beardsell *et al.* 1972). Because of stomatal sensitivity to ambient humidity (see below), it is preferable to operate with the chamber humidity close to ambient, whichever mode of operation is used.

Continuous-flow porometers provide the best method currently available for rapid but accurate studies of stomatal conductance in the field. They can vary in sophistication from simple instruments having only an air supply, a leaf chamber and a humidity sensor (Fig. 6.5) to instruments that include temperature measurement and control and readout in conductance units. One advantage over transit-time instruments is that calibration only involves calibration of the humidity sensor and flow meter and one does not need to take account of actual temperature (though see Parkinson & Day

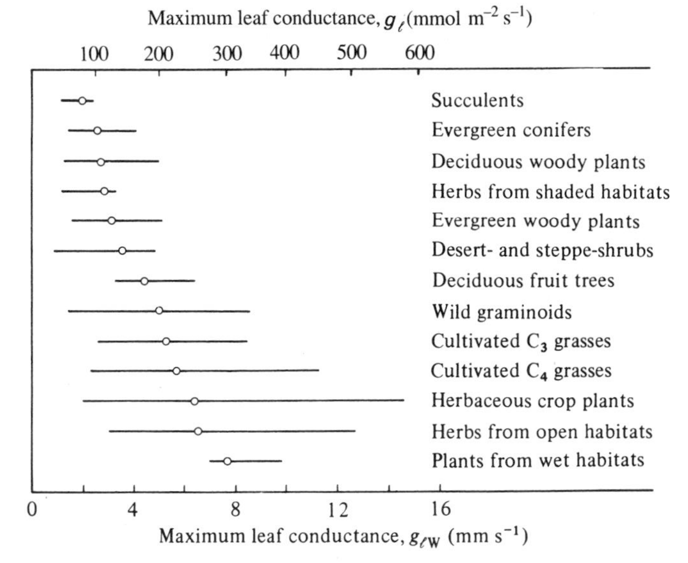

Fig. 6.6. Maximum leaf conductance ($g_{\ell W}$) in different groups of plants. The lines cover about 90% of individual values reported. The open circles represent group average conductances. (Adapted from Körner *et al.* 1979).

1980) so long as a relative humidity sensor is used. Equally important is the fact that continuous-flow porometers have similar sensitivity over a wide range of g_ℓ. This is in contrast to transit-time instruments that are relatively insensitive at the physiologically important high values of g_ℓ.

Stomatal response to environment

Maximum conductance

The great variability in stomatal frequency and size that exists between different species, leaf position or growth condition (Table 6.1) leads us to expect corresponding differences in stomatal conductance. Figure 6.6 summarises a large number of reported measurements of maximum leaf conductance that have been reported for different groups of plants. Although the range of values found within any group is very wide, there are some clear-cut differences with maximum conductances ($g_{\ell W}$ on a total leaf surface area basis) averaging less than 80 mmol m^{-2} s^{-1} (2 mm s^{-1}) for succulents and evergreen conifers and four times that value for plants from wet habitats.

In addition to genotypic differences, maximum leaf conductance is strongly affected by growth conditions and changes with leaf age.

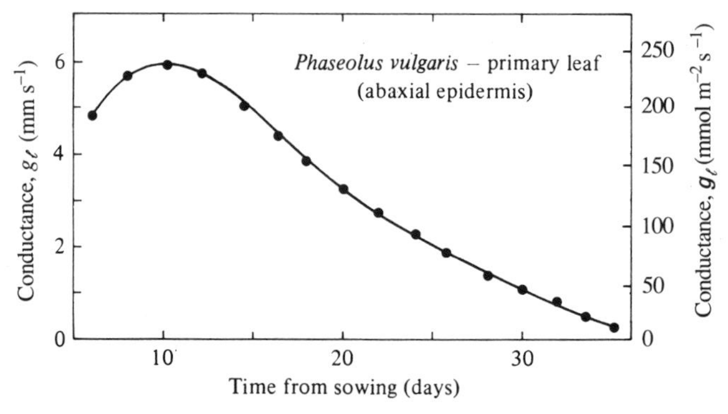

Fig. 6.7. Typical trend of leaf conductance (at 1200 μmol m^{-2} s^{-1}) for a single leaf from early expansion to final senescence. (After Solárová 1980.)

Characteristically, maximum conductance does not attain a peak value until several days after leaf emergence, it may then stay close to this value for a time that is characteristic of the species, finally declining to a very low value as the leaf senesces (Fig. 6.7).

Stomatal response to environment

The effects of individual factors such as radiation, temperature, humidity or leaf water status on stomatal conductance can be studied best in controlled environments or leaf chambers where each factor may be varied independently. However, the use of this information to predict the stomatal conductance in a natural environment is complicated by various factors including (i) interactions between the responses (that is any response depends on the level of other factors), (ii) variability of the natural environment, (iii) the fact that the stomatal time response is frequently of the same order or longer than that of changes in the environment (therefore stomata rarely reach the appropriate steady-state aperture), (iv) in species with amphistomatous leaves the stomata on the upper surface tend to be more responsive than those on the lower surface, and (v) endogenous rhythms tend to affect stomatal aperture independently of the current environment (e.g. night-time closure tends to occur even in continuous light).

Environmental factors affecting stomatal aperture include the following:

(*a*) *Light*. Perhaps the most consistent and well-documented stomatal response is the opening that occurs in most species as irradiance increases (Fig. 6.8*a*). Maximum aperture is usually achieved with irradiances greater than about a quarter of full summer sun (i.e. about 200 W m^{-2} (total

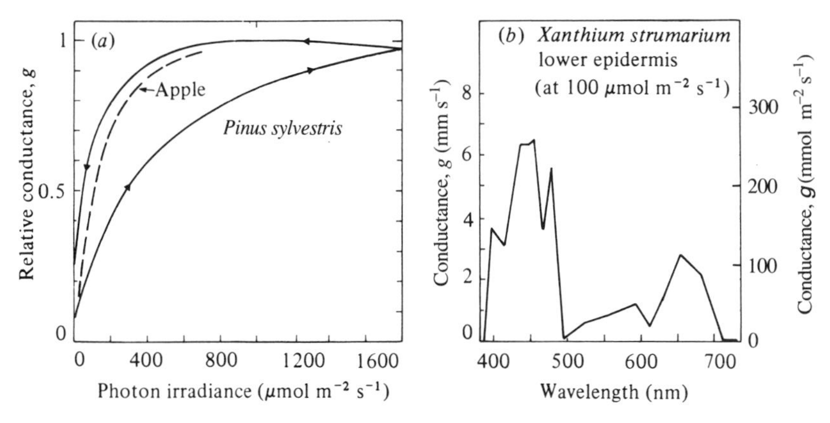

Fig. 6.8. (*a*) Examples of stomatal light response curves. The curve for apple (data from Warrit *et al.* 1980) is typical for many species, approaching a maximum at about a quarter full sunlight. The curves for *Pinus sylvestris* (after Ng & Jarvis 1980) illustrate the hysteresis that can occur in certain conditions. (*b*) Action spectrum of stomatal opening in *Xanthium strumarium* (calculated from Sharkey & Raschke 1981). Action is represented as the conductance achieved at each wavelength for a photon irradiance of 100 μmol m^{-2} s^{-1}.

shortwave) or 400 μmol m^{-2} s^{-1} (PAR)), though this value depends on species and on the natural radiation environment: stomata on shade-grown leaves open at lower light levels than do those on sun-adapted leaves. The conductance–irradiance relationship often shows hysteresis (Fig. 6.8*a*), particularly if time is not allowed for complete equilibrium when the light is altered. Although some of the stomatal response to light may be indirect and attributable to the decrease in intercellular CO_2 concentration that occurs on illumination (see below), there is strong evidence for a direct light response involving two independent photoreceptors. Stomata are particularly sensitive to blue light (Fig. 6.8*b*), a response that probably involves a flavin photoreceptor. The rather smaller response to red light is important at high irradiances and probably involves chlorophyll. Evidence for this comes from the observation that sensitivity to blue light but not to red is found in the white part of variegated leaves.

The rate of stomatal response to changing light is variable, though closing responses tend to be more rapid than opening. Half-times are generally of the order of 2–5 min, which are of the same order as environmental changes. There is some evidence that stomatal closure in response to decreasing light can be potentiated by water stress. An example is shown for sorghum leaves in Fig. 6.9 where mild water deficits decreased the half-time for closure to less than 1 min. The kinetics of response are

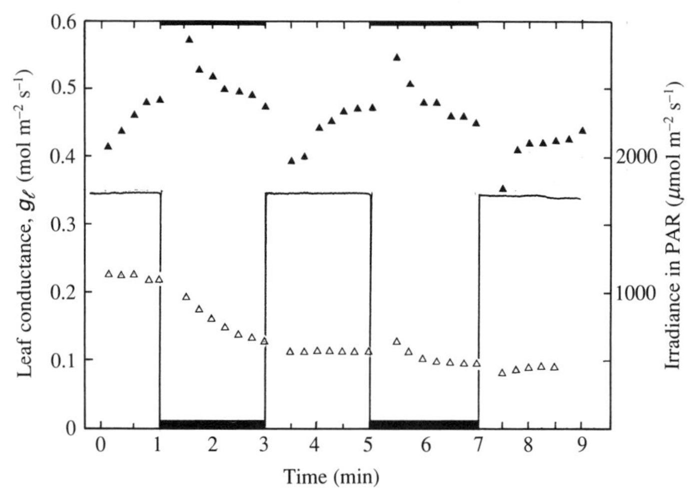

Fig. 6.9. The time course of stomatal conductance for leaves of field-grown *Sorghum bicolor* in response to sudden darkening for well-watered plants (▲) or for plants subject to mild drought (△); irradiance in the PAR is indicated by the continuous line (H. G. Jones, D. O. Hall & J. E. Corlett, unpublished data). The transients obtained on changing irradiance are artefacts related to time for the porometer system to reach equilibrium.

likely to be particularly important when light is changing rapidly, as in sunflecks.

Although in most plants stomata open in response to light and close in the dark, the reverse is true in plants having the Crassulacean acid metabolism (CAM) pathway of photosynthesis (see Chapter 7). In these plants maximal opening is in the dark, particularly in the early part of the night period.

(*b*) *Carbon dioxide*. Although the CO_2 concentration in the natural environment is relatively constant (see Chapter 11), stomata are sensitive to CO_2, responding to the CO_2 mole fraction in the intercellular spaces (x_i'). In general stomata tend to open as x_i' decreases, with the sensitivity to CO_2 being strongly species and environment dependent, being greatest in C_4 species, and at concentrations below about 300 vpm (Morison 1987). Stomata respond to CO_2 in both light and dark, so the response cannot depend only on photosynthesis. The value of x_i' is maintained surprisingly constant (at approximately 130 vpm in C_3 species and 230 vpm in C_4 species), over a wide range of conditions and rates of photosynthesis (Wong *et al.* 1979). This would occur if stomatal conductance varied in proportion to assimilation rate, and has led to the suggestion that a signal from the

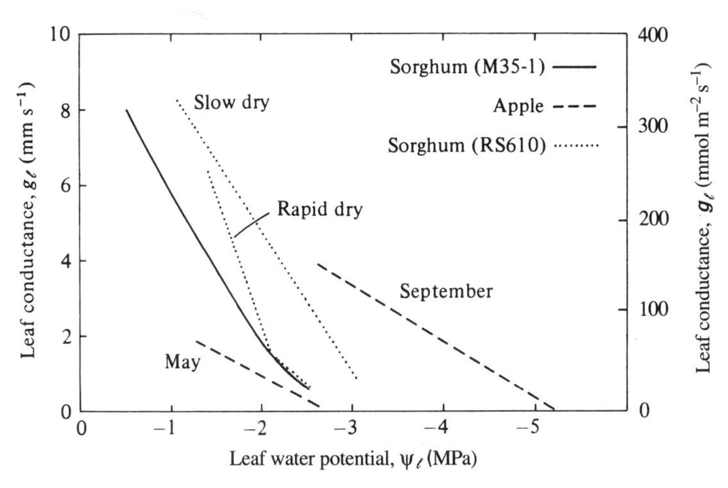

Fig. 6.10. Relationships between g_ℓ and Ψ_ℓ for apple (after Lakso 1979), and sorghum (data from Henzell *et al.* 1976, and Jones & Rawson 1979). Slow drying at 0.15 MPa day^{-1}, fast drying at 1.2 MPa day^{-1}.

mesophyll controls stomatal aperture, though mechanistic evidence for such a hypothesis is lacking and it seems more likely that this results more from a close matching of g_s and assimilation rate.

(*c*) *Water status.* Stomata are sensitive to leaf water status, tending to close with decreasing leaf water potential (Fig. 6.10). Closure occurs over a wide range of Ψ_ℓ, and this relationship can be modified by exposure to previous stress, or by the rate of desiccation. The effects of growth conditions on the Ψ_ℓ at which g_ℓ tends to zero are summarised for several species in Table 6.2. The degree of adjustment ranged from zero for *Hibiscus* to 3.6 MPa for *Heteropogon.* At the leaf water potentials that occur normally for well-watered plants during the course of a day, however, stomatal conductance is relatively insensitive to Ψ_ℓ, and may even *increase* with decreasing leaf water potential (see Fig. 6.14*d* and Jones 1985*b*); this response is what one would expect if stomatal conductance were controlling leaf water potential (through an altered transpiration rate), rather than the reverse and is discussed further below.

Although it is clear from studies with detached leaves that stomata can respond to leaf water status through locally mediated effects on active solute accumulation in the guard cells, there are several lines of evidence that stomatal closure in response to soil drying may often be controlled by other factors (see also Chapter 4). For example, when the leaf water status of *Helianthus annuus* and *Nerium oleander* plants was manipulated by modifying the whole-plant transpiration by changes in atmospheric

Table 6.2. *Some examples of stomatal adaptation to stress in different species*

		Ψ_ℓ at which g_ℓ tends to zero (MPa)			
Species	Condition	Maximum	Minimum	Adjust-ment	Ref.
Apple	Seasonal change, field	−2.7	−5.2	2.5	1
Heteropogon contortus	CE vs field	−1.4	−5.3	3.6	2
Eucalyptus socialis	Outdoor hardening	−2.5	−3.8	1.3	3
Cotton	CE vs field	−1.6	−2.7	1.1	4
Cotton	Drought cycles in CE	−2.8	−4.0	1.2	5
Sorghum	Field	−1.9	−2.3	0.4	6
Wheat	CE, solution culture	−1.4	−1.9	0.5	7
Hibiscus cannabinus	CE vs field	−2.1	−2.0	0	2

CE refers to controlled environment.
References: 1. Lakso 1979; 2. Ludlow 1980; 3. Collatz *et al.* 1976; 4. Jordan & Ritchie 1971; 5. Brown *et al.* 1976; 6. Turner *et al.* 1978; 7. Simmelsgaard 1976.

humidity, while maintaining fixed environmental conditions for an individual leaf whose conductance was monitored, it was found that leaf conductance of the monitored leaf was much more closely related to soil water status than to leaf water status (Fig. 6.11). The hypothesis that leaf conductance could respond directly to soil water status was tested further by pressurising the root system to maintain pressure in the xylem at zero (thus maintaining the leaf cells fully turgid), as soil was allowed to dry (Gollan *et al.* 1986). Again leaf conductance appeared to respond to soil water content, rather than to leaf turgor (Fig. 6.12).

Further evidence that soil water status may, on occasions, be more important than the water status of the leaf itself in controlling g_ℓ has been discussed in Chapter 4. It is commonly observed that stomatal conductance is better related to Ψ_p than to Ψ_ℓ, with at least some of the change in response to growth conditions (Table 6.2) being related to osmotic adjustment (see Chapter 10) where a given turgor pressure is achieved at a lower water potential as a result of solute accumulation in the cell sap. It is likely that the close relationship to bulk leaf Ψ_p, where it is observed, is indirect, acting through an effect on ion pumping at the guard cells.

When a plant is re-watered after a period of drought, the stomata may

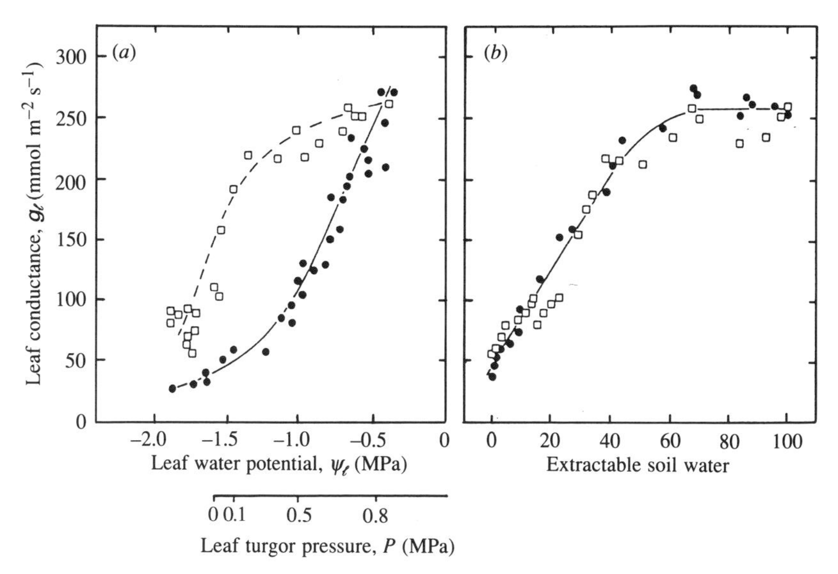

Fig. 6.11. (*a*) Relationship between leaf conductance and Ψ_ℓ in *Nerium oleander* as soil water content decreased and the leaf-to-air vapour concentration difference ($\delta x_{w\ell}$) was maintained at 10 Pa kPa^{-1} while the whole plant was maintained either in an atmosphere with a constant vapour concentration difference of 10 Pa kPa^{-1} (●) or where the concentration difference was as great as 30 Pa kPa^{-1} (□). (*b*) The data from (*a*) replotted as a function of extractable soil water. (Redrawn from Schulze 1986 with data from Gollan *et al.* 1985.)

take some days to recover (depending on the severity and duration of the stress), even though leaf water potential may recover rapidly.

(*d*) *Humidity*. Until the early 1970s it had been thought that stomata were insensitive to ambient humidity. For example Meidner & Mansfield (1968) in their text on stomatal physiology state '... stomatal behaviour ... is comparatively unaffected by changes in relative humidity of the ambient air'. It is now clear, however, that the stomata in many species close in response to increased leaf-to-air vapour pressure difference (δe_ℓ), as shown in Fig. 6.13. The magnitude of this response is dependent on species, growing conditions and particularly plant water status, the response being smaller at high temperature (Fig. 6.13) or in stressed plants.

Although the humidity response frequently takes minutes to reach full expression, it has been reported, at least for apple leaves, that as much as 90 % of the total stomatal response to an increased δe_ℓ can occur within 20 s of the change (Fanjul & Jones 1982). Opening responses tend to be rather slower.

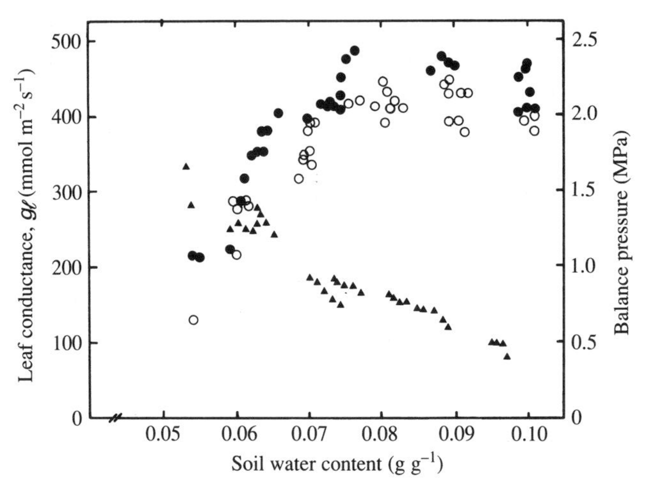

Fig. 6.12. Relationship between leaf conductance (g_ℓ) and soil water content for wheat leaves, when the soil was dried either without maintaining leaf turgor by pressurising the system (○) or with maintenance of leaf turgor by applying a balancing pressure (●). The corresponding values of the balancing pressure (P) required to maintain turgor at each soil water content are indicated by (▲). (Data from Gollan *et al.* 1986.)

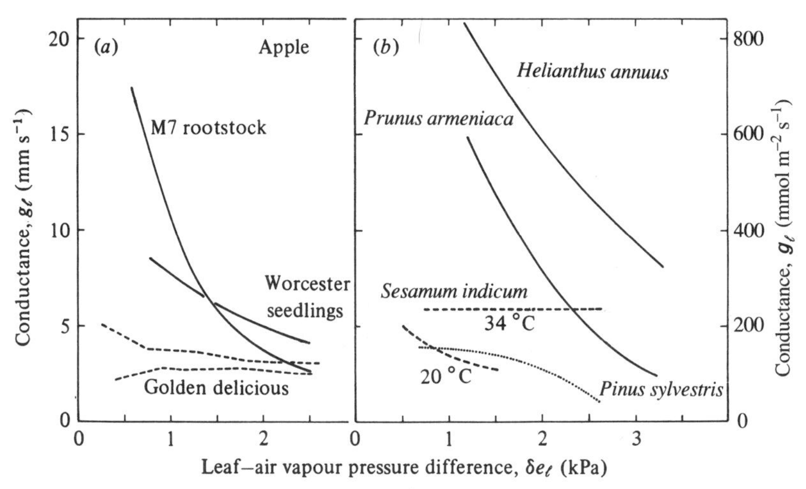

Fig. 6.13. Examples of stomatal humidity responses. (Data for apple from Warrit *et al.* 1980 and Fanjul & Jones 1982; *Helianthus* and *Sesamum* from Hall & Kaufmann 1975; *Prunus* from Schulze *et al.* 1972; and *Pinus* from Jarvis & Morison 1981.)

(*e*) *Temperature*. Many studies of stomatal temperature responses have yielded contradictory results. Unfortunately, in many earlier studies temperature was confounded with variation in leaf–air vapour pressure difference. It is necessary to conduct temperature response studies at constant values of δe_ℓ (under which conditions absolute and relative humidity increase with temperature). In general, stomata tend to open as temperature increases over the normally encountered range, though an optimum is sometimes reached (see Hall *et al.* 1976). The magnitude of the temperature response does, however, depend on the vapour pressure.

(*f*) *Other factors*. Stomatal aperture is also affected by many gaseous pollutants such as O_3, SO_2 and nitrogen oxides (Mansfield 1976; Unsworth & Black 1981). Many of these responses are probably related to the toxic effects of these substances on membrane integrity. Stomatal aperture is also dependent on many other factors such as leaf age, nutrition and disease.

Stomatal behaviour in natural environments

As pointed out earlier, it is difficult to determine stomatal responses from measurements in the field. Typical results for leaf conductance of extension leaves of apple trees during the course of one day are presented in Fig. 6.14, together with the irradiance on a horizontal surface. These results are also plotted against irradiance and Ψ_ℓ in Fig. 6.14*b* and *d*. The variability in g_ℓ is a result partly of the fluctuations in irradiance and partly of leaf-to-leaf variability and differences in orientation. Much of the scatter in the plot of g_ℓ against $\mathbf{I}$ (Fig. 6.14*b*) results from the stomatal time constant being longer than that of the fluctuations of irradiance.

Several approaches can be used to determine the stomatal response to individual factors or to predict g_ℓ for given conditions, using results such as those in Fig. 6.14. One approach is to use boundary-line analysis. In Fig. 6.14*b* a hypothetical ideal boundary-line relationship between g_ℓ and $\mathbf{I}$ is shown as the dashed line, which may approximate the response when no other factors are limiting. It is assumed that many points fall below this line as a result of factors such as lowered Ψ_ℓ. Unfortunately this approach is difficult to quantify statistically as the upper points that define the boundary line are measured with some degree of error.

A widely used approach is that of multiple regression, where g_ℓ is regressed against various independent variables to give an equation of the form

$$g_\ell = a + b\mathbf{I} + c\delta e_\ell + d\Psi_\ell + \ldots \qquad (6.13)$$

where a, b, c, d, etc. are regression coefficients. Non-significant terms can

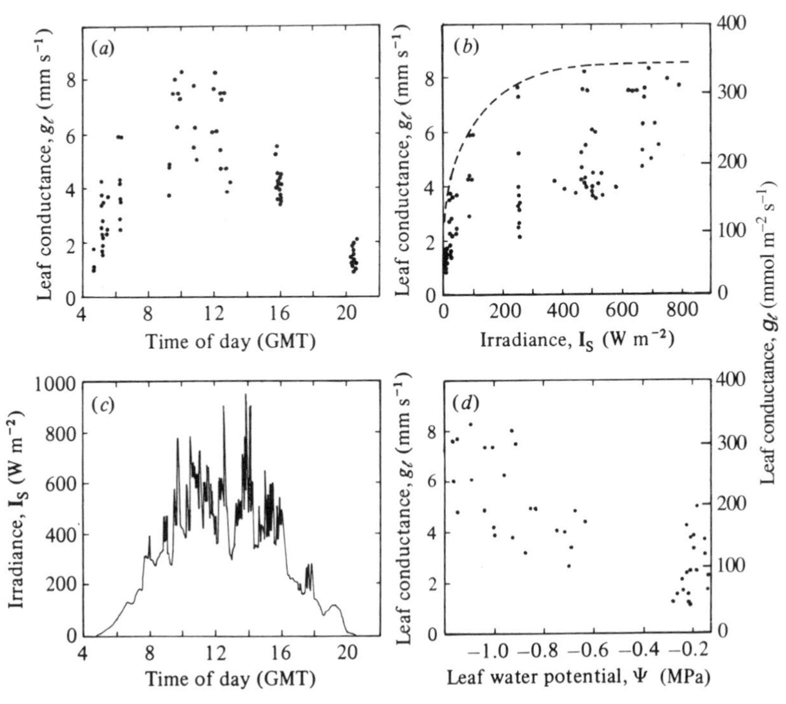

Fig. 6.14. Typical variation during one day (26 July 1980) of g_ℓ of mid-extension leaves of Bramley apples (*a*), together with irradiance above the canopy (*c*), and g_ℓ as a function of **I** (*b*) or Ψ_ℓ (*d*). For details see text. (H. G. Jones, unpublished data.)

be omitted and non-linear relations fitted by the use of higher-order polynomials (e.g. quadratic). This technique can give useful summaries of large amounts of data, but it is entirely empirical and thus has rather little predictive value for untried combinations of environmental conditions. It is frequently necessary to include time of day as an independent variate.

It is worth noting that the 'important' variates found in multiple regression models are often not the same as the individual variates most closely related to g_ℓ when regressed singly. For example, in an analysis of several data sets for apple in England, Jones & Higgs (1989) showed that the most useful three variates in a multiple regression were δe_ℓ, $\mathbf{I}_{50}$ (a derived variate $= \mathbf{I}/(50+\mathbf{I})$), and T_a, even though the best individual variates were either e_s or h. This latter observation is of particular interest in relation to the recent suggestion that g_ℓ tends to vary directly in proportion to assimilation rate scaled by the relative humidity at the leaf surface (Ball *et al.* 1986).

In order to get a robust model for predicting g_ℓ Jones & Higgs (1989) proposed the replacement of the environmental variates in equation 6.13 by differences (Δ) from typical values so that equation 6.13 becomes

$$g_\ell = g_o + \beta 1\,\Delta \mathbf{I} + \beta 2\,\Delta \delta e + \beta 3\,\Delta \Psi_\ell + \ldots \qquad (6.14)$$

where g_o is a reference value (at typical environmental conditions) and the βi are multiple regression coefficients. For prediction of g_ℓ for new data sets, the sizes of the coefficients can be normalised by dividing through by g_o to give

$$g_\ell = g_o(1 + \mathrm{b}1\,\Delta \mathbf{I} + \mathrm{b}2\,\Delta \delta e + \mathrm{b}3\,\Delta \Psi_\ell + \ldots) \qquad (6.15)$$

where the bi $= \beta\mathrm{i}/g_o$. The model is scaled to different data sets by using an appropriate g_o; absolute errors in predicted g_ℓ that result from incorrect regression coefficients are minimised by this technique. Using this approach it was shown that a model derived for one data set could fit data sets obtained for other orchards in different years as successfully as a freely-fitted model. The model using δe_ℓ, T_a and $\mathbf{I}_{50}$ explained between 32 and 62% of the variance in g_ℓ for different sets of apple data (Jones & Higgs 1989).

Perhaps the best method for analysing stomatal conductance (see Jarvis 1976) is to use a multiplicative model (rather than the additive model in equation 6.13) with appropriate non-linear components of the form

$$g_\ell = g_o \cdot \mathrm{f}(\mathbf{I}) \cdot \mathrm{g}(\delta e_\ell) \cdot \mathrm{h}(\Psi_\ell) \cdot \ldots \qquad (6.14)$$

where the forms of the individual functions (f(**I**), g(δe_ℓ), etc.) are obtained from controlled environment studies. Some particularly useful functions are summarised in Fig. 6.15. It is worth noting that for well-watered plants the stomata do not usually close in response to the normal diurnal fall in Ψ_ℓ. In fact an opposite effect is sometimes observed (Fig. 6.14*d*). This somewhat paradoxical result (compare with Fig. 6.10) arises because Ψ_ℓ is falling as a result of the increased evaporation rate as stomata open, rather than control operating the other way around.

Stomatal resistance in relation to other resistances

Typical values for the components of leaf and boundary layer resistances to water vapour loss from single leaves (see Fig. 6.3) are summarised in Table 6.3. Although the cuticular resistance tends to be by far the largest, the dominant component of the total resistance is stomatal, because the resistance of two parallel resistors is determined primarily by the smaller.

The high cuticular resistance results from the low water permeability (liquid and vapour) of the hydrophobic cuticle and the overlying wax layer. The thickness, composition and morphology of the cuticle and wax layers

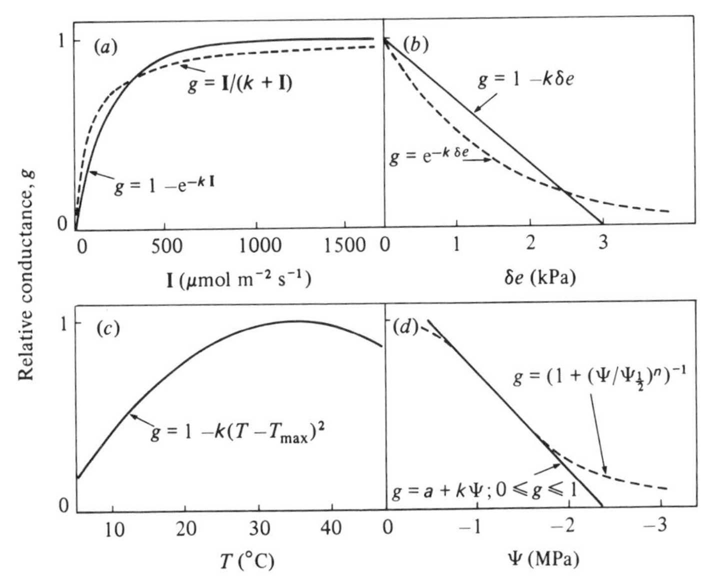

Fig. 6.15. (*a*)–(*d*) The most useful functions for describing stomatal response to environment (solid lines), and other useful functions (dashed lines). a, k, n are constants, as are T_{max} (T at which g is maximal) and $\Psi_{\frac{1}{2}}$ (the value of Ψ when g is half maximum). All functions give g as a fraction of a reference value. Details of the Ψ functions are discussed by Fisher *et al.* (1981).

is very dependent on species and growth conditions (see Martin & Juniper 1970), being most developed (with highest r_c) in plants from arid sites where water conservation is most crucial. The waxes may vary in form from smooth layers or platelets to rods or filaments several micrometres long. Some epidermal wax structures are shown in Fig. 6.2*e*.

Mathematical and physical models of gas diffusion in the intercellular spaces suggest that most transpired water originates from cell walls close to the stomatal pore, so that r_{iW} is small (probably less than 0.5 m² s mol⁻¹) (see Tyree & Yianoulis 1980). Because the sites of CO_2 assimilation are more evenly distributed throughout the leaf, r_{iC} may be significant and perhaps as much as 2.5 m² s mol⁻¹.

There has been controversy about the magnitude of the 'wall' resistance, r_w, since Livingston & Brown (1912) reported evidence for non-stomatal control of water loss. They envisaged that the evaporation sites retreated into the cell walls thus increasing the gas-phase diffusion path, a process they called 'incipient drying'. An alternative possibility is that the effective water vapour concentration at the liquid–air interface is significantly below saturation at leaf temperature. From equation 5.11, the relative lowering of

Table 6.3. *Relative values of the different resistances (see Fig. 6.3) and the corresponding conductances for single leaves. The leaf resistance is dominated by the stomatal component*

		Resistance r^m ($m^2\ s\ mol^{-1}$)	Conductance g^m ($mmol\ m^{-2}\ s^{-1}$)	Corresponding conductance g ($mm\ s^{-1}$)
Intercellular space and wall resistance	$(r_i + r_w)$	< 1	> 1000	> 25
Cuticular resistance	(r_c)	50– > 250	4–20	< 0.1–0.5
Stomatal resistance	(r_s)			
– minimum for many succulents, xerophytes and conifers		5–25	40–200	1–5
– minimum for mesophytes		2–6	170–500	4–13
– maximum when closed		> 125	8	< 0.2
Boundary layer resistance	(r_a)	0.25–2.5	400–4000	10–100

the partial pressure of water vapour (e) compared with that at saturation (e_s) is related to water potential by

$$e/e_s = \exp(\Psi \bar{V}_W/\mathscr{R}T) \tag{6.15}$$

Application of this equation predicts that at 20 °C, e/e_s would be 0.99 at -1.36 MPa, and only falls to 0.95 at -6.92 MPa, a figure found only in extremely severely stressed leaves. For a relative humidity of 0.50 in the ambient air, even this latter figure would introduce less than a 10% error in the driving force for evaporation (or equivalently r_w would be less than 10% of the total resistance).

In addition to effects of bulk leaf water potential on Ψ at the evaporating surface, Ψ could also be lowered by accumulation of solutes, or because of the presence of a large internal hydraulic resistance. Experimental results for several species and theoretical calculations (Jarvis & Slatyer 1970; Jones & Higgs 1980) all suggest that r_w is small ($\ll 1\ m^2\ s\ mol^{-1}$) at physiological water contents.

Table 6.3 shows that, for single leaves, r_ℓ is normally at least an order of magnitude greater than r_a. In canopies, however, the relative importance of

Table 6.4. *Canopy leaf or physiological resistance to water vapour* (r_L) *and canopy boundary layer* (r_A) *resistance for different types of vegetation* (*from Jarvis* 1981). *Approximate equivalents in molar units* (m^2 *s* mol^{-1}) *are given in parentheses*

	r_L(s m^{-1}) per unit leaf area	r_L (s m^{-1}) per unit ground area	r_A (s m^{-1}) per unit ground area
Grassland/heathland	100 (2.5)	50 (1.25)	50–200 (1.25–5)
Agricultural crops	50 (1.25)	20 (0.5)	20–50 (0.5–1.25)
Plantation forest	167 (4.2)	50 (1.25)	3–10 (0.08–0.25)

the boundary layer resistance increases. This is because all the individual leaf resistances are in parallel and therefore the total leaf resistance decreases as leaf area index increases. In addition, for a crop, the boundary layers for the individual leaves must be added to an overall crop boundary layer. Some typical ranges for the values of the canopy boundary layer resistance (r_A) and the canopy leaf resistance (r_L) are given in mass units (because these are more commonly used for canopy studies) in Table 6.4 (the equivalents in molar units are given in parentheses). The values shown indicate that the ratio between r_A and r_L can vary widely between different plant communities. Some implications of differences in this ratio between grassland and aerodynamically rough canopies such as tall forests, were discussed in Chapter 5.

Stomatal function and the control loops

The two main control systems involved in the regulation of stomatal aperture are related to the fluxes of water vapour and CO_2 that need to be controlled (Fig. 6.1):

(*a*) *Water control loop.* The main function of stomata is the control of water loss. In general they respond to factors that lower Ψ_ℓ in such a way as to minimise further increases in stress. The responses involve either feedback via leaf water status itself, or else direct feedforward control (Fig. 6.1). The distinction between feedback and feedforward can be illustrated by the possible effects of altered δe_ℓ (see Fig. 6.16).

Negative feedback is illustrated by the control pathway in the lower part of Fig. 6.16. Here a decrease in air humidity (increased δe_ℓ) increases **E**, since

$$\mathbf{E} = (2.17/T) g_\ell \delta e_\ell \qquad (6.16)$$

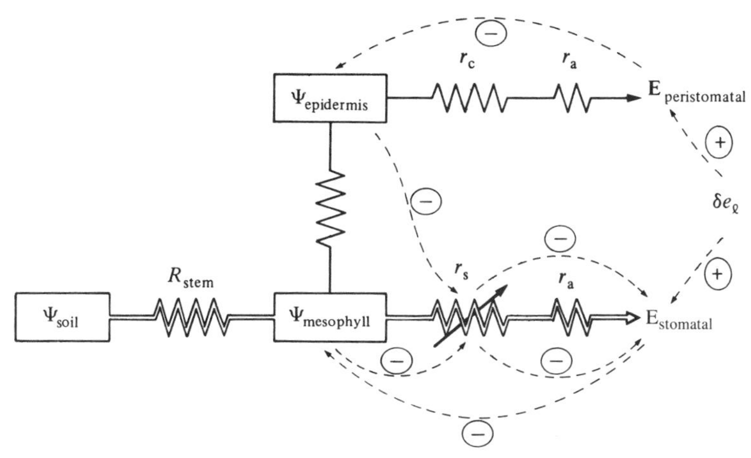

Fig. 6.16. A simplified diagram of the various feedback and feedforward control loops involved in humidity effects on stomatal aperture. The solid lines represent the flows of water with the majority of the flow occurring through (and controlled by) the stomatal pores (double lines) and with only a small proportion of peristomatal evaporation directly through the epidermis (single lines). The dotted lines indicate control processes with the signs indicating the effect of an increase in the first factor on the second (e.g. the negative sign relating $\mathbf{E}_{stomatal}$ to $\Psi_{mesophyll}$ indicates that an increase of $\mathbf{E}_{stomatal}$ results in a decrease of $\Psi_{mesophyll}$).

This in turn tends to lower bulk Ψ_ℓ because, from equation 4.25 (representing the sum of all the soil and plant resistances by R_{sp}):

$$\Psi_\ell = \Psi_s - R_{sp}\mathbf{E} \tag{6.17}$$

This lowered Ψ_ℓ then leads to stomatal closure which finally has a negative feedback effect on **E**. Assuming a linear relationship (Fig. 6.15*d*):

$$g_\ell = a + b\Psi_\ell \tag{6.18}$$

and, combining this with equation 6.17 gives

$$g_\ell = a + b(\Psi_s - R_{sp}\mathbf{E}) = c - d\mathbf{E} \tag{6.19}$$

where a, b, c and d are constants. From equations 6.16 and 6.19 the relationships describing this negative feedback are

$$\mathbf{E} = c\delta e_\ell/(1 + d\delta e_\ell) \tag{6.20}$$

$$g_\ell = c/(1 + d\delta e_\ell) \tag{6.21}$$

and are illustrated by the solid lines in Fig. 6.17. Note that equation 6.20 is a saturation-type curve and that negative feedback cannot cause a steady-state reduction in transpiration (dotted curve in Fig. 6.17) as δe_ℓ increases. (Any reduction in **E** implies increases in Ψ_ℓ and hence g_ℓ which would immediately restore **E**.) A feedback loop with a high *gain* would tend to maintain **E** relatively constant. However, a high gain can lead to instability and regular stomatal oscillations (see Cowan 1977) where there are time delays in the responses.

Feedforward is illustrated in the top half of Fig. 6.16. In this case the environment affects the controller (the stomata) directly *without* depending on changes in the flux that is being controlled (i.e. evaporation through the stomata). An increased δe_ℓ increases the rate of peristomatal transpiration (water loss not passing through the stomatal pore). This leads to lowered guard-cell water potential and consequent stomatal closure, as long as the hydraulic flow resistance in the pathway to the evaporating sites on the outer surface of the guard cells is great enough. (Evidence for a significant flow resistance in this pathway is discussed by Meidner & Sheriff (1976).) Using an argument similar to that used to derive equation 6.21, it is possible to show that in this case (cf. Fig. 6.15)

$$g_\ell = e - f\delta e_\ell \tag{6.22}$$

where e and f are constants. Combining with equation 6.16 and rearranging gives

$$\mathbf{E} = e'\delta e_\ell - f'(\delta e_\ell)^2 \tag{6.23}$$

Depending on the relative values of the new constants e' and f', **E** may even fall with increased evaporative demand (dotted curve in Fig. 6.17). Such responses have been observed in several species. In practice, both feedback and feedforward responses usually occur together. Further discussion of feedback and feedforward in this context may be found in articles by Cowan (1977) and Farquhar (1978). Figure 6.17 also shows the expected linear response of **E** to δe_ℓ when there is no stomatal response (dashed curve).

The general processes involved in stomatal response to guard cell or mesophyll water status were outlined above, and it was pointed out that although hydropassive movements can occur, and that they may be responsible for the very rapid movements observed on detaching leaves, active responses, probably mediated by ABA, are much the more important.

(*b*) *Carbon dioxide control loop.* The central role of the stomata in the control of photosynthesis depends on sensing the photosynthetic rate. Most evidence indicates that the photosynthetic feedback control (Fig. 6.1) depends on sensing the intercellular space CO_2 concentration. For example, as light increases, the rate of CO_2 fixation increases, thus lowering the intercellular CO_2 concentration, and the stomata open to compensate.

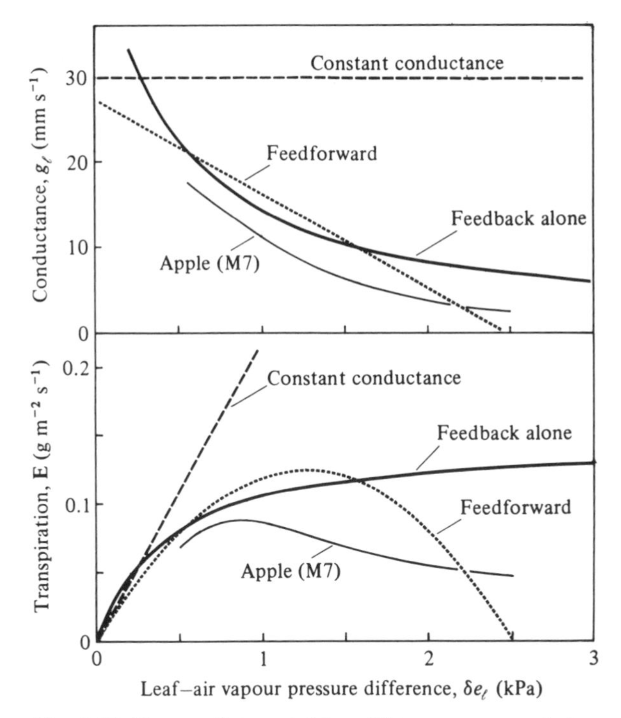

Fig. 6.17. Types of stomatal humidity response and the consequent relation between **E** and δe_ℓ. The thick solid lines show negative feedback (equations 6.20 and 6.21), the dotted lines show feedforward (equations 6.22 and 6.23) and the dashed line constant g_ℓ. Data for M7 apple from Fig. 6.13 are shown for comparison.

Other changes that reduce the intercellular space CO_2 have a similar effect (though see Jarvis 1980).

There is a strong feedback control that tends to maintain intercellular CO_2 remarkably constant (at a mole fraction of around 130×10^{-6} in C_4 species and 230×10^{-6} in C_3 species – see Chapter 7). The night-time stomatal opening in CAM plants can also be explained on the same basis: dark CO_2 fixation at night lowers intercellular CO_2 and stomata open.

Sample problems

6.1 A leaf detached from a plant and suspended in a moving airstream ($h = 0.2$) initially loses weight at a rate of 80 mg m^{-2} s^{-1}, falling to a constant rate of 2 mg m^{-2} s^{-1} after about 20 min. A piece of wet blotting paper suspended alongside the leaf loses water at a rate of 230 mg m^{-2} s^{-1}. Assuming that leaf, air and blotting paper are all at 20 °C, calculate (i) the boundary layer resistance, r_{aW}, (ii) the cuticular resistance, r_{cW}, (iii) the initial value of the stomatal resistance, r_{sW}, if the leaf has equal numbers of stomata on each surface.

6.2 Calculate (i) the stomatal diffusion resistance (r_{sW}) and (ii) the corresponding conductance (g_{sW}) for a leaf with 200 stomata mm^{-2} on each surface, each pore being 10 μm deep and circular in cross-section ($d = 5$ μm).

6.3 A continuous-flow porometer with a chamber area of 1.5 cm^2 is attached to the lower surface of a leaf. If the flow rate is 2 cm^3 s^{-1}, and the relative humidity of the outlet air is 35% (inlet air is dry), calculate $g_{\ell W}$ assuming (i) that $T_\ell = T_a$ or (ii) that $T_\ell = 25$ °C and $T_a = 27$ °C.

6.4 Assuming that g_ℓ for a particular species decreases linearly from 10 mm s^{-1} at $\delta e = 0$ kPa to 0 at $\delta e = 3$ kPa: (i) plot a graph of the relationship between **E** and δe; what is **E** when $\delta e = 1$ kPa? (ii) If g_ℓ is also sensitive to Ψ_ℓ, falling by 50% per MPa below zero, plot the dependence of **E** on δe if Ψ_ℓ falls linearly at a rate of 1 MPa for each 0.1 g m^{-2} s^{-1} increase in evaporation; what is **E** when $\delta e = 1$ kPa?

7

Photosynthesis and respiration

The most important characteristic of plants is their ability to harness energy from the sun to 'fix' atmospheric carbon dioxide into a range of more complex organic molecules. This process of photosynthesis provides the major input of free energy into the biosphere; some of the free energy stored in these photosynthetic assimilates is then transferred in the process of respiration to high energy compounds that can be used for synthetic and maintenance processes. The net rate of photosynthetic CO_2 fixation by a photosynthesising plant (net photosynthesis, $\mathbf{P}_n$) is the difference between the gross rate of CO_2 fixation ($\mathbf{P}_g$) and the rate of respiratory CO_2 loss (**R**). Because of the central nature of these two processes to all aspects of plant growth and adaptation, they will be discussed in some detail in this chapter.

Photosynthesis

The overall reaction of photosynthesis can be represented by

$$CO_2 + 2H_2O \xrightarrow{\text{light}} CO_2 + 4H + O_2 \rightarrow (CH_2O) + H_2O + O_2 \qquad (7.1)$$

The net effect is the removal of one mole of water and the production of one mole of O_2 for every mole of CO_2 that is reduced to the level of sugar ($(CH_2O)_6$). Many of the individual reactions take place in specialised organelles called chloroplasts, within the leaf mesophyll cells (Fig. 7.1). These are bounded by a double membrane and contain a network of vesicles called thylakoids arranged either as single lamellae or stacked up to form characteristic granal stacks (Fig. 7.1). The ground material of the chloroplast is called the stroma. Details of the biochemistry and physiology of photosynthesis can be found in appropriate texts such as those by Clayton (1981), Hatch & Boardman (1981), Foyer (1984), Lawlor (1987) and Briggs (1989).

Photosynthesis can be conveniently treated as three separate components: (i) light reactions, in which radiant energy is absorbed and used to generate the high energy compounds ATP and NADPH; (ii) dark reactions,

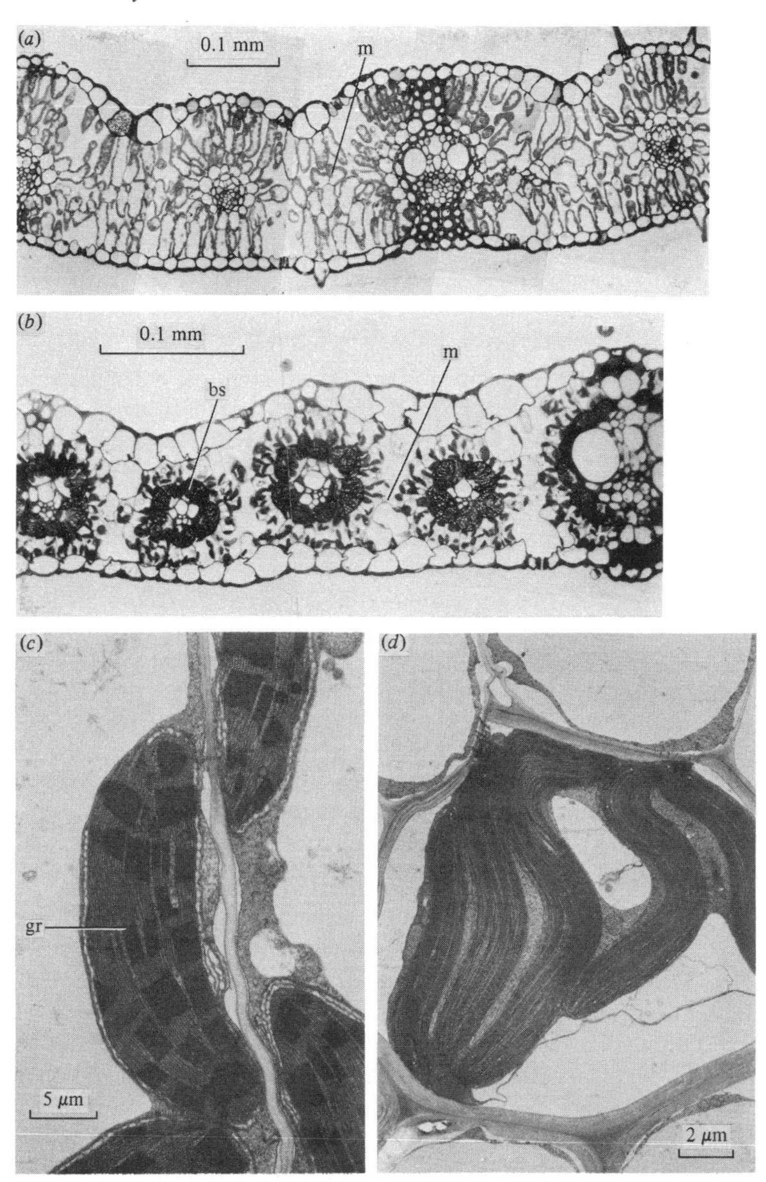

Fig. 7.1. (*a*) Transverse section of the leaf of a C_3 grass, the wheat species (*Triticum urartu* Tum) showing the photosynthetic mesophyll cells (m); (*b*) transverse section of a leaf of the C_4 grass millet (*Pennisetum americanum*) showing the distinct mesophyll (m) and bundle-sheath (bs) cells; (*c*) electron micrograph of mesophyll chloroplast from (*b*) showing the photosynthetic lamellae and the granal stacks (gr); (*d*) an agranal chloroplast from the bundle-sheath of (*b*). Courtesy of Dr M. L. Parker, Plant Breeding Institute, Cambridge.

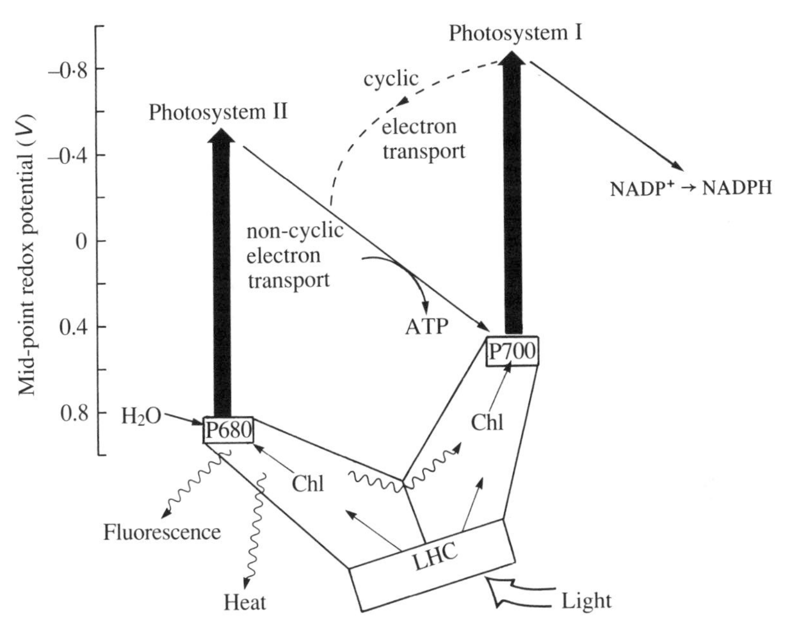

Fig. 7.2. Schematic representation of the photosynthetic electron transport chain showing light trapping by the major chlorophyll-protein antenna complexes and energy transfer to the reaction centres of the two photosystems with fluorescence from PS II, energy transfer to PS I, and dissipation as heat competing with photochemistry at PS II. (LHC = light harvesting complex, Chl = chlorophyll.)

which include the biochemical reduction of CO_2 to sugars using the high energy compounds generated in the light reactions; and (iii) supply of CO_2 from the ambient air to the site of reduction in the chloroplast.

Light reactions

The primary process in photosynthesis is the absorption of incoming solar radiation by the pigments located on the grana and stroma lamella membranes of the chloroplasts. The main pigments, the chlorophylls, are most effective at absorbing in the red and blue (Fig. 2.4, p. 17) while carotenoids and other accessory pigments permit the absorption of other wavelengths in the PAR (400–700 nm). In green plants the chlorophylls are embedded in three chlorophyll–protein complexes: the light harvesting complex (LHC), the photosystem I antenna complex (PS I) and the photosystem II antenna complex (PS II). The complete electron transport pathway is represented schematically in Fig. 7.2.

Radiation absorption causes excitation of electrons in the pigment molecules, with excitation being funnelled to one of the two specialised 'reaction centres' (P_{680} in PS II and P_{700} in PS I) by resonance transfer, which can occur if the energy available corresponds to that required for excitation in the receptor. Blue light is about 1.5 times as energetic as red (i.e. 4.62×10^{-19} J quantum^{-1} at 430 nm compared with 3.00×10^{-19} J quantum^{-1} at 662 nm), but because the upper singlet state of chlorophyll (excited by blue light) is very unstable, decaying rapidly to the lower excited state with the generation of heat, blue quanta tend to be no more effective than red (Fig. 2.4). As well as being lost as heat, the energy released when excited electrons return to the ground state can either be harnessed at the reaction centres to cause charge separation and hence to drive electron transport and the generation of ATP and NADPH, or else it may be lost by re-radiation (a process called fluorescence). At room temperature *in vivo* the majority of fluorescence arises from chlorophyll *a* in PS II, and this fluorescence can be used as a powerful probe of photosynthetic functioning. As fluorescence of photosystem II is only one of a number of competing processes for de-excitation of excited chlorophyll molecules, the probability that an excited chlorophyll molecule will de-excite by fluorescence (equal to the quantum yield for fluorescence, ϕF = number of quanta emitted as fluorescence (F)/number of quanta absorbed (**I**)) is given by the ratio of the rate constant for fluorescence to that of all competing processes

$$\phi F = \frac{k_F}{k_F + k_D + k_T + k_P} \qquad (7.2)$$

where k_F, k_D, k_T and k_P, respectively, are the rate constants for de-excitation through fluorescence, thermal dissipation as heat, energy transfer to photosystem I, and PS II photochemistry with all reaction centres open. Normally ϕF is of the order 0.01–0.02.

Proper operation of the photosynthetic system requires balanced excitation of the two photosystems and energy transfer between the two is regulated by the degree of phosphorylation of LHC protein, with increased phosphorylation increasing the transfer of excitation to PS I. Charge separation at the PS II reaction centre leads to the reduction of the primary quinone acceptor (Q_A), and hence by transport through a series of further electron acceptors (Q_B, plastoquinone, etc.), drives the phosphorylation of ADP or the reduction of $NADP^+$. The oxidised P_{680} is regenerated by electrons derived from the splitting of water.

It is thought that the major pathway of electron flow is the non-cyclic pathway from H_2O to $NADP^+$. In this pathway one molecule of $NADP^+$ is reduced to NADPH for each two electrons that flow. The coupling of this electron flow to the generation of ATP probably involves Mitchell's chemi-osmotic mechanism where the reaction centres and redox carriers in the electron transport pathway are asymmetrically arranged in the membranes such that electron transport gives rise to charge separation across the thylakoid membrane, and transport of H^+ from the stroma to inside the thylakoid. This increases the pH of the stroma. The return flow of H^+ is then thought to drive the production of ATP by its linkage to a membrane-bound ATPase. It is not certain how many ATP molecules can be generated by non-cyclic electron flow, but it is thought to be somewhere between 1 and 2 ATP per $2e^-$. The electron pathway is somewhat flexible, with cyclic and pseudocyclic pathways that lead to ATP production without reduction of $NADP^+$ being possible, as is the direct reduction of oxygen (the Mehler reaction).

Dark reactions

Plants can be classified into at least three major groups (C_3, C_4 and CAM) on the basis of the biochemical pathway by which they fix CO_2. The characteristic anatomical differences between C_3 and C_4 species are illustrated in Fig. 7.1, while the characteristics of each pathway are listed in Table 7.1 and will be discussed in detail in succeeding sections. The essential features of the three pathways are summarised in Fig. 7.3.

(*a*) *C_3 pathway*. C_3 plants use the enzyme ribulose bisphosphate carboxylase (RuBP carboxylase or Rubisco) for the primary fixation of CO_2 in the chloroplast to form the 3-carbon compound 3-phosphoglyceric acid (PGA) which is then converted to triose phosphate using ATP and NADPH. Most of the triose phosphate then takes part in a complex reaction sequence (the photosynthetic carbon reduction (PCR) or Calvin cycle) that requires further ATP to generate the substrate for the initial carboxylation reaction (RuBP). Some of the triose is siphoned off as the net product of photosynthesis to form sugar phosphate (fructose-1,6-bisphosphate) and sugars. This cycle requires 3 ATP and 2 NADPH per CO_2 converted to sugar phosphate. It is the dominant photosynthetic pathway of species from cool, temperate or moist habitats and is the only pathway found in trees (with very few exceptions), or lower plants. The majority of crop plants use the C_3 pathway, including all the temperate cereals (wheat, barley, etc.), root crops (e.g. potato and sugar beet) and leguminous species (beans, etc.).

(*b*) *C_4 pathway*. In this pathway, the initial carboxylation reaction involves phosphoenolpyruvate carboxylase (PEP carboxylase) producing oxaloacetate (OAA) as the first product of fixation, with other 4-carbon compounds (especially malate and aspartate) also being formed very rapidly. The next stage in these plants is the transfer of these 4-carbon compounds to specialised 'bundle sheath cells' where they are de-

Table 7.1. *Characteristics of the main photosynthetic pathways*
Data collated from a number of sources.

	C_3	C_4	CAM Day	CAM Night
Anatomy				
'Kranz' anatomy (distinct bundle sheath	No	Yes	Succulent	
Frequency of leaf bundles	Low	High	Low	
Leaf air space volume (%)				
monocots.	10–35 %	< 10 %	Low	
dicots.	20–55 %	< 30 %	Low	
Biochemistry				
Early products of ^{14}C fixation	PGA	C_4 acids	PGA	C_4 acids
Primary carboxylase	RuBP	PEP	RuBP	PEP
Discrimination against ^{13}C ($\delta^{13}C$, ‰)	−22 to −40	−9 to −19	C_3-like	C_4-like
Absolute sodium requirement	No	Yes	Only for night fixation	
Physiology				
CO_2 compensation point (vpm)	30–80	< 10	*c.* 50	< 5
Post illumination burst of CO_2	Yes	Slight	Yes	—
Enhancement of $\mathbf{P}_n$ in low O_2	Yes	No	Yes	No
Quantum requirement	15–22	19	—	—
Mesophyll resistance				
r'_m ($m^2\ s\ mol^{-1}$)	7–15	1.2–5	*c.* 20	?
r'_m ($s\ cm^{-1}$)	3–6	0.5–2.0	*c.* 8	?
Relative stomatal sensitivity to environment	Insensitive	Sensitive	Reversed cycle	
Intercellular space CO_2 partial pressure	$\sim 0.7p'_a$	$\sim 0.4p'_a$	$0.5p'_a$	?
Maximum photosynthetic				
rate ($\mu mol\ m^{-2}\ s^{-1}$)	14–40	18–55	6	8
($mg\ CO_2\ m^{-2}\ s^{-1}$)	0.6–1.7	0.8–2.4	0.25	0.3
Optimum day temp (°C)	*c.* 15–30 (Wide acclimation)	25–40	*c.* 35 (Needs low night temperature)	
Light response saturating well below full sunlight	Usually	Rarely	Usually	—

Table 7.1 (*cont.*)

	C_3	C_4	CAM Day	CAM Night
Ecology				
Regions where commonest	Temperate	Tropical, arid	Arid	
Transpiration ratio	High	Low	Medium	Very low
(g H_2O lost per g CO_2 fixed)	450–950	250–350	50–600	< 50
Max. growth rate (g m^{-2} day^{-1})	34–39	51–54	7	
Average productivity (tonne ha^{-1} yr^{-1})	*c.* 40	60–80	Low	

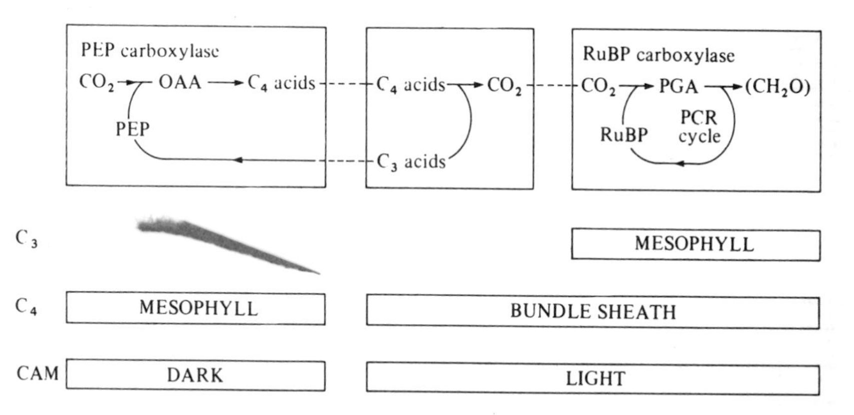

Fig. 7.3. The essential features of C_3, C_4 and CAM photosynthetic pathways, showing the spatial separation of the two carboxylations in C_4 and the temporal separation in CAM.

carboxylated (see Fig. 7.3). The CO_2 released is then refixed using the normal C_3 enzymes of the PCR cycle located within the bundle sheath cells. The initial fixation by PEP carboxylase in the mesophyll cells acts as a CO_2 'concentrating' mechanism because PEP carboxylase has a much higher affinity for CO_2 than does Rubisco.

C_4 plants can be subdivided into three groups on the basis of the decarboxylation enzyme employed, but as these three types are physiologically and ecologically similar, they will be treated together. As will be discussed in later sections, the C_4 pathway has particular adaptive value in hot dry environments and is common in species from tropical and semi-arid habitats including the cereals maize, millet and sorghum.

A number of species from seven genera that include *Moricandia*, *Flaveria*

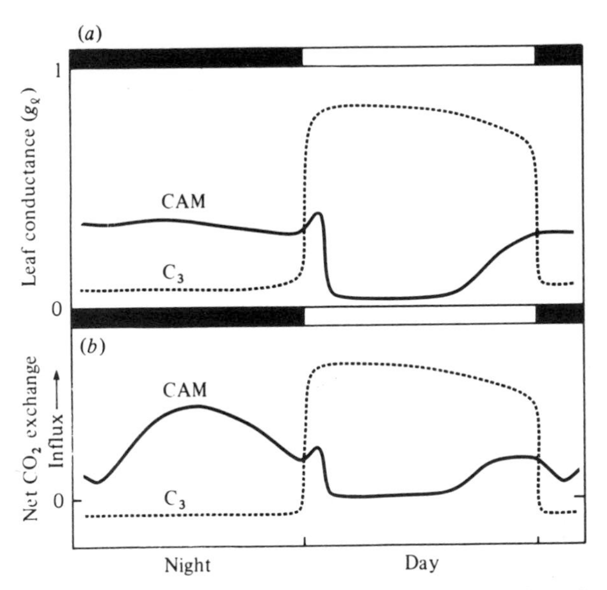

Fig. 7.4. Characteristic night–day cycles of (*a*) leaf conductance and (*b*) net CO_2 exchange in C_3 and CAM plants. The early morning and late afternoon stomatal opening and net photosynthesis in the CAM plant disappears when water stressed.

and *Panicum* are known which are intermediate between C_3 and C_4 plants in both physiological and anatomical characters (Edwards & Ku 1987; Monson & Moore 1989). In these plants the CO_2 compensation concentration tends to be intermediate between C_3 and C_4 species. Although the capacity for photorespiration appears to be similar in C_3 and C_3–C_4 intermediate species, there is a difference in the distribution of the enzyme (glycine decarboxylase) which catalyses the release of CO_2 in photorespiration. In the C_3–C_4 intermediates (e.g. *Moricandia arvensis*) that have been studied it is only present in the mitochondria of bundle sheath cells, while in C_3 species it is present in mitochondria of all chloroplast-containing cells (Hylton *et al.* 1988). This localisation of the photorespiratory CO_2 release at the inner wall of the bundle sheath cells probably explains the effective light-dependent recapture of photorespiratory CO_2 in these species because any CO_2 released must diffuse out through the overlying chloroplasts. This may have been an early step in the evolution of the C_4 pathway.

(*c*) *Crassulacean acid metabolism* (*CAM pathway*). In many ways the CAM pathway (see Kluge & Ting 1978; Osmond 1978) resembles the C_4 in that CO_2 is initially fixed into C_4 compounds using PEP carboxylase and is subsequently decarboxylated and refixed by Rubisco. In this case, however, the initial carboxylation occurs during the dark night period when large quantities of C_4 acids accumulate in the mesophyll cell vacuoles. During the day, malic enzyme acts to decarboxylate the stored malate thus providing CO_2 as a substrate for the normal C_3 enzymes. In CAM plants, the two carboxylation systems occur in the same cell but are separated temporally (Fig. 7.3), while in C_4 plants the two carboxylation systems operate at the same time, but are spatially separated in different cells. A consequence of the night-time activity of the primary carboxylase is that in CAM plants the stomata tend to be open during the night and closed during the day (Fig. 7.4). This is clearly advantageous in terms of water conservation. The CAM pathway is usually found in succulent plants, such as the cacti, that occur in arid areas. The few economically important CAM plants include pineapple and sisal.

Although some plants such as *Opuntia basilaris* always use the CAM pathway, a number of species (e.g. *Agave deserti*) are facultative CAM plants in that they can operate either as C_3 plants when water supply is adequate or may have a varying expression of CAM activity depending on the environment. This may range from full CAM expression to an intermediate form called 'idling', where there is no net night fixation of CO_2 but some vacuolar acidification occurs. The switch between C_3 and CAM activity appears to be controlled by water stress (von Willert *et al.* 1985).

CO_2 supply

The CO_2 that is fixed in photosynthesis must diffuse from the ambient air (where the partial pressure averages nearly 34 Pa – equivalent to a volume fraction $\simeq$ 340 parts per million) through a series of resistances to the carboxylation site. The first part of the pathway through the leaf boundary layer, the stomata and the intercellular spaces to the mesophyll cell walls is in the gas phase and has been discussed in detail in Chapter 6. The remainder of the transport pathway through the cell wall to the carboxylation site in the chloroplasts is in the liquid phase. This will be discussed in detail below.

Respiration

The oxidation of carbohydrate to CO_2 and H_2O in living cells is generally termed respiration. In photosynthetic organisms there are two main types of respiration. The first is called dark respiration ($\mathbf{R}_d$) and includes various pathways of substrate oxidation such as glycolysis, the oxidative pentose phosphate pathway and the tricarboxylic acid (TCA or Krebs) cycle (Fig.

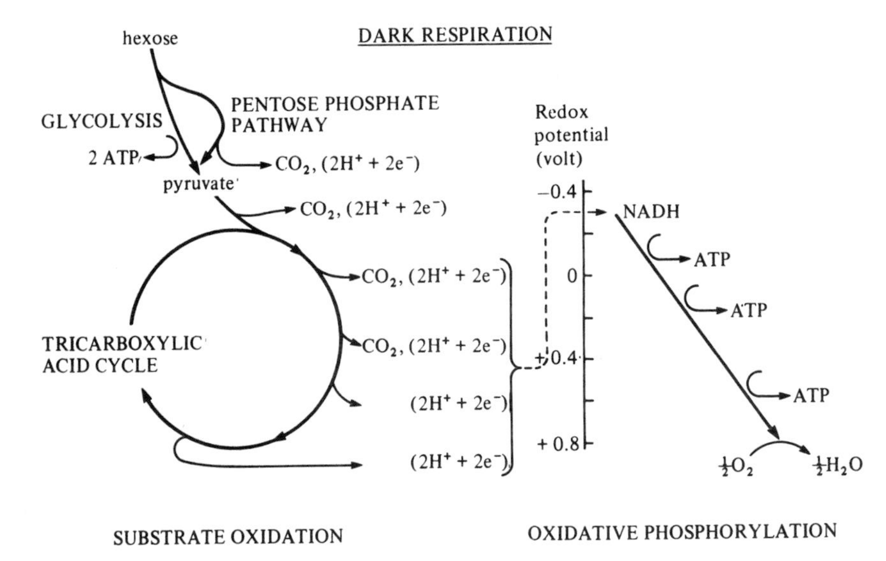

Fig. 7.5. Summary of the main pathways involved in dark respiration. The $2H^+ + 2e^-$ represent reductant that ultimately goes to reduce oxygen.

7.5) that conserve some of the free energy in carbohydrate in the high energy bonds of ATP, reduced pyridine nucleotide (NADH) and $FADH_2$. The term dark respiration also covers the further oxidation of NADH and $FADH_2$ by transfer of electrons through the various electron transfer complexes of the mitochondrial electron transport pathway in the mitochondrial membranes to O_2 as the final electron acceptor. There are a number of routes for this electron transfer, all of which feed through ubiquinone. As only three of the six main complexes transport protons and hence allow the energy to be harnessed in ATP synthesis (by a mechanism similar to that in chloroplasts), up to three protons may be transported per electron depending on the exact pathway used. Normally efficient mitochondrial respiration leads to phosphorylation of close to three ADP per oxygen (a P/O ratio of up to three), but this can be greatly reduced if significant electron flow occurs through the so-called alternate oxidase system which bypasses cytochrome *c*. The alternate oxidase is not inhibited by cyanide and is especially important in thermogenesis (e.g. in the *Arum* spadix) because its low efficiency at generating ATP leads to high heat output.

The second type of respiration in plants is called photorespiration (see review by Ogren 1984). This is the pathway of CO_2 production via the photorespiratory carbon oxidation (PCO) cycle (also known as the

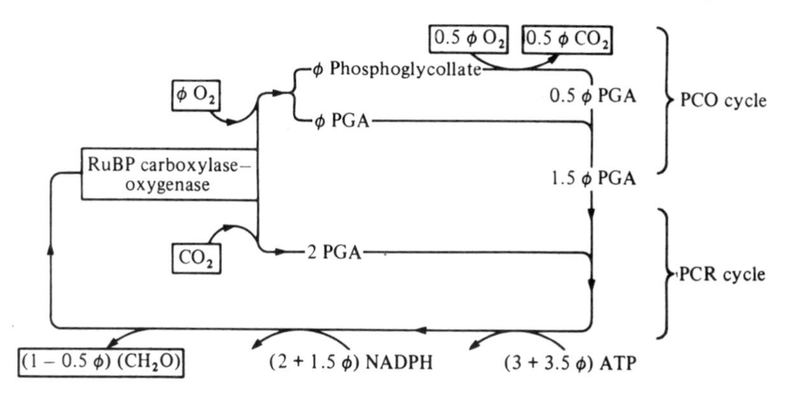

Fig. 7.6. Simplified diagram of the photorespiratory (PCO) and Calvin (PCR) cycles showing the stoichiometry where ϕ molecules of RuBP are oxygenated to every one carboxylated. (Modified from Farquhar *et al.* 1980*b*.)

glycollate pathway) (Fig. 7.6). The same enzyme (Rubisco) that catalyses the carboxylation of RuBP as the first step of the PCR cycle can also catalyse the oxygenation of RuBP to phosphoglycollate as the first step of the PCO cycle.

Details of respiratory metabolism in plants are reviewed in Davies (1980) and Moore & Beechey (1987) while a more general review of respiration in relation to productivity may be found in Amthor (1989).

There are a number of physiologically important differences between photorespiration and dark respiration:

(*a*) True photorespiration is obligatorily coupled to operation of the PCR cycle for generation of its substrate (RuBP), so it only occurs in photosynthetic cells in the light. 'Dark' respiration, on the other hand, can occur in all cells in the dark or light, and probably even continues in photosynthetic cells in the light at 5–15% of the rate of net photosynthesis (Raven 1972; Graham 1980). The term light respiration ($\mathbf{R}_\ell$) will be used to refer to the total CO_2 production of leaves by photorespiration ($\mathbf{R}_p$) and dark respiration pathways in light.

(*b*) Photorespiration is affected by the concentrations of both O_2 and CO_2. This is because of the competitive nature of oxygenation versus carboxylation of RuBP. Increases in CO_2 concentration increase the proportion of RuBP that is carboxylated (and hence net photosynthesis), but increases in O_2 concentration increase oxygenase activity and the proportion of carbon lost in photorespiration. In contrast dark respiration is not affected by CO_2 or by O_2 concentration in excess of 2–3%.

(*c*) In dark respiration, about 35–40% of the energy available from the oxidation of sugars is conserved in the form of ATP. The energetic yield for respiration of glucose, for example, is made up as follows: 6 ATP per glucose for glycolysis (2 via substrate level phosphorylation, 4 via mitochondrial electron transport from NADH (up to a P/O ratio of 2 via this route)), plus 2 ATP via the TCA cycle substrate level phosphorylation, plus 24 via mitochondrial phosphorylation with a P/O of 3 and two ATP for each of the $FADH_2$ from the TCA cycle – a total of 36 ATP per glucose (though this is likely to be an overestimate). From these figures about 1.10 MJ mol^{-1} of the 2.87 MJ mol^{-1} released by complete oxidation of glucose is potentially trapped in ATP, though this is only an approximation as it assumes standard conditions which do not pertain in plant cells. In contrast to dark respiration, however, the PCO cycle requires a net input of energy to drive it (28 ATP equivalents per CO_2 released – see Fig. 7.6 and Lorimer & Andrews 1981).

Respiratory quotients

The ratio of the numbers of moles of CO_2 released to the number of moles of O_2 absorbed in respiration is known as the respiratory quotient. Respiration of glucose and other hexoses gives rise to a respiratory quotient of unity, but oxidation of reduced compounds such as fats or proteins yields a respiratory quotient of less than 1 (0.7 for many lipids, 0.8 for some proteins), while oxidised compounds such as organic acids yield respiratory quotients greater than unity (about 1.33 for citric acid). Overall a value of about 1 is usually applicable.

Respiratory efficiency

There is evidence for genetic variation in respiratory efficiency. For example, genotypes of ryegrass with slow or fast respiration by young fully expanded leaves have been selected by Wilson (1982). Dark CO_2 efflux and apparent net photosynthesis were negatively correlated, possibly because the slow respiring lines had a lower rate of maintenance respiration, and also faster growth; these differences were reflected in higher herbage yields. Low alternative pathway activity should also be beneficial if it can be selected for (some results are summarised by Amthor 1989).

Calculations of the theoretical efficiency of conversion of substrate to biomass can be made on the basis of known biochemical pathways (Penning de Vries *et al.* 1983). The amount of CO_2 released during the synthesis of different compounds (mg CO_2 per g of compound formed) ranges from −11 for organic acids, through 170 for carbohydrates, 544 for proteins to as much as 1720 for lipids. Overall the average for leaf tissue is 333, while a peanut seed could be over 1000.

Photorespiration in C_4 plants

Many of the physiological characters of C_4 plants listed in Table 7.1 indicate a lack of external symptoms of photorespiration in these plants. These include a CO_2 compensation point near zero (indicating little or no respiratory release of CO_2), a low mesophyll resistance (see below), a lack of photosynthetic response to O_2 partial pressure, high rates of photosynthesis and even a high temperature optimum (the oxygenase to carboxylase ratio tends to increase with temperature). Many of these can be explained in terms of efficient refixation of any respired CO_2, but this cannot explain the lack of any O_2 effect on the efficiency of radiation use (quantum efficiency), as this would be expected to change with photorespiration rate, but such effects have not yet been detected. In spite of this, there is good evidence that the appropriate enzymes are present in bundle sheath cells (albeit at fairly low activities) so that the lack of photorespiration probably results from high internal CO_2 concentrations that shift the path of RuBP metabolism almost entirely to carboxylation.

Function of respiration

Dark respiration acts as a source of the ATP needed for biosynthetic and maintenance processes and also provides the essential carbon skeletons. The oxidative pentose phosphate pathway may also be an important source of NADPH for reactions in the cytoplasm. Dark respiration is often separated into growth ($\mathbf{R}_g$) and maintenance ($\mathbf{R}_m$) components, on the basis that $\mathbf{R}_g$ provides energy for growth and synthesis of new cell constituents and $\mathbf{R}_m$ is used for maintenance of existing cell structure. Although there may be no biochemical distinction between these components, $\mathbf{R}_m$ is assumed to be proportional to dry weight and strongly temperature sensitive, while $\mathbf{R}_g$ is directly dependent on photosynthesis and insensitive to temperature.

The function of photorespiration is more controversial since its operation only seems to lower plant productivity, which hardly seems evolutionarily advantageous! One suggestion is that oxygenation is an inevitable consequence of the chemical mechanism of carboxylation of RuBP, so that the PCO cycle acts to recycle any phosphoglycollate formed. Alternatively photorespiration may be an evolutionary 'hangover' serving no useful purpose. This suggestion has led to attempts (so far unsuccessful) by plant physiologists and breeders to increase net photosynthesis by inhibiting photorespiration or by selecting lines with reduced photorespiration. The most favoured hypothesis is that photorespiration acts to dissipate excess reductant produced under conditions of high irradiance but low potential

CO_2 fixation (e.g. water stress when the stomata are closed), and so avoid damage to the photosynthetic apparatus (termed photoinhibition – see below). Although there is good evidence for this hypothesis (e.g. Powles & Osmond 1978), there are a number of other protective mechanisms which may explain why C_4 plants are apparently not more sensitive to photoinhibition than are C_3 plants.

Measurement and analysis of CO_2 exchange

The main techniques for measuring the carbon balance and gas exchange of plants are reviewed by Šesták *et al.* (1971) and include growth analysis, radioactive tracers, and net CO_2 or O_2 exchange.

Growth analysis

Growth analysis (see also Evans 1972; Hunt 1990) is a powerful method for estimating long-term net photosynthetic production (photosynthesis minus respiration). It is based on readily obtainable primary measurements of plant dry weight and leaf dimensions made at intervals on growing plants or plant stands. It is also useful for analysing physiological adaptations of different species in terms of their partititioning of carbohydrate into leaves and other organs such as roots or seeds. This partitioning is at least as important as photosynthetic activity per unit area in determining productivity of different plant stands. The growth rate or rate of change of total plant dry weight ($\mathrm{d}W/\mathrm{d}t$) is obtained from a series of destructive harvests. It can be calculated for single plants or for plant stands and either expressed per unit total dry weight as a relative growth rate (RGR)

$$\mathrm{RGR} = (1/W)(\mathrm{d}W/\mathrm{d}t) \tag{7.3}$$

or else it is expressed per unit ground area as a crop growth rate (CGR). It is possible to derive the net photosynthetic rate per unit leaf area (A), called unit leaf rate or net assimilation rate (NAR)

$$\mathrm{NAR} = (1/A)(\mathrm{d}W/\mathrm{d}t) \tag{7.4}$$

from either of the following relations

$$\mathrm{NAR} = \mathrm{RGR}/\mathrm{LAR} \tag{7.5}$$

or

$$\mathrm{NAR} = \mathrm{CGR}/L \tag{7.6}$$

where LAR is the leaf area ratio (leaf area divided by total dry weight) and L is the leaf area index (leaf area per unit ground area). Note that NAR allows for respiratory losses at night and from non-photosynthesising tissues so is not quite equivalent to $\mathbf{P}_n$ measured on single leaves.

Use of radio-tracers

Photosynthesis can be measured using radioactive tracers for carbon or oxygen. One approach is to monitor the rate of loss of activity in the air in a closed chamber containing a photosynthesising leaf and initially supplied with a known specific activity of $^{14}CO_2$. More usually the leaf is allowed to fix $^{14}CO_2$ for a short period and then it is killed and the amount of ^{14}C incorporated into the leaf is determined in a scintillation counter. For short exposures (less than about 20 s), ^{14}C incorporation is a measure of *gross* photosynthesis (though it is an underestimate because the ^{12}C released by respiration may dilute the ^{14}C supplied). As the exposure time increases, the proportion of ^{14}C initially fixed that is re-released during the exposure increases so that with long exposures the technique may tend to estimate *net* photosynthesis. Another method that is useful for studying photosynthesis with stomatal control eliminated is to use thin slices of leaf tissue (see e.g. Jones & Osmond 1973), or isolated protoplasts or chloroplasts incubated with $H^{14}CO_3^-$ in solution.

Net gas-exchange

Micrometeorological measurements. Net CO_2 exchange of large areas of vegetation or complex plant communities is best estimated from micro-meterological measurements, as they interfere little with the environment and provide effective averages. In general, net CO_2 flux density per unit area of ground ($\mathbf{J}_C$) is negative during the day (representing absorption by the surface, $\mathbf{P}_n$) and positive at night (representing net respiratory loss) and may be obtained using the familiar transport equation (see equation 3.34, p. 66)

$$\begin{aligned} \mathbf{J}_C &= -\mathbf{P}_n = -\boldsymbol{K}_C(M_C/\mathcal{R}T)(\mathrm{d}p'/\mathrm{d}z) \\ &\simeq -\boldsymbol{K}_C(\mathrm{d}c'/\mathrm{d}z) \end{aligned} \tag{7.7}$$

from measurements of the gradient of CO_2 partial pressure ($\mathrm{d}p'/\mathrm{d}z$) within the crop boundary layer. Note that a prime (′) will be used throughout this chapter to distinguish various measures of CO_2 concentration (c', m', p' and x') and resistances (r') and conductances (g') to CO_2 diffusion. The appropriate transfer coefficient ($\boldsymbol{K}_C$) may be obtained from aerodynamic or heat balance measurements. Other techniques such as eddy correlation can also be used.

Cuvette measurements. The commonest technique involves measurements of net gas-exchange in cuvettes that can range in size from small (*c.* 1 cm^2) single-leaf chambers for portable porometers (cf. Chapter 6) to large (> 10 m^3) chambers that can enclose whole plants or areas of canopy (see e.g. Pearcy *et al.* 1989). The degree of environmental control in these cuvettes can vary in sophistication up to completely independent control of temperature, light, humidity and of CO_2 and O_2 concentrations in the best laboratory systems.

In a closed system the molar flux density of CO_2 ($\mathbf{J}_C^m$, mol m^{-2} s^{-1}) can be obtained from the rate of change of CO_2 concentration (dx'/dt, in units of (mol CO_2 per mol air) s^{-1}) as

$$\mathbf{J}_C^m = -\mathbf{P}_n^m = (PV/\mathcal{R}TA)(dx'/dt) \tag{7.8}$$

where A is the reference area (usually the leaf area in the chamber, m^2) and V is the volume of the chamber (m^3). [The corresponding mass flux density ($\mathbf{P}_n$) is given by V/A multiplied by the rate of change of CO_2 concentration (dc'/dt).]

In the more common open or semi-open gas-exchange systems, the flux of CO_2 is obtained from the difference in the flow of CO_2 into the cuvette ($u_e x'_e$) and the flow of CO_2 out ($u_o x'_o$) as

$$\mathbf{P}_n^m = (u_e x'_e - u_o x'_o)/A \simeq u_e(x'_e - x'_o)/A \tag{7.9}$$

where u is the molar flow rate (mol s^{-1}). This corresponds to equation 6.4 for the estimation of evaporation. For precise work it is necessary to allow for the small difference between the flow entering the cuvette (u_e) and the outflow (u_o) that results from water vapour efflux from leaves (any CO_2 flux can be neglected as it is balanced by an O_2 flux). For a typical ($e_o - e_e$) of 1 kPa the relative increase in u is only about 1% (i.e. $(e_o - e_e)/P$).

Measurement of CO_2 concentration – IRGAs. The most usual method for CO_2 detection is by means of infra-red gas analysers (IRGAs) that make use of the strong CO_2 absorption in the IR (especially at 4.26 μm: see Fig. 2.5, p. 22). The radiation from an IR source passes through a fixed-volume analysis cell containing the sample gas to a detector, and the absorptance depends on the number of molecules of CO_2 in the optical path. Cross sensitivity to other atmospheric gases (especially water vapour, which has a common absorption band with CO_2 at 2.7 μm) can be eliminated by use of interference filters to restrict sensitivity to the 4.26 μm absorption band.

IRGAs are usually calibrated in terms of the volume fraction of CO_2 (obtained for example by mixing known volumes of CO_2 and CO_2-free air or nitrogen with precision mixing pumps). In the atmosphere, the mole fraction of CO_2 is now nearly 350×10^{-6} or 350 volume parts per million (vpm), which is a concentration of 632 mg m^{-3} at 100 kPa and 20 °C (see Chapter 11). For gases the volume fraction is identical to the mole fraction (x). Other measures of CO_2 concentration can be obtained using the conversions described in Chapter 3 (i.e. $p' = x'P$; and $c' = x'PM_C/\mathcal{R}T = p'M_C/\mathcal{R}T$). Although the IRGA actually measures the molar concentration ($= c'/M_C$) in the analysis cell it is more convenient to use the more conservative quantities of mole fraction or partial pressure. For example, as the analysis cell is usually maintained at a constant temperature, the IRGA reading is constant for a given mole fraction in the cuvette whatever the cuvette temperature, though it is necessary to correct for any pressure difference between measurement and calibration, i.e.

$$x'_{true} = x'_{observed} P^o/P \quad (7.10)$$

where P^o is the calibration pressure.

Oxygen exchange. Sensitive polarographic electrodes for the study of O_2 evolution in the gas phase (Delieu & Walker 1983) are particularly useful for the measurement of maximum O_2 evolution rates. Although stomatal closure limits the value of measurements at normal CO_2 concentrations, it is possible to make measurements at higher CO_2 concentrations than are possible using IRGAs, so that it is possible to estimate the photosynthetic capacity. For stressed tissue with stomatal closure it may be necessary to make measurements with as much as 15% CO_2 to obtain a true $\mathbf{P}_{max}$ (though beware possible inhibition of photosynthesis in C_4 plants).

Estimation of photorespiration. Although many methods have been proposed for estimating photorespiration (see Ludlow & Jarvis 1971*a*), none is altogether suitable because of our incomplete understanding of the gas-exchange processes within leaves. Most methods suffer from the difficulty of separating photorespiration from any dark respiration continuing in the light (often assumed to equal the rate in the dark, but see Graham 1980), and inter- or intracellular reassimilation of CO_2 is difficult to quantify. One useful estimate is the magnitude of the enhancement of $\mathbf{P}_n$ in low O_2 (see below), but this is an overestimate because it probably includes some stimulation of carboxylase resulting from relief of competitive inhibition by the oxygenase. Similarly the **CO_2 compensation concentration**, Γ (or CO_2 release into CO_2-free air) can be used to estimate photorespiration (see e.g. Farquhar *et al.* 1980*b*). Other methods include the size of the transient burst of CO_2 release observed on darkening a leaf, the rate of $^{14}CO_2$ output by a leaf supplied with [^{14}C]glucose, or studies using simultaneous measurements of exchange of different O or C isotopes.

Chlorophyll fluorescence

As indicated above, the analysis of chlorophyll *a* fluorescence provides a powerful probe of the functioning of the intact photosynthetic system. Two main techniques are available for fluorescence analysis: (i) analysis of the rapid (ms timescale) and slow (s–min timescale) components of the changes in chlorophyll fluorescence (the 'Kautsky' effect) that are observed on

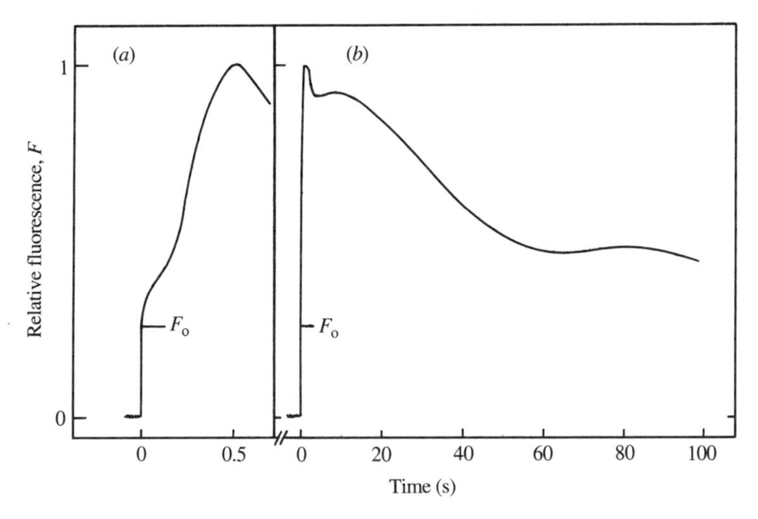

Fig. 7.7. Characteristic fluorescence induction kinetics (Kautsky curve) on illuminating a dark-adapted leaf, showing the very rapid rise to F_0 followed by fairly fast rise to a maximum, followed by slower decay to a steady state as photosynthesis starts up: (*a*) fast kinetics, (*b*) slow kinetics.

illuminating a dark-adapted leaf (Fig. 7.7), and (ii) use of modulated fluorescence to detect the fluorescence from a normally photosynthesising leaf in white light. When combined with the use of high intensity (saturating) pulses of light, this latter technique permits analysis of the processes that quench (i.e. decrease) fluorescence and provides a non-destructive measure of the relationship between the light and dark reactions of photosynthesis.

In a traditional Kautsky system, the leaf would be illuminated with an exciting light with wavelengths less than *c*. 620 nm, and fluorescence from PS II (peak emission at 695 nm) measured with a sensitive photodetector having a narrow band filter (centred on 695 nm). This arrangement ensures that the small fluorescence signal can be separated from the much larger amount of reflected exciting light.

On first illuminating a dark-adapted leaf, where all components of the electron transport chain would be fully oxidised, fluorescence (F) immediately rises to a level (F_0) that is characteristic of open PS II reaction centres and fully oxidised primary electron acceptor (Q_A). The quantum yield of fluorescence in this situation is given by equation 7.2. As light is absorbed and charge separation occurs, thus closing reaction centres and reducing Q_A, fluorescence rises in a complex manner (Fig. 7.7) reaching a peak (F_m) when all of Q_A is fully reduced. At this point, because all the reaction centres are closed, chlorophyll excitation can no longer decay through photochemistry (i.e. k_P falls to zero), so the quantum yield of fluorescence decreases (is quenched) from the value given by equation 7.2 to

$$\phi F = \frac{k_F}{k_F + k_D + k_T} \tag{7.11}$$

As electron transport starts, and photosynthesis increases, fluorescence slowly declines through a number of transients to a steady state.

Unfortunately the Kautsky system cannot be used to study photosynthesis in white light in the field because of the need to separate the actinic and fluorescent wavelengths. The development of modulated fluorescence systems overcomes this limitation and enables one to distinguish the various quenching processes that cause the decline in fluorescence at time t ($F(t)$) from F_m. The principle of modulated systems is that a very weak ($\sim$ 1–5 μmol m^{-2} s^{-1}) light source is switched on and off rapidly (modulated), and the measuring system is set to detect only the fluorescence signal that corresponds to the modulated exciting light. By the use of sensitive electronics it is possible to distinguish this signal from the very much larger signal arising from the illuminating white light. As long as the modulated light is so weak that it cannot drive electron transport, the fluorescence obtained with illumination by the modulated source alone is equivalent to F_0. It is common to normalise measurements using F_0, so that differences in leaf chlorophyll content or sensor geometry that give rise to different absolute levels of fluorescence can be neglected.

Quenching analysis

A lot of information about the functioning of the photosynthetic system can be obtained from an analysis of fluorescence quenching. The various fluorescence parameters that can be obtained using a modulated system and the method of calculation of the different quenching coefficients is illustrated in Fig. 7.8 and summarised in Table 7.2. (As a word of warning it is worth noting here that although some standardisation of terminology has been attempted, there are still important differences in the precise meaning of similar terms as used by different authors. In particular, some authors use F_m or a similar term only to refer to the maximum fluorescence after a long dark adaptation period, while others (as here) use it to describe the maximum obtainable with a saturating flash at any time.)

Three main forms of fluorescence quenching are distinguished:

(*a*) *Photochemical quenching* (q_P sometimes written as q_Q). This is a measure of the redox state of Q_A and ranges from zero when Q_A is fully reduced (i.e. at F_m where fluorescence is maximal and the maximum amount

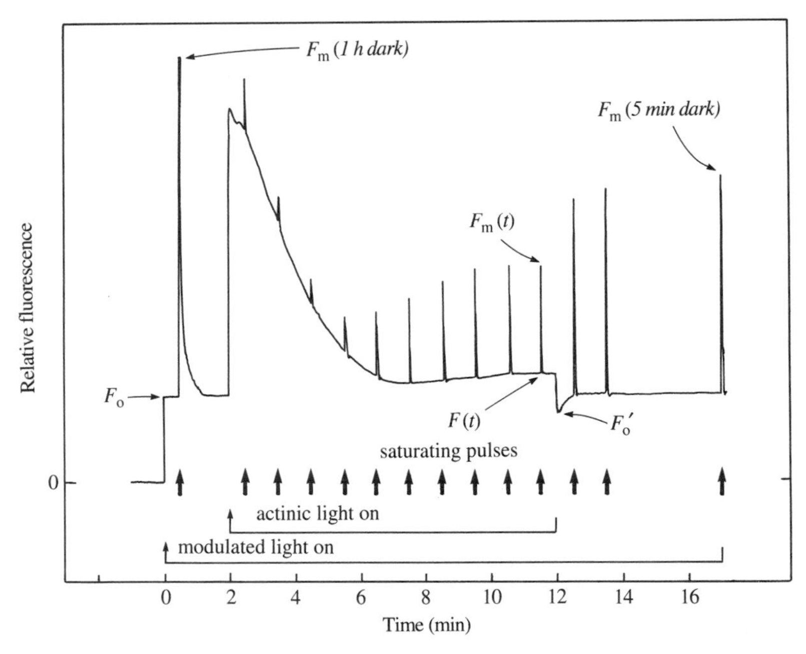

Fig. 7.8. Use of a modulated fluorescence system to analyse quenching, showing the variables measured. $F(t)$ is the fluorescence level at any time t, F_o is the fluorescence level with only the low irradiance ($\sim 1\ \mu$mol m^{-2} s^{-1}) modulated light source, F_m is the fluorescence with a saturating light pulse (annotated as necessary – i.e. F_m (1 *h dark*) refers to F_m after a 1 h dark pretreatment), F_o' is a quenched value of F_o.

of incident energy is re-emitted as fluorescence because all reaction centres are closed) to unity when Q_A is oxidised and all PS II reaction centres are open to accept excitation energy (i.e. a minimum amount of excitation is re-emitted as fluorescence). The value of q_P at any instant t may be obtained from the fluorescence at that time ($F(t)$) together with corresponding values of F_m (obtained at the same time with a saturating flash of light) and F_o according to

$$q_P = (F_m(t) - F(t))/(F_m(t) - F_o) \quad (7.12)$$

(*b*) *Non-photochemical quenching* (q_N). This is used to describe a wide range of mechanisms that can cause F_m at any time t ($F_m(t)$) obtained with a saturating flash to fall below the maximum value obtained after a long (e.g. one hour) period in the dark (e.g. F_m(1 *h dark*)). These are normally lumped together in q_N, which is calculated from (see Fig. 7.8)

$$q_N = (F_m(1\ h\ dark) - F_m(t))/(F_m(1\ h\ dark) - F_o) \quad (7.13)$$

Table 7.2. *Calculation of various fluorescence parameters*

Efficiency of PS II:	
$F_v/F_m = (F_m(t) - F_o)/F_m(t)$	(7.17)
Quenching coefficients:	
$q_P = (F_m(t) - F(t))/(F_m(t) - F_o)$	(7.12)
$q_N = (F_m(1\ h\ dark) - F_m(t))/(F_m(1\ h\ dark) - F_o)$	(7.13)
Where there is F_o quenching, so that the fluorescence (F'_o) in the presence of far red light (or immediately after switching off the actinic light) is lower than F_o, q_o quenching is defined as:	
$q_o = (F_o - F'_o)/F_o = 1 - (F'_o/F_o)$	
Since it is necessary to correct for the fact that $F_m(1\ h\ dark)$ is measured without any q_o (Bilger & Schreiber 1986), the other quenching coefficients become:	
$q_P = (F_m(t) - F(t))/(F_m(t) - F'_o)$	(7.12a)
$q_N = [F_m(1\ h\ dark) - (F_m(t) \cdot F_o/F'_o)]/(F_m(1\ h\ dark) - F_o)$	(7.13a)

The main components of q_N are:

(i) Energy-dependent quenching (q_E). This depends on the state of energisation of the thylakoids or the pH gradient (ΔpH) across the thylakoid membrane. Although the mechanism is not clear, it seems that an increase in ΔpH causes an increase in the radiationless dissipation of absorbed radiant energy as heat (i.e. k_D in equations 7.2 and 7.11 increases) and thus decreases the amount of fluorescence. This presumably helps to dissipate the excess energy that is available for causing damage (photoinhibition) to the photosynthetic machinery when the energy cannot be utilised in photosynthesis (e.g. at low CO_2 levels or in water stress).

(ii) State transitions (q_T). The maximum fluorescence from PS II can also be decreased by a transfer of some of the excitation energy from PS II to PS I (increasing k_T); this transfer is under control of the phosphorylation state of the LHC protein.

(iii) It is also likely that carotenoids within the chloroplast can provide a further non-photochemical quenching mechanism. Excess light leads to an increase in the level of zeaxanthin (by de-epoxidation of violaxanthin) which acts as a particularly efficient quencher of excitation in the antenna chlorophyll.

(*c*) *Photoinhibition* (q_I). A particularly important type of quenching is termed photoinhibition (q_I) and is generally taken to include irreversible, or only slowly reversible, damage to the photosynthetic system that lowers F_m(1 *h dark*) below the value found in a healthy plant. Although often lumped in with other forms of non-photochemical quenching, the present definition clearly indicates the distinction in terms of rate of relaxation. Useful reviews of various aspects of photoinhibition may be found in Kyle *et al.* (1988). All types of non-photochemical quenching act to dissipate excess energy absorbed by the chloroplasts in high light conditions when it cannot be used in photosynthesis (e.g. when photosynthesis is inhibited by cold, water stress or low CO_2 concentrations). In the absence of these dissipation mechanisms, photoinhibitory and photobleaching damage would be much greater.

In some situations, especially in stressed plants, it is also possible for F_o to be quenched (Fig. 7.8). This must be allowed for in calculating the various quenching coefficients (see Table 7.2).

Efficiency of photochemistry

The efficiency of excitation energy capture by open PS II reaction centres can be derived from consideration of equation 7.2. With all reaction centres open, $F = F_o = \mathbf{I}\,\phi F$, so this equation gives F_o as

$$F_o = \mathbf{I}\left\{\frac{k_F}{k_F + k_D + k_T + k_P}\right\} \tag{7.14}$$

Assuming no change in the non-photochemical dissipation processes, the variable fluorescence (F_v, the difference between the maximal fluorescence, F_m (given by equation 7.11) and F_o) is given by

$$F_v = F_m - F_o = \mathbf{I}\left\{\frac{k_F}{k_F + k_D + k_T} - \frac{k_F}{k_F + k_D + k_T + k_P}\right\} \tag{7.15}$$

Dividing through by F_m and rearranging gives

$$F_v/F_m = \left\{1 - \frac{k_F + k_D + k_T}{k_F + k_D + k_T + k_P}\right\} = \frac{k_P}{k_F + k_D + k_T + k_P} \tag{7.16}$$

which shows that the proportion of absorbed energy used in photochemistry is given by F_v/F_m, which is therefore a measure of the efficiency of photochemistry of open reaction centres of PS II.

The ratio F_v/F_m for dark-adapted healthy plants is normally close to 0.83 (Björkman & Demmig, 1987), showing that with open reaction centres $k_P \gg (k_F + k_D + k_T)$, though slightly different values are obtained with different measuring instruments. Not only does photoinhibition lower F_v/F_m from this optimal level, but so do all other non-photochemical quenching mechanisms. Although q_E (and ΔpH) is normally thought to disappear within a few seconds of darkening in isolated chloroplasts, it may take a minute or so to disappear in an intact leaf. For leaves, this rapidly decaying component of q_N is usually ascribed to q_E, while more slowly decaying quenching can be attributable to state transitions or the rather slow reconversion of zeaxanthin back to violaxanthin in the xanthophyll cycle. Any q_N that takes longer than an hour or so to relax is often operationally defined as photoinhibition, so that F_v/F_m measured on a dark-adapted leaf is a measure of photoinhibition.

Because q_P measures the proportion of PS II reaction centres that are oxidised or 'open', and F_v/F_m measures the efficiency of these open centres for electron transport, it has been argued by Genty *et al.* (1989) that the product of these two variables gives the quantum yield of non-cyclic electron transport through PS II (ϕ_{PSII}). Substituting from equation 7.12 and cancelling terms gives

$$\begin{aligned}\phi_{PSII} &= (F_v/F_m)\,q_P = ((F_m - F_o)/F_m)(F_m - F(t))/(F_m - F_o)\\ &= \Delta F/F_m \end{aligned} \qquad (7.17)$$

where ΔF is the difference between steady state and maximal fluorescence at time *t* under the same conditions of non-photochemical quenching. It is therefore very simple to estimate ϕ_{PSII} using measurements of only the steady-state fluorescence (F) and the F_m obtained with a saturating flash.

Control of photosynthesis

It is of general interest to be able to determine the relative importance of different processes, such as the light reactions, the dark reactions, diffusion through the boundary layer or the stomata, or even supply of phosphate, in controlling the rate of net assimilation. For example, workers concerned to improve crop photosynthesis need to be able to identify those processes that are most important in restricting the rate of assimilation. Traditionally this has been achieved by the use of the electrical analogues that were introduced in Chapter 3. Because of the series nature of the CO_2 uptake pathway it is most convenient to use resistances rather than conductances for this purpose. In this approach the whole photosynthetic system is treated as a linear process to which the transport equation applies, with the overall resistance to CO_2 uptake ($\Sigma r'$) being partitioned into, for example, gas-phase (r'_g) and liquid-phase or mesophyll (r'_m) components, according to:

$$\mathbf{P}_n^m = \frac{x'_a - x'_x}{\Sigma r'} = \frac{x'_a - x'_i}{r'_g} = \frac{x'_i - x'_x}{r'_m} \qquad (7.18)$$

where x'_a, x'_i and x'_x, respectively, are the mole fractions of CO_2 in the ambient air, in the air at the surface of the mesophyll cells, and 'internal' to the carboxylation site (see below). [Instead of using mole fraction one can express this equation in terms of the partial pressure of CO_2 (p_C or p') replacing x' by p'/P).]

Calculation of resistances. An essential step in calculation of the resistances in equation 7.18 is the estimation of CO_2 mole fraction in the intercellular spaces (x'_i). The usual approach is first to determine the gas-phase resistance to water loss ($r_{gW} = r_{aW} + r_{\ell W}$) using the methods outlined in Chapter 6; this is then converted to the corresponding gas-phase CO_2 transfer resistance (r'_g) assuming that the paths for CO_2 and water vapour are similar (see Table 3.2):

$$\begin{aligned} r'_g &= r'_a + r'_\ell = (\mathbf{D}_W/\mathbf{D}_C)^{\frac{2}{3}} r_{aW} + (\mathbf{D}_W/\mathbf{D}_C) r_{\ell W} \\ &= 1.39 r_{aW} + 1.64 r_{\ell W} \simeq 1.6 r_{gW} \end{aligned} \quad (7.19)$$

The approximate conversion factor of 1.6 is normally used where r_{aW} and $r_{\ell W}$ are not measured separately. Significant cuticular water loss can cause equation 7.19 to underestimate r'_g because the large liquid-phase component of the cuticular pathway would lead to a large extra resistance to CO_2 transport. The third step is calculation of x'_i from a rearranged equation 7.18 as

$$x'_i = x'_a - \mathbf{P}^m_n r'_g \quad (7.20)$$

Although the approximation represented by equation 7.20 is adequate for many purposes, in precise studies it is better to take account of the ternary interactions between air, H_2O and CO_2. Von Caemmerer & Farquhar (1981) derived the following improved approximation:

$$x'_i = \frac{\{(1/r'_g) - (\mathbf{E}/2)]x'_a\}}{[(1/r'_g) + (\mathbf{E}/2)]} - \mathbf{P}^m_n \quad (7.21)$$

A very useful graphical device for investigating the relative contribution of gas- and liquid-phase resistances was introduced by Jones (1973*a*) and is illustrated in Fig. 7.9. Figure 7.9*a* shows a response curve relating $\mathbf{P}^m_n$ to x'_i, obtained by varying ambient CO_2 concentration and plotting $\mathbf{P}^m_n$ obtained against the calculated x'_i; this curve represents the photosynthetic 'demand' function. Figure 7.9*b* shows the relationship between $\mathbf{P}^m_n$ and x'_i calculated according to equation 7.20 for two different values of r'_g; these straight lines represent photosynthetic 'supply' funtions, and indicate the drop in mole fraction across the gas phase, showing how x'_i would fall below x'_a as $\mathbf{P}^m_n$ increases. The actual value of x'_i and $\mathbf{P}^m_n$ in any particular conditions is obtained from the intersection of the supply and demand functions (Fig. 7.9*c*).

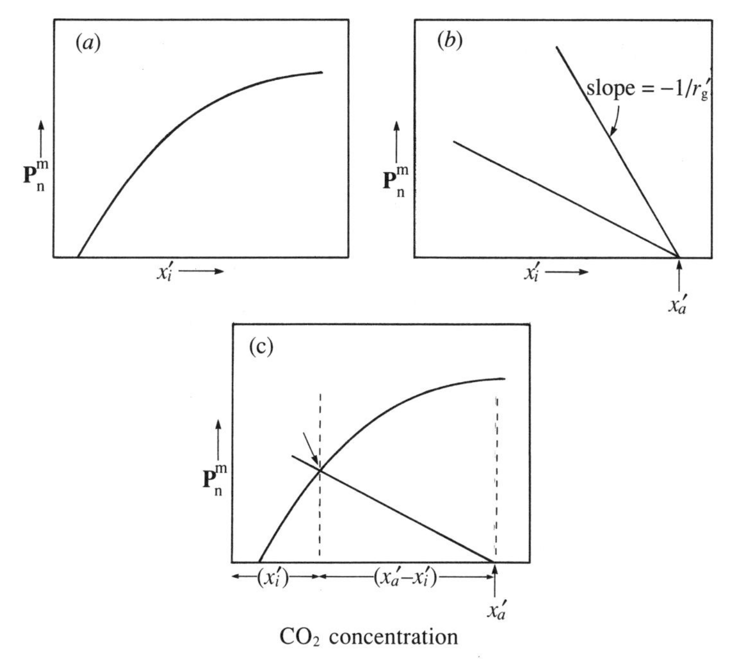

Fig. 7.9 (*a*) Response curve or 'demand' function relating $\mathbf{P}_n^m$ to x_i', (*b*) two 'supply' functions relating $\mathbf{P}_n^m$ to x_i' for different gas-phase resistances, with the slopes of the lines equal to $1/r_g'$, and (*c*) simultaneous solution of supply and demand functions to obtain x_i' at the operating point indicated by the arrow.

Effect of stomatal heterogeneity. A critical assumption in gas-exchange studies and especially in the calculation of the CO_2 concentration (expressed in terms of mole fraction) at the cell wall and the mesophyll resistance is that the stomata are equally open over the whole area of leaf whose gas-exchange is being measured. If that assumption does not hold, as for example where the stomata close in 'patches', $\mathbf{P}_n^m$, r_g' and the calculated values of x_i' and r_m' only represent average values; this can give very misleading indications of physiological responses to environment. The reason for this is illustrated in Fig. 7.10. All leaves are to some extent heterogeneous and evidence is accumulating that there can be variation in stomatal aperture over the leaf surface, especially when abscisic acid is applied exogenously (Terashima *et al.* 1988), but also in other situations (Laisk *et al.* 1980). The implications of heterogeneity are most important where its degree changes as a result of an environmental change. It is therefore essential to confirm whether stomatal heterogeneity exists for any material being studied before applying resistance analysis.

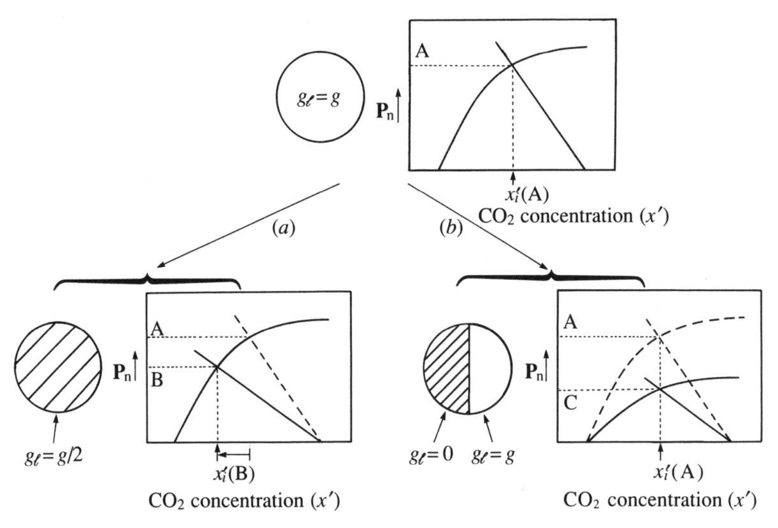

Fig. 7.10. An illustration of the effect of patchy stomatal closure on calculated x_i' and on the estimated mesophyll properties. The top figure represents a healthy leaf with the photosynthesis demand function given by the solid line, resulting in a photosynthetic rate A. Assuming no change in the photosynthetic properties of the leaf we can compare what happens if (a) all the stomata close by 50 %, or (b) if half the stomata (in discrete patches unconnected via the intercellular spaces) close completely. In (a) stomatal conductance per unit area falls to $g/2$, assimilation decreases according to the response curve, falling to B (> A/2), and the calculated x_i' decreases significantly (implying increased stomatal control of photosynthesis as one would expect). In (b) on the other hand, although the average stomatal conductance over the large area also falls to $g/2$, assimilation would now decrease to C (= A/2) as only half the area is now photosynthesising, but because both A and g_ℓ have changed similarly it follows from equation 7.20 that x_i' is apparently unchanged. In addition there is a large apparent shift in the photosynthetic demand function with the photosynthetic capacity apparently reduced by 50 % even though there is no real change in the mesophyll properties.

Mesophyll resistance. The liquid-phase (or mesophyll) resistance is a complex term that includes a liquid-phase transport component and an 'enyme' component that depends on the activities of biochemical and photochemical processes including respiration. The major component of r_m' is probably a carboxylation resistance, so it is appropriate that its reciprocal is often termed 'carboxylation efficiency'. The definition of r_m' using equation 7.18 involves an important extension of the transport equation to biochemical processes. An assumption implicit in the use of resistance analysis is that the relative importance of the component processes in limiting photosynthesis (their relative 'limitations') is proportional to their respective resistances. This is discussed further below.

There are two main difficulties with equation 7.18 for defining the mesophyll resistance. The first difficulty arises because the calculated r'_m increases as CO_2 saturation approaches, so that unlike the situation for the gas-phase resistance (a true transport resistance), the concentration drop across the mesophyll does not vary in direct proportion to the flux. The second problem is the need to have an estimate of x'_x; although it is sometimes assumed equal to 0 (e.g. Gaastra 1959), probably with some validity for C_4 plants, in other cases this assumption leads to the calculated r'_m being variable for all x'_i. Both these problems limit the general predictive value of equation 7.18. The best approach is to limit application of the term mesophyll resistance (r'_m) to the linear part of the CO_2 response curve and to assume that x'_x equals the compensation concentration (Γ), that is the CO_2 concentration at which $\mathbf{P}_n^m$ is zero. This gives a constant r'_m that can be obtained from either

$$r'_m = 1/g'_m = (x'_i - \Gamma)/\mathbf{P}_n^m \tag{7.22}$$

or

$$r'_m = 1/g'_m = \mathrm{d}x'_i/\mathrm{d}\mathbf{P}_n^m \tag{7.23}$$

That is, r'_m is given by the inverse of the *initial* slope of the $\mathbf{P}_n^m \cdot x'_i$ curve. This restricted definition of r'_m is the most useful as it is always valid. In all other cases (e.g. at non-limiting CO_2 or when x'_x is not equal to Γ) it is more appropriate to call the mesophyll component a 'residual' resistance.

Some typical values for $\mathbf{P}_n$ and $\mathbf{R}_d$, and for r'_m and r'_ℓ are given in Table 7.3. This shows that the ratio r'_ℓ/r'_m tends to be greater in C_4 than in C_3 plants.

Photosynthetic limitations

There have been many attempts to quantify the relative control of photosynthesis exerted by different factors (e.g. CO_2 supply, carboxylase activity, etc.). These efforts were stimulated particularly by Blackman's (1905) paper enunciating the *Principle of Limiting Factors*: 'When a process is conditioned as to its rapidity by a number of separate factors, the rate of the process is limited by the pace of the "slowest" factor'. Although the concept that the rate is limited by the slowest factor is useful to a first approximation, it has led to a widespread failure to recognise that in well adapted plants several factors (e.g. gas-phase and liquid-phase processes) may be evenly balanced and all contribute to the overall limitation. Before discussing methods for estimating the relative limitation imposed by

Table 7.3. *Some representative values for various CO_2 exchange parameters for different species*

$\mathbf{P}_n$ at saturating light and normal ambient CO_2 (μmol m^{-2} s^{-1}), $\mathbf{R}_d$ (μmol m^{-2} s^{-1}), Γ (μmol mol^{-1}), minimum r'_m (m^2 s mol^{-1}), and corresponding values of r'_ℓ (m^2 s mol^{-1}) and r'_ℓ/r'_m. Values converted from original units assuming a temperature of 20 °C.

	$\mathbf{P}_n$	$\mathbf{R}_d$	Γ	r'_m	r'_ℓ	r'_ℓ/r'_m	Ref.
C_3 plants							
Atriplex hastata (young plants)	25	2.5	50	6.8	1.8	0.26	1
Wheat (9 species)	24	—	41	7.5	2.5[a]	0.33	2
Sitka spruce (field)	11	0.7	50	12.5	10.8	0.86	3
4 Tropical legumes (30 °C)	18	1.8	35	8.3	1.8	0.22	4
Larrea divaricata	23–27	3–4	—	6.5[b]	2.5	0.38	6
C_4 plants							
Atriplex spongiosa	40	2.3	0	1.5	4.3	2.8	1
6 Pasture grasses (30 °C)	36	2.5	0	2.5	2.1	0.85	4
Zea mays	14–55	1.4	0	2.0	6.0	3.0	7
CAM plants							
Kalanchoe diagremontiana (light)	6	—	51	20	21	1.06	5
Kalanchoe diagremontiana (dark)	5	—	—	—	25	—	5

[a] = r'_g.
[b] = estimated.

References: 1. Slatyer 1970; 2. Dunstone *et al.* 1973; 3. Ludlow & Jarvis 1971*b*; 4. Ludlow & Wilson 1971*a*, 1972; 5. Allaway *et al.* 1974; 6. Mooney *et al.* 1978; 7. Gifford & Musgrave 1973; El-Sharkawy & Hesketh 1965.

stomatal and mesophyll processes it is useful to introduce a method for the analysis of control in biochemical pathways.

Control analysis. A widely applicable quantitative approach that can be applied to the analysis of any complex metabolic pathway, where changes in one component affect the performance of other steps (for example, by altering substrate concentration or by feedback regulation), as occurs in photosynthesis, has been introduced by Kacser & Burns (1973). They developed a general theory of 'control' in relation to biochemical pathways that can be used to predict system response to *small* changes in one of the components, from a knowledge of the properties of the components in isolation. Central to the approach is the calculation of **Control** or

Sensitivity coefficients (C) which reflect the response of steady-state fluxes ($\mathbf{J}$) through the *whole* system to small changes in certain parameters (b_i) such as individual enzyme activities. In order to make the control coefficients independent of the units chosen for $\mathbf{J}$ and b, they are defined in terms of fractional changes in these two quantities:

$$C_{bi} = (\partial \mathbf{J}/(\mathbf{J})/(\partial b_i/b_i) \qquad (7.24)$$

Defined in this way the control coefficient varies from 0 (when the component being studied has no effect on the overall process) to 1 (when that step exerts complete control). The relative size of the different control coefficients in a pathway is a measure of their relative importance in controlling flux. An important property of the C_{bi} is that all the control coefficients in one pathway sum to 1.

A particular advantage of control analysis is that it provides a technique for estimating system behaviour (that is control coefficients) from a knowledge of the **elasticities** (ε_i) of all the individual reactions when isolated from the whole pathway, where the individual ε_i are defined as the fractional changes in the *local* reaction rate for a fractional change in substrate concentration at that step (with all other components of the overall sequence being held constant). Further details of the calculation of elasticities and control coefficients have been outlined by Kacser & Burns (1973). So far control analysis has been applied to photosynthesis in a few cases only, though some of the problems and techniques required are discussed by Giersch *et al.* (1990).

Relative limitation due to stomata and mesophyll. A detailed discussion of methods for partitioning limitations may be found in Jones (1985a). Implicit in the use of resistance analogues, as described above, is the assumption that they provide a means for quantifying the relative contributions of different processes to the control of $\mathbf{P}_n$. In principle, the ratio of the resistances of different components (e.g. r'_g and r'_m), or the concentration drops across them, is a measure of their relative control of $\mathbf{P}_n$ (see Fig. 7.11). On this basis the gas-phase limitation (ℓ'_g) can be defined as the relative contribution of gas-phase diffusion to the overall limitation as in

$$\ell'_g = r'_g/(r'_g + r'_m) = r'_g/\Sigma r' \qquad (7.25)$$

Unfortunately this definition of limitation is only useful under CO_2-limiting conditions and fails as CO_2 saturation is approached. The reason may be illustrated by taking an extreme case: consider a plant where the photosynthetic system is operating at its maximum rate ($\mathbf{P}_{max}$) so that diffusion is non-limiting. In this case the calculated value of ℓ'_g may be finite because r'_m and r'_g remain finite, but clearly it is not reasonable to say that

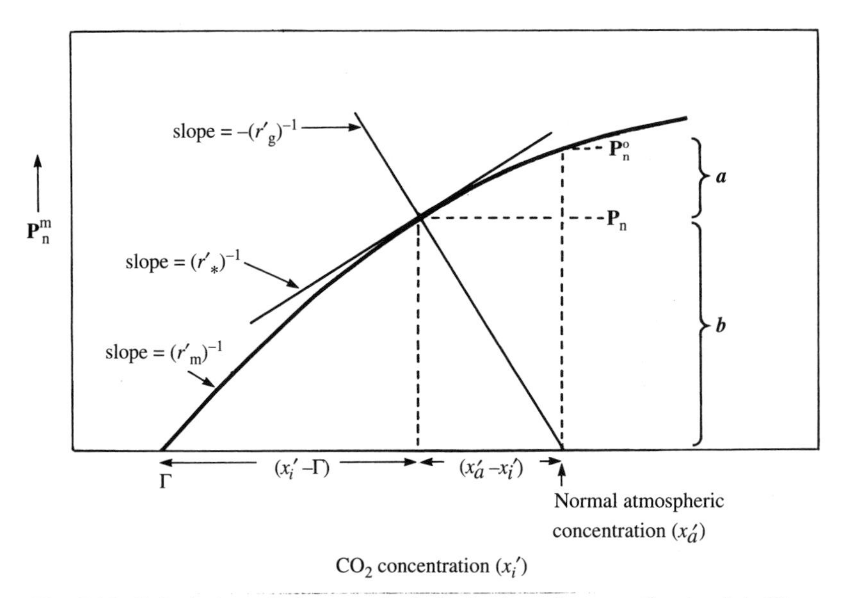

Fig. 7.11. Calculation of photosynthetic limitations (see text for details). The heavy line represents the demand function relating $\mathbf{P}_n^m$ to x'_i for a hypothetical leaf. The CO_2 concentration drop across the mesophyll is represented by $(x'_i - \Gamma)$, and that across the gas-phase by $(x'_a - x'_i)$. Farquhar & Sharkey's (1982) definition of ℓ'_g is given by $\boldsymbol{a}/(\boldsymbol{a}+\boldsymbol{b})$; the recommended definition (equation 7.27) is $r'_g/(r'_g + r'_*)$.

there is a finite gas-phase limitation as a small change in stomatal conductance would have no effect on $\mathbf{P}_n$.

As an alternative, Farquhar & Sharkey (1982) suggested calculation of the gas-phase limitation as the relative reduction in actual $\mathbf{P}_n$ below the potential rate ($\mathbf{P}_n^o$) at infinite gas-phase conductance (see Fig. 7.11). Unfortunately the assumption of infinite gas-phase conductance as used in this method is unrealistic and little better (and no more likely to be achieved in practice) than the logical alternative of deriving $\mathbf{P}_n^o$ at infinite mesophyll conductance. It seems more appropriate to use a method such as that given below that is generally valid and does not involve unrealistic extrapolation.

Returning to control analysis, it is clear that one option, by analogy with the definition of control coefficient, would be to define the gas-phase limitation (ℓ'_g) as the relative sensitivity of $\mathbf{P}_n$ to a small change in r'_g:

$$\ell'_g = (\partial \mathbf{P}_n/\mathbf{P}_n)/(\partial r'_g/r'_g) = (\partial \mathbf{P}_n/\partial r'_g)(r'_g/\mathbf{P}_n) \qquad (7.26)$$

It can be shown (see Jones 1973*a*) that this is equivalent to

$$\ell'_g = r'_g/(r'_g + r'_*) \tag{7.27}$$

where r'_* is the slope $dx'_i/d\mathbf{P}_n$ at the operating point (Fig. 7.11). Advantages of this approach include: (i) it does not involve extrapolation, and (ii) it is simple to apply. It is particularly useful in plant breeding or in studies of plant response to environment where one is concerned with small changes in stomatal resistance.

So far only the relative limitation by gas-phase and mesophyll processes under one set of conditions has been considered. In practice, it is useful to be able to determine the relative contributions to a *change* in $\mathbf{P}_n$ (see Jones 1985*a* for a detailed discussion of possible methods).

Carbon isotope discrimination

The two stable isotopes of carbon (^{13}C and ^{12}C) occur in the molar ratio of 1:89 in the atmosphere, but the relative abundance of ^{13}C in plant material is commonly less than its abundance in atmospheric carbon dioxide. The value of the molar abundance ratio, $^{13}C/^{12}C$, in plant material provides information on the physical and chemical processes involved in its synthesis because processes such as diffusion or carboxylation discriminate against the heavier isotope by different amounts (see Farquhar *et al.* 1989 for a useful review). These 'isotope effects' result in an amount of discrimination (Δ), usually defined by

$$\Delta = \frac{(^{13}C/^{12}C)_{reactants}}{(^{13}C/^{12}C)_{products}} - 1 \tag{7.28}$$

In practice Δ is not measured directly, and isotopic abundances (determined using a mass spectrometer) are usually expressed as deviations ($\delta^{13}C$ or δ) from the abundance ratio in a fossil belemnite formation (the PDB standard, where $^{13}C/^{12}C = 0.01124$), as

$$\delta = \frac{(^{13}C/^{12}C)_{sample}}{(^{13}C/^{12}C)_{standard}} - 1 \tag{7.29}$$

By combining equations 7.28 and 7.29 it follows that

$$\Delta = \frac{\delta_a - \delta_p}{1 + \delta_p} \tag{7.30}$$

Table 7.4. *Typical values for the molar abundance ratio* ($^{13}C/^{12}C$), *the deviation of isotopic composition from that of the PDB standard* ($\delta^{13}C$ *or* δ), *and the isotopic discrimination* (Δ) (*from data quoted by Ehleringer & Osmond* 1989; *Farquhar* et al. 1989)

	$^{13}C/^{12}C$	$\delta^{13}C$ (‰)	Δ (‰)
PDB standard	0.011237	0	—
free atmospheric CO_2 (in 1988)	0.01115	−7.7	—
C_3 plant material	0.01085–0.01102	−20 to −35	13 to 28
C_4 plant material	0.01107–0.01116	−7 to −15	−1 to 7
CAM plant material	0.01099–0.01112	−10 to −22	2 to 15
Coal	0.01087	−32.5	—

where the subscripts 'a' and 'p' refer to the air and plant material respectively. Some typical values of Δ and δ are presented in Table 7.4, which shows that although δ tends to be negative when using the PDB standard, the value of the discrimination, Δ, is positive. Because both Δ and δ are small numbers they are conventionally presented as parts per thousand (‰) so that using ‰ as equivalent to 10^{-3} we can write the typical Δ for C_3 plants ($0.02 = 20 \times 10^{-3}$) as 20‰.

The value of δ_a is declining as a result of burning fossil fuels (from −6.7‰ in 1956 at 314 ppm to −7.9‰ in 1982 at 342 ppm: see Farquhar *et al.* 1989), and can also vary seasonally (by about 0.2‰) and diurnally. In large conurbations these fluctuations may be as large as 2‰ as a result of man's activities. This variation in the isotopic composition of the source air provides another reason for the general use in physiological studies, where possible, of Δ rather than δ. In general δ is expected to decline as the atmospheric CO_2 concentration increases. For example, at the bottom of a dense tropical forest where respiratory processes release fixed C (with a smaller proportion of ^{13}C than in air) and raise the concentration of CO_2 to as much as 390 vpm, δ_a can be as low as −11.4‰.

Discrimination can result from both equilibrium effects, such as the different relative concentrations that occur in the gas phase and liquid phase when in equilibrium, and kinetic effects such as those involved in enzyme reactions or transport processes. The main factors determining Δ in C_3 plants are diffusion in the air (including in the boundary layer and through the stomata) where Δ is about 4.4‰, and carboxylation at Rubisco

Table 7.5. *Typical maximum rates of net photosynthesis at saturating light and normal ambient CO_2 for leaves and crops and short-term crop growth rates*

Data from Cooper 1970; Slatyer 1970; Allaway *et al.* 1974; Hartsock & Nobel 1976; Monteith 1976, 1978; Kluge & Ting 1978; Milthorpe & Moorby 1979; Nobel 1988.

	C_3	C_4	CAM
$\mathbf{P}_n$ (μmol m^{-2} s^{-1})			
Single leaves	14–40	18–55	8
Crops	14–64	64	—
Crop growth rate (g m^{-2} day^{-1})			
Typical non-stressed	15–30	15–50	3–5
Maximum	34–39	51–54	7

where Δ is about 30‰. Farquhar *et al.* (1982) showed that the net effect is approximated by

$$\begin{aligned}\Delta &= 0.0044(x'_a - x'_i)/x'_a + 0.030\,x'_i/x'_a \\ &= 0.0044 + (0.0256)x'_i/x'_a \qquad (7.31)\end{aligned}$$

This equation shows that the value of Δ depends on x'_i/x'_a, which itself depends on the stomatal aperture: when stomata are closed x'_i declines and Δ also falls. (For precise work it should be noted that equation 7.31 may overestimate Δ by an amount that depends on assimilation rate; Evans *et al.* 1982). The use of Δ in studies of water use efficiency is discussed in Chapter 10. In C_4 plants, Δ is relatively insensitive to x'_i/x'_a as the large discrimination by Rubisco (*c.* 30‰) is replaced by a much smaller or negative effective discrimination by PEP carboxylase (*c.* 5.7‰). This net discrimination by PEP carboxylase is the result of a 2‰ Δ for fixation of HCO_3^-, combined with factors dependent on the 'leakiness' of the bundle sheath and the equilibrium between gaseous CO_2 and HCO_3^- in solution.

Response to environment

Maximum values of $\mathbf{P}_n^m$ at normal CO_2 concentrations and saturating light are usually 14–40 μmol m^{-2} s^{-1} for C_3 leaves, 18–55 μmol m^{-2} s^{-1} for C_4 and up to about 10 μmol m^{-2} s^{-1} for CAM plants (see Table 7.5). Leaf respiration rates in the dark are often in the range 0.5–3 μmol m^{-2} s^{-1}. There are large variations in respiration rates of other tissues: for example,

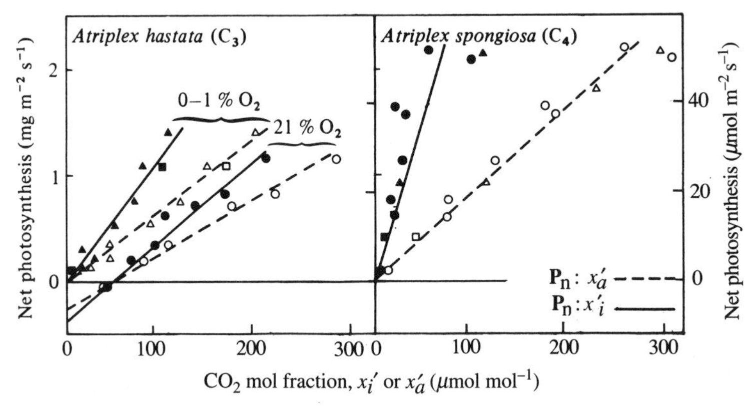

Fig. 7.12. The initial linear portions of CO_2 response curves for typical C_3 and C_4 species, showing a response to O_2 concentration in C_3 but not C_4 leaves. Zero O_2 – triangles, 1% O_2 – squares, 21% O_2 – circles; solid lines and closed symbols give $\mathbf{P}_n : x'_i$ response curves, dashed lines and open symbols refer to $\mathbf{P}_n : x'_a$ response. (After Slatyer 1970.)

fruits tend to have high respiration rates during early cell division and also to have another peak called the 'climacteric rise' near maturity. The rate of photorespiratory carbon loss in photosynthesising C_3 leaves is probably between 20 and 25% of $\mathbf{P}_n$.

CO_2 and O_2 concentration

The main reason for determining the photosynthetic CO_2 response is to provide mechanistic information for interpreting photosynthetic responses to environment. It is only in controlled environments and glasshouses where x'_a can be artificially manipulated, and in relation to long-term climatic changes in atmospheric CO_2 (see Chapter 11), that the CO_2 response curves are of direct interest.

The main physiological differences between photosynthetic pathways are illustrated by the CO_2 responses of C_3 and C_4 leaves in normal air (21% O_2) or low oxygen (Fig. 7.12). Reduction of O_2 concentration only enhances $\mathbf{P}_n$ in C_3 plants (and in CAM plants during the light period). The value of Γ is normally close to zero for C_4 (or night-time CAM) where the primary fixation is by PEP carboxylase, as it is for C_3 plants when photorespiration is inhibited by low O_2. Plotting these results as a function of x'_i shows the smaller r'_m in C_4 leaves (averaging 2–5 $m^2\ s\ mol^{-1}$) compared with 5–12.5 $m^2\ s\ mol^{-1}$ in C_3 leaves.

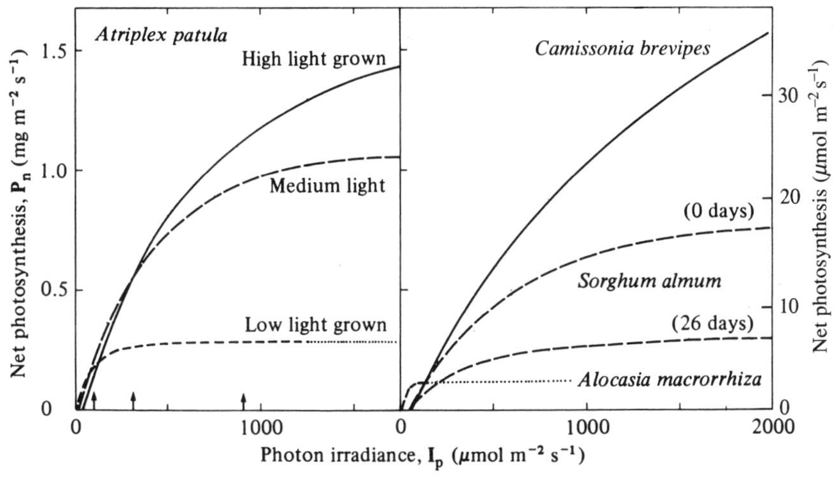

Fig. 7.13. Photosynthetic light response curves for (*a*) *Atriplex patula* grown at three different irradiances as shown by the arrows (after Björkman *et al.* 1972*a*), and (*b*) *Alocasia macrorrhiza* from extreme shade (after Björkman *et al.* 1972*b*), newly unfolded or 26 day old *Sorghum almum* leaves (after Ludlow & Wilson 1971*b*) and *Camissonia brevipes* (after Armond & Mooney 1978).

In C_3 plants x_i' is often maintained at between 0.6 and $0.7\,x_a'$ (200–230 vpm at normal ambient concentrations) compared with 0.3–$0.4\,x_a'$ (about 100–130 vpm) in C_4 plants (see Wong *et al.* 1979). However, because of the different time constants of intracellular and stomatal responses to factors such as irradiance, x_i' is likely to be rather more variable in the fluctuating environment in the field.

Light

Photosynthetic light responses have been reviewed by Boardman (1977) and Patterson (1980). Typical light response curves for several different species, and for one species grown at a range of irradiances, are shown in Fig. 7.13. Although there is a slight tendency for $\mathbf{P}_n$ in C_4 species to continue increasing at higher irradiances than in C_3 species, there are greater differences between sun and shade species, or between leaves of one species grown at different irradiances (Fig. 7.13). In shade species or in shade-grown leaves, $\mathbf{P}_n$ may be saturated at less than 100 μmol m^{-2} s^{-1} (PAR), which is approximately 5% full sunlight. Sun leaves, on the other hand, often continue to respond up to typical values for full sunlight. The

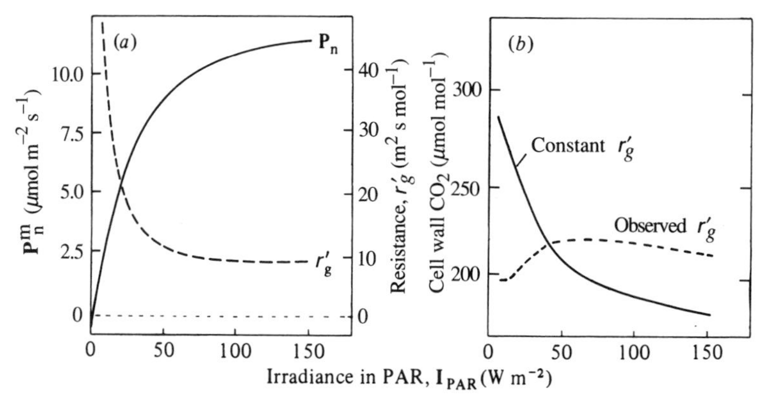

Fig. 7.14. (*a*) Light response curves for $\mathbf{P}_n$ and r'_g in Sitka spruce (data of Ludlow & Jarvis 1971*b*). (*b*) The same data plotted to show calculated variation of x'_i with constant r'_g and with the observed r'_g.

light compensation point also varies from as low as 0.5–2 μmol m^{-2} s^{-1} in extreme shade species such as *Alocasia macrorrhiza* growing in the Queensland rainforest to as much as 40 μmol m^{-2} s^{-1} in sun leaves.

A difficulty with interpreting light response curves is that the CO_2 concentration at the cell wall or photosynthetic site is not constant with changing light as it is a function of $\mathbf{P}_n$ and r'_g (see equation 7.20), as shown in Fig. 7.14, though the effect of the stomatal response to light is to minimise these changes in x'_i.

Several factors contribute to the differences in the photosynthetic behaviour of sun and shade leaves. There is good evidence that all components of the photosynthetic system adapt together: for example, high light leaves tend to be thicker with a greater internal surface area than shade leaves and to have more chlorophyll and very much more carboxylase per unit area. In addition, although there is evidence that r'_m per unit cell surface area is approximately constant, mesophyll resistances tend to decrease with increasing irradiance as a result of changes in A_{mes}/A (the ratio of mesophyll cell surface area to leaf surface area: Holmgren 1968; Nobel *et al.* 1975), as do stomatal resistances. The light reactions are also affected by growth irradiance; there is more extensive granal development in shade leaves but the capacity for electron transport is markedly reduced. For example, electron transport through both photosystems (when expressed on a chlorophyll basis) may be as much as 14 times higher in chloroplasts extracted from sun plants than in those from shade plants (Boardman *et al.* 1975). Part of the effect may result from a slightly smaller photosynthetic unit size (the ratio of collector chlorophyll to reaction centres – e.g. Malkin

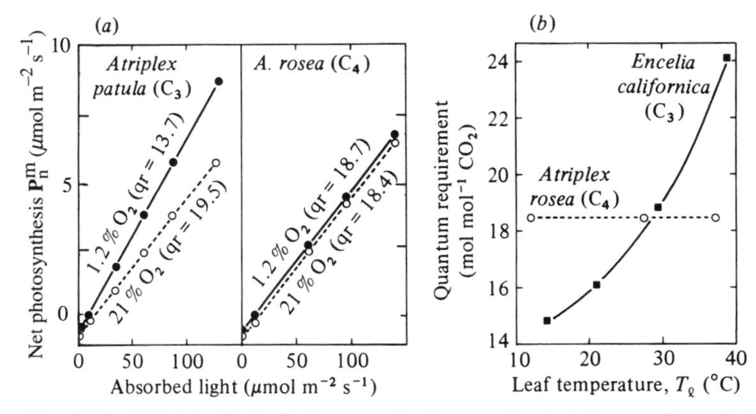

Fig. 7.15. (*a*) Light-limited photosynthetic rates in *Atriplex patula* and *A. rosea* leaves at 1.2% and 21% O_2, showing effects on the quantum requirement (qr) (after Björkman *et al.* 1970). (*b*) Temperature dependence of quantum requirement in C_3 and C_4 plants (after Ehleringer & Björkman 1977).

& Fork 1981) but at least for photosystem I, the ratio of chlorophyll to the reaction centre (P_{700}) is often quite constant. It is the cytochrome f and cytochrome b components of the transport chain that are particularly reduced in low light plants. The various adaptations in response to altered irradiance can occur within days of the change.

In spite of large differences in light-saturated $\mathbf{P}_n$, only small differences have been observed in the initial slope of the photosynthetic light response curves in healthy leaves. The reciprocal of this slope, the quantum requirement (quanta per CO_2 fixed) is a measure of the efficiency of photosynthesis and is relatively constant at a value of about 19 for C_4 plants, but is strongly dependent on temperature and oxygen concentration in C_3 plants (Fig. 7.15).

Very high irradiances can damage the photosynthetic system, particularly in shade-adapted leaves, or in leaves where photosynthetic metabolism has been inhibited by other stresses such as extreme temperature or water stress. The damage can be a result of photooxidation where bleaching of the chlorophyll occurs. Where no bleaching is observed the damage is usually termed photoinhibition.

Dark respiration rates are also influenced by growth irradiance, being as low as 4 $\mu g\ m^{-2}\ s^{-1}$ in shade plants compared with 50–150 $\mu g\ m^{-2}\ s^{-1}$ in sun leaves. This difference can contribute to the net photosynthetic advantage at low light that is often exhibited by shade leaves.

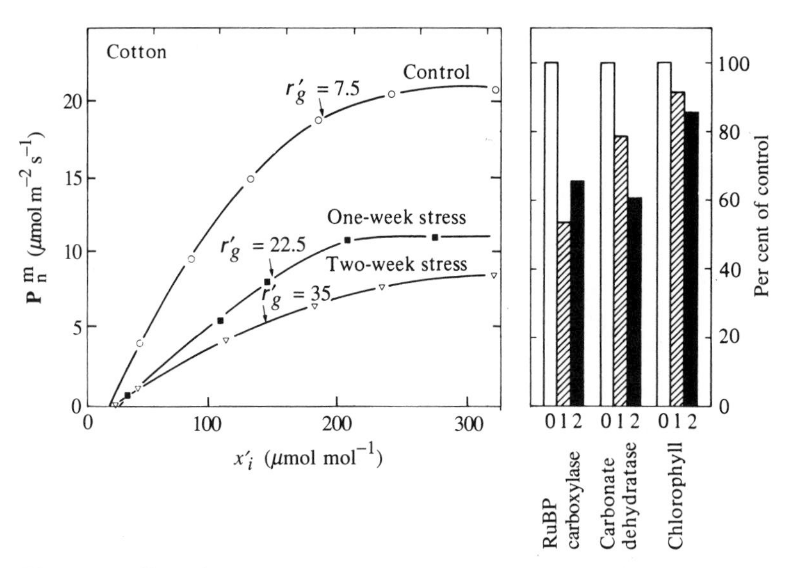

Fig. 7.16. Effect of moderate-term water stress (zero, one- or two-week stress) on various photosynthetic parameters in cotton (resistances, $m^2\ s\ mol^{-1}$). (Data from Jones 1973*b*.)

Water status

The effects of water deficits on assimilation have been the subject of many recent papers (see e.g. Kaiser 1987), but there is still uncertainty concerning the relative importance of stomatal closure and changes in mesophyll capacity in causing the observed decreases in assimilation. Although many pot-studies have suggested that changes in stomatal conductance are the main cause of the decreased photosynthesis (see Boyer 1976), many field studies using more realistic longer-term drying cycles have indicated that the capacities of most photosynthetic components are modified in parallel, such that their relative limitations remain nearly constant (e.g. Fig. 7.16 and Table 7.6). Recent evidence from chlorophyll fluorescence, however, suggests that electron transport is relatively little affected by low water potential, though high irradiance during the water stress can lead to photoinhibitory damage at PS II. The possibility exists that non-uniform stomatal closure has been the cause of some reports of lowered mesophyll photosynthetic capacity and constant x'_i, but this is unlikely to be the case for data such as those in Fig. 7.16 where enzymes were extracted from the whole leaf. The use of O_2 evolution measurements at saturating CO_2 can

Table 7.6. *Effect of medium-term water stress on respiration and photosynthesis of potted apple 'mini-trees' (unpublished data of L. Fanjul)*

	$\Psi\ell$	$\mathbf{P}_n$	$\mathbf{R}_p^a$	$\mathbf{R}_d$	$\Gamma_{(21\%O_2)}$	$\Gamma_{(1\%O_2)}$	$(\mathbf{R}_d+\mathbf{R}_p)/\mathbf{P}_n$
	(MPa)	(μmol m^{-2} s^{-1})			(μmol mol^{-1})		
cv. James Grieve (24 days stress)							
Control	−1.3	27.3	14.1	1.6	49	7	0.47
Moderate stress	−2.0	17.3	7.0	2.3	32	1	0.53
Severe stress	−2.8	9.5	3.9	1.6	71	28	0.56
cv. Egremont Russet (14 days stress)							
Control	−1.0	14.1	5.7	2.0	74	32[b]	0.55
Moderate stress	−1.8	10.0	3.2	2.7	78	37[b]	0.59
Severe stress	−3.6	2.0	1.4	1.8	136	83[b]	1.56

[a] $\mathbf{R}_p$ was estimated from the difference in $\mathbf{P}_n$ at 1% and 21% O_2, so is an overestimate of true photorespiration.
[b] 2% O_2.

provide a useful independent check, and has shown that although C_4 photosynthesis may be fairly sensitive to water deficits, some of the reports that C_4 photosynthetic *capacity* is reduced by moderate water deficits (less than about 25% relative water content) arise because the high CO_2 concentration used in the O_2 electrode (*c.* 5%) was probably not adequate to saturate photosynthesis when the stomata are closed. For example, it has been found that CO_2 concentrations as high as 15% can be inadequate to saturate photosynthesis in water stressed C_4 plants (G. Cornic and A. Massacci, unpublished data), even though 2% CO_2 can be saturating with stressed leaves of spinach and other C_3 plants.

Respiration is also sensitive to water deficits, though both increases and decreases have been reported (e.g. Bunce & Miller 1976). Because respiration tends to be rather less sensitive than $\mathbf{P}_n$, $\mathbf{R}$ often increases in relation to gross photosynthesis, thus explaining some of the decline in $\mathbf{P}_n$ and the tendency for Γ to increase with stress (Table 7.6).

Temperature and other factors

Over the normal physiological temperature range, $\mathbf{P}_n$ shows a broad optimum response. The decline at high temperatures results partly from the more rapid increase of respiration with temperature and partly from a time-dependent photosynthetic inactivation at high temperatures. The fact that Γ increases markedly with temperature (see Bykov *et al.* 1981) is a manifestation of this changing balance between CO_2 fixation and release including enhanced photorespiration at high temperatures (see Fig. 7.15). On the basis that dark respiration can be partitioned into a temperature-sensitive maintenance component ($\mathbf{R}_m$) and a temperature-insensitive growth component ($\mathbf{R}_g$), McCree (1970) proposed the following expression for total $\mathbf{R}_d$ over a 24 h period:

$$\mathbf{R}_d = \mathbf{R}_g + \mathbf{R}_m = a\mathbf{P}_g + bW \qquad (7.32)$$

where $\mathbf{P}_g$ (g m^{-2} day^{-1}) is total photosynthesis (excluding that lost in photorespiration), W (g m^{-2}) is the leaf weight in CO_2 equivalents ($44/12 \times$ carbon content of the leaf $\simeq 44/30 \times$ dry weight), and a (dimensionless) and b (day^{-1}) are constants. The value of a is usually 0.25–0.34 and is relatively independent of species or temperature though it depends on the length of the light period. The value of b ranges from 0.007 day^{-1} to 0.015 day^{-1} at 20 °C, approximately doubling for every 10 °C rise in temperature (see Legg 1981), so that

$$b_{(T)} = b_{(20)} 2^{0.1(T-20)} \qquad (7.33)$$

where $b_{(20)}$ is the value at 20 °C and T is in °C. Temperature responses are discussed in more detail in Chapter 9.

Many other factors such as ambient humidity (Chapter 6) and windspeed (Chapter 11), which both primarily affect the gas-phase resistance, affect $\mathbf{P}_n$. In addition, factors such as leaf age (responses parallel the stomatal response shown in Fig. 6.7 (p. 146); see also Fig. 7.13) altered nutrition, disease, etc. may affect any of the component processes. An important factor is the internal control of $\mathbf{P}_n$ exercised by the demand for photosynthate. $\mathbf{P}_n$ can be inhibited as much as 50% by removal of sinks such as developing grain in cereals (e.g. King *et al.* 1967), and there are many reports that $\mathbf{P}_n$ is higher in fruiting than in non-fruiting plants (e.g. Hansen 1970; Lenz & Daunicht 1971), with at least part of the difference being attributable to differences in stomatal aperture. Evidence for and against direct feedback inhibition and hormonal control of photosynthesis has been reviewed by Neales & Incoll (1968) and Guinn & Mauney (1980).

As well as leaves and leaf-like structures, several other plant organs can make important photosynthetic contributions (see e.g. Biscoe *et al.* 1975*a*; Pate 1975). For example, ears of cereals, especially the awns, and the pods and stems of many species can have high rates of photosynthesis. Many green stems and fruits have relatively impermeable 'skins' with low conductances. In such cases their photosynthetic capacity acts to refix the respired CO_2 rather than to perform net photosynthesis (see Jones 1981*b*).

Photosynthetic models, efficiency and productivity

Leaf photosynthesis models

Many more or less mechanistic models have been proposed for the biochemical reactions of photosynthesis following Maskell's (1928) and Rabinowitch's (1951) pioneering applications of Michaelis–Menten kinetics (e.g. Hall 1979; Farquhar *et al.* 1980*b*). A useful summary of models that incorporate present understanding of the photosynthetic system is given by Farquhar & von Caemmerer (1982), while some more empirical models are outlined by Thornley & Johnson (1990).

Although a rectangular hyperbola is sometimes used to simulate the photosynthetic response to x'_i, in most cases the actual response tends to reach saturation much more rapidly, sometimes approaching the so-called 'Blackman response' of two straight lines (a constant initial slope where photosynthesis is entirely CO_2-limited, switching suddenly to a horizontal light-limited portion). The most useful general form of equation to simulate photosynthetic responses to x'_i is the non-rectangular hyperbola, which can be written in the form (e.g. Jones & Slatyer 1972)

$$\mathbf{P}_n = \mathbf{P}_{max}(x'_i - \Gamma - b\mathbf{P}_n)/(a + x'_i - \Gamma - b\mathbf{P}_n) \qquad (7.34)$$

where a and b are constants. The detailed mechanistic interpretation of the parameters in this quadratic equation is unresolved, but a determines the sharpness of the transition from the linear phase to saturation and b determines the initial slope of the response. The initial slope ($d\mathbf{P}_n/dx'_i$), sometimes called the 'carboxylation efficiency', is largely determined by the carboxylation process, while the maximum rate is mainly determined by the light reactions in that they limit the rate of regeneration of the substrate for the carboxylase (RuBP). Light responses may be incorporated by assuming that light only affects $\mathbf{P}_{max}$, according to

$$\mathbf{P}_{max} = \mathbf{P}^{max}_{max}(\mathbf{I} - \mathbf{I}_c)/(c + \mathbf{I} - \mathbf{I}_c) \qquad (7.35)$$

where $\mathbf{P}^{max}_{max}$ is an absolute maximum photosynthetic rate at saturating light and CO_2 and is an appropriate function of temperature (see Chapter 9), $\mathbf{I}_c$

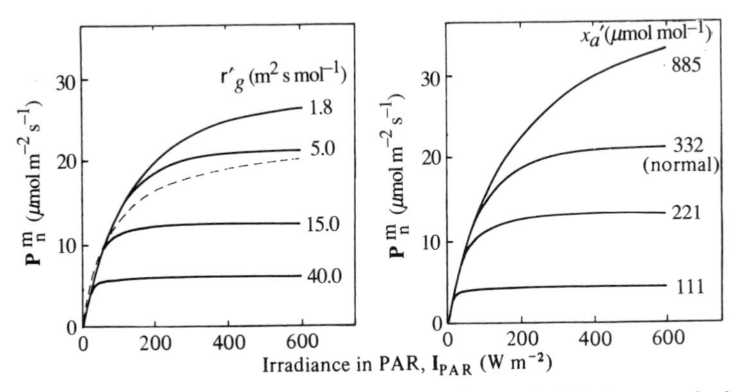

Fig. 7.17. Photosynthesis–irradiance curves for typical C_3 leaves calculated using equation 7.34 and 7.35 (see Jones & Slatyer 1972 for details) for a range of values of r'_g and x'_a (parameter values unless stated otherwise: x'_a = 332 μmol mol^{-1}, Γ = 55 μmol mol^{-1}, $K_m^C = 11.1$, K_m^I = 200 W m^{-2}, $\mathbf{P}_{max}^{max}$ = 45.5 μmol m^{-2} s^{-1}, r'_t = 7.5 m^2 s mol^{-1}, r'_g = 5 m^2 s mol^{-1}). The dashed line indicates a rectangular hyperbola.

is a light compensation point and c is a constant. Combination of equations 7.34 and 7.35 gives a useful general expression for photosynthesis, whose behaviour is illustrated in Fig. 7.17. This figure shows a close fit to observed responses (e.g. Fig. 7.13) and more rapid saturation than can be achieved with a rectangular hyperbola.

Photosynthetic efficiency of single leaves

The photosynthetic efficiency (ε_p) may be defined thermodynamically as the available energy stored in plant dry matter expressed as a fraction of incoming radiant energy. The free energy content of the immediate product of photosynthesis (sucrose) is about 480 kJ (mol C)$^{-1}$ or 16 kJ g^{-1}. However, plant material also contains a range of other compounds (proteins, fats, etc.) that lead to an average energy content for dry matter of about 17.5 kg g^{-1} (see Monteith 1972), this is approximately equivalent to 525 kJ (mol C)$^{-1}$.

Assuming an average energy content in the PAR for solar radiation of 220 kJ mol^{-1}, the theoretical minimum quantum requirement for gross photosynthesis of 8 absorbed quanta per CO_2 fixed, equals an efficiency of about 27% (i.e. $100 \times 480/8 \times 220$). In practice minimum quantum requirements are about 19 for C_4 leaves and between 15 and 22 (depending on temperature) for C_3 leaves (Fig. 7.15) and represent efficiencies (in terms of absorbed PAR) of 12.6% and 10.8–16%, respectively. These may be converted to an incident PAR basis by multiplying by the leaf absorptance in the PAR ($\simeq 0.85$) and to an incident total solar basis by further multiplication by the ratio of PAR/total solar ($\simeq 0.5$). A typical maximum photosynthetic efficiency in terms of incident solar radiation is therefore 5.4%.

Typical minimum quantum requirements are of the order of 19, rather than 8, largely as a result of respiratory CO_2 losses concurrent with photosynthesis. Inhibition of photorespiration in C_3 plants by low O_2 (see Fig. 7.12) reduces the minimum quantum requirement to about 12.5. Some of the remaining difference may be related to dark respiration continuing in the light. The rather higher quantum requirement of C_4 plants (unaffected by O_2 concentration, indicating no photorespiratory contribution) may result from the additional requirement of at least 2 ATP per CO_2 inherent in the C_4 pathway.

For a bright day ($\mathbf{I}_S = 900$ W m^{-2}) even a high photosynthetic rate of 2.4 mg CO_2 m^{-2} s^{-1} (which is equivalent to a sucrose accumulation rate of 1.8 mg m^{-2} s^{-1} ($2.4 \times 30/44$)) represents an efficiency of only 3.2%. This low efficiency at high irradiance results from light saturation. In practice most single leaves do not achieve even this high efficiency because of non-optimal nutrition, water status or temperature, or because of internal factors such as senescence or sink limitation.

Canopy photosynthesis

An important application of photosynthesis models is in the prediction of productivity in different environments of canopies with different structure (e.g. erect or horizontal leaves) or photosynthetic characteristics (e.g. C_3 or C_4). This is commonly done by means of complex simulation models that combine leaf photosynthesis models (such as equations 7.34 and 7.35) with other submodels for predicting (*a*) light incident on the canopy and its distribution within the canopy (see Chapter 2 for details), (*b*) stomatal resistance as a function of water status (see Chapter 6) and (*c*) respiratory CO_2 loss. Details of particular models may be found elsewhere (e.g. Monteith 1965*a*; Duncan *et al.* 1967; Allen *et al.* 1974; Hesketh & Jones 1980; Thornley & Johnson 1990).

Canopy photosynthesis can then be calculated by summing the contributions of each type of organ as illustrated in Fig. 7.18, but using a light penetration model to determine the light profile. Unfortunately it is necessary to estimate a large number of parameters for this approach and

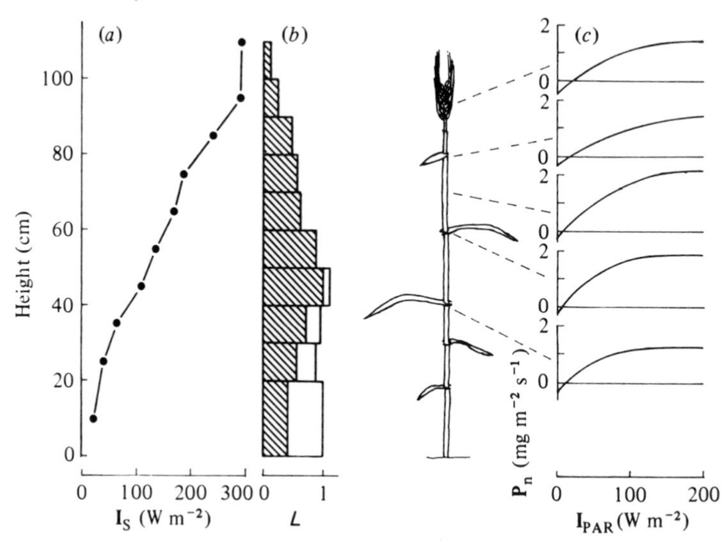

Fig. 7.18 The method for determining canopy photosynthesis from (*a*) irradiance at different levels in the crop, (*b*) corresponding leaf area index (L) (green leaf area shown hatched), and (*c*) photosynthesis–light response curves for each organ. Measurements are for a barley crop in central England on 28 June. (After Biscoe *et al.* 1975*a*.)

calculations can become very tedious. For many purposes, simple models can be adequate. For example, since leaves only reflect or transmit between 10 and 20% of incident PAR, in photosynthesis models it is only necessary to take account of the first or second interception of any ray of light. Various limiting cases can be considered.

(*a*) *All leaves are at an acute angle to any direct radiation.* In this case direct irradiance at the surface of each leaf ($\mathbf{I}_o \cos\theta$) is low, so that each leaf is light-limited and $\mathbf{P}_n \propto \mathbf{I}$ or $\mathbf{P}_n = \varepsilon_p \mathbf{I}$. Assuming that any diffuse irradiance is also within the light-limited range, total canopy photosynthesis is therefore proportional to light interception with a proportionality constant ε_p. At a high leaf area index, therefore, where all radiation is intercepted, canopy photosynthesis is simply $\varepsilon_p \mathbf{I}_o$.

(*b*) *Very low leaf area index canopies.* For homogeneous canopies where $L < 1$, one only needs to take account of the first interception of any radiation. For a horizontal-leaved canopy with no mutual shading, $\mathbf{I} = \mathbf{I}_o$ for all leaves. Therefore single-leaf photosynthetic models can be applied successfully.

In practice most canopies are between these two extremes with their light

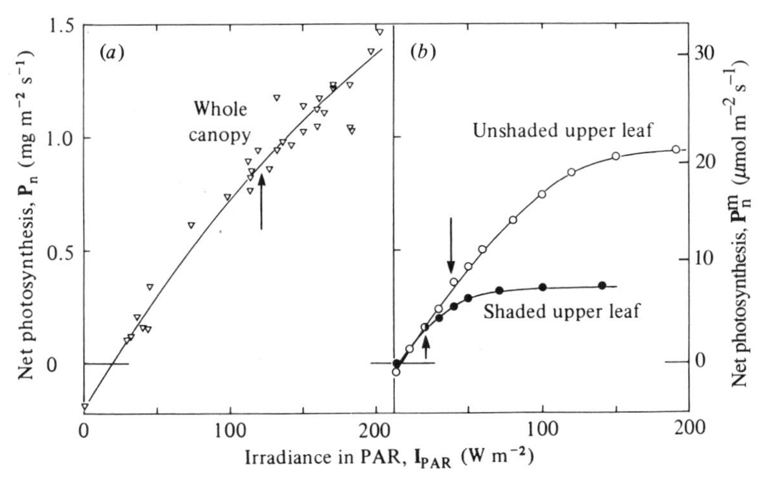

Fig. 7.19. Photosynthetic light response curves for tomato plants in daylit cabinets: (*a*) whole canopy and (*b*) individual leaves. The irradiance giving the highest photosynthetic efficiency is the point on each curve where a tangent to the curve passes through the origin, these points are indicated by arrows. (Data from Acock *et al.* 1978).

response showing less tendency to light saturation than single leaves (Fig. 7.19). As a result, canopy photosynthesis is greater than for single leaves, up to about 2.8 mg m^{-2} s^{-1} (64 μmol m^{-2} s^{-1}: Table 7.5). Another important consequence is that the irradiance giving maximal photosynthetic efficiency (ε_p) is higher for a crop canopy than for single leaves (Fig. 7.19). In this example the optimal irradiance (giving maximal $\mathbf{P}_n/\mathbf{I}$) was between 15 and 25 W m^{-2} for single leaves and over 100 W m^{-2} for the whole canopy.

One application of canopy photosynthesis models has been the prediction of effects of leaf angle on productivity. A simple model to study these effects can be obtained by combining (*a*) the single-leaf photosynthesis model (assuming $K_m^I = 55\ \mu$mol mol^{-1} and $r'_g = 5$ m^2 s mol^{-1}, with other parameters as for Fig. 7.17) with (*b*) a model for predicting shortwave irradiance on a horizontal surface (equation 2.10 (p. 23) with β obtained according to Appendix 7) and (*c*) the simple Beer's Law model for predicting sunlit leaf area and irradiance per unit sunlit area (equation 2.18 (p. 35) where $\ell = 1$ for a horizontal-leaved canopy and $\ell = 2/\pi \tan\beta$ for a vertical-leaved canopy). The results of some simulations for low, medium and high leaf area indices, for clear days at tropical and mid-temperate latitudes are shown in Fig. 7.20.

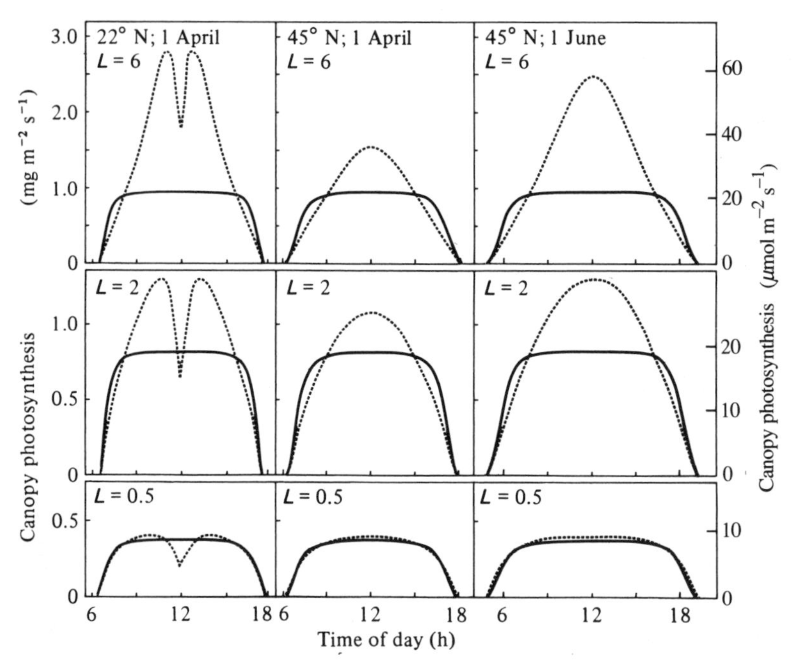

Fig. 7.20. Simulations of photosynthesis by erect leaved (..................) and horizontal-leaved (————) canopies of high (6), medium (2) and low (0.5) leaf area index (see text).

These curves show that, as expected, erect leaves are only advantageous at high solar elevations and where L is moderately high. At the low solar elevations reached in temperate latitudes in the spring, the difference between total daily photosynthesis of erect- and horizontal-leaved canopies can be very small, even at high leaf area index. The midday depression of photosynthesis predicted for vertical leaves in the tropics arises when the sun is immediately overhead. Alteration of the model to include diffuse radiation would tend to reduce the effect of leaf orientation on daily photosynthesis.

Productivity

In calculating the productivity of vegetation, it is necessary to allow for two additional loss factors: dark respiration and incomplete interception of incident radiation. The single-leaf calculations only allowed for respiration occurring in leaves in the light. In a crop, however, there is an additional

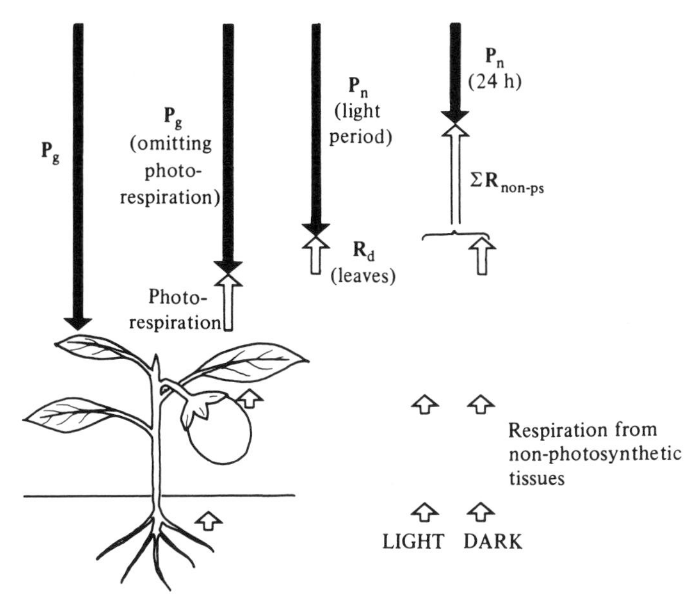

Fig. 7.21. Diagrammatic representation of the CO_2 exchanges showing how net photosynthesis over a 24 h period is derived.

respiratory loss from other organs (e.g. stems and roots) during the light and from the whole plant during the dark.

The various components of the CO_2 balance of a crop are illustrated in Fig. 7.21 and summarised for a barley crop in Table 7.7. Although gross photosynthesis changed relatively little over the season, respiratory losses (especially the maintenance component) increased markedly. The respiratory loss factor represents the loss by non-photosynthesising tissues ($\Sigma \mathbf{R}_{non\text{-}ps}$), that is the sum of the loss from non-photosynthetic tissues in the light and all tissues in the dark. On this basis the respiratory loss factor can be defined as $(\Sigma \mathbf{P}_n - \Sigma \mathbf{R}_{non\text{-}ps})/\Sigma \mathbf{P}_n$, where $\Sigma \mathbf{P}_n$ is the total net photosynthesis during the light period. If one assumes for the barley data that half the respiratory losses in Table 7.7 come from photosynthetic tissues, in May, $\Sigma \mathbf{R}_{non\text{-}ps} = 43 + (78/2) = 82$ mg m^{-2} week^{-1} and $\Sigma \mathbf{P}_n = 167 + 82 = 249$ mg m^{-2} week^{-1}. Therefore the respiratory loss factor is 0.68 in May and is 0.43 in July. These results are comparable to other results (see Monteith 1972) and imply that between about 40 and 70% of net photosynthesis is conserved as dry matter growth. Assuming a typical respiratory loss factor of 0.6 would give an expected maximum efficiency of use of solar radiation of $0.6 \times 5.4 = 3.2\%$.

Table 7.7 *Components of CO_2 balance (see Fig. 7.21) of a barley crop (per unit ground A) at two stages of development. (Data from Biscoe* et al. 1975b). *Dark respiration by leaves assumed to be half total* $\mathbf{R}_d$

		Mid-May (vegetative)	Mid-July (grain filling)
a	Standing dry matter in CO_2 equivalents (g m^{-2})	526	1605
Respiration (g CO_2 m^{-2} $week^{-1}$)			
b	Soil microbial respiration	23	18
c	Measured night-time $\mathbf{R}_d$ losses	43	64
d	Day-time $\mathbf{R}_d$, estimated from *c* by adjusting for temperature differences	78	138
e	$\Sigma\mathbf{R}_{non\text{-}ps}$ ($= c + 0.5 \times d$)	82	133
f	Maintenance respiration ($\mathbf{R}_m$ as % total)	30	64
g	Growth respiration ($\mathbf{R}_g$ as % total)	70	36
h	Root respiration (% total)	14	5
Photosynthesis (g CO_2 m^{-2} $week^{-1}$)			
i	Net photosynthesis for 24 h	167	102
j	Net photosynthesis in light ($\Sigma\mathbf{P}_n = e + i$)	249	235
k	Gross photosynthesis in light (omitting photorespiratory losses) ($= c + d + i$)	288	304

The maximum short-term crop growth rates that have been observed for C_3 crops of approximately 36 g m^{-2} day^{-1} and for C_4 crops of 52 g m^{-2} day^{-1} (Table 7.5) correspond to efficiencies of 3.1% for rice (20 MJ m^{-2} day^{-1} insolation) and between 3.1 and 4.5% for maize (see Monteith 1978). The difference between C_3 and C_4 species is maintained in spite of their similar quantum requirement at low light, probably as a result of the smaller tendency for light saturation in C_4 plants.

There is increasing evidence that dry matter production, particularly during the vegetative phase of plant growth, is a linear function of the amount of radiation *intercepted* (as expected for limiting case (*a*) above). Figure 7.22 shows that total dry matter production for a wide range of different crops was approximately 1.4 g dry matter per MJ intercepted solar radiation (an efficiency of 2.5%, which is not far from the theoretical value).

Nutrition and water stress apparently exert a large proportion of their effects on yield by altering leaf area index and consequent light interception, though in one water-stressed barley crop ε_p was reduced by 20% (Legg

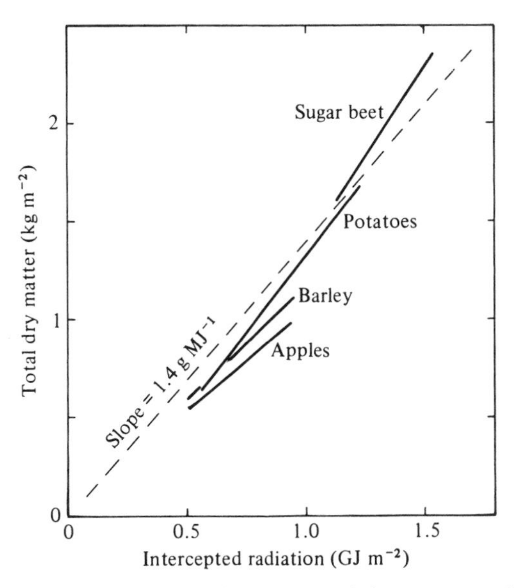

Fig. 7.22. Relation between total dry matter at harvest and total radiation intercepted by the foliage. (After Monteith 1977.)

et al. 1979), with even more extreme deviations having been observed in some other cases. Much of the limitation to the yield of annual crops arises from poor light interception in the early stages of crop growth when leaf area index is small. Low temperatures during winter also inhibit growth in temperate climates and contribute much towards the different productivity achieved in temperate and tropical environments. Actual productivities of different plant communities and estimated efficiencies of utilisation of incident solar radiation are presented in Table 7.8. This shows that in extreme environments photosynthetic efficiency can be several orders of magnitude less than potential.

Evolutionary and ecological aspects

In general, the C_4 pathway is most advantageous in high light, high temperature environments where water is limited. In these conditions leaf photosynthetic rates of C_4 species generally exceed those of C_3 species, though there are C_3 species that have photosynthesis rates equal to the highest known in C_4 species (Mooney *et al.* 1976) and also C_3 species that can achieve high rates of net photosynthesis at temperatures above 45 °C

Table 7.8. *Typical values for net primary production and efficiency of conversion of incident solar radiation for different ecosystems*

Data collected from Cooper 1970, 1975; Milthorpe & Moorby 1979; Trewartha 1968.

	Net primary production	Incident solar radiation, $\mathbf{I}_S$	Efficiency ε_p
Over whole year	($g\ m^{-2}\ yr^{-1}$)	($GJ\ m^{-2}\ yr^{-1}$)	(%)
Dry desert	3–(250)	7	0.0007
Arctic tundra	100–600	3	0.06–0.36
Temperate grassland	500–2000	3.3	0.25–1.1
Sugar cane	2500–7500	6	0.7–2.2
120-day growing season	($g\ m^{-2}$)	($GJ\ m^{-2}$)	(%)
Potatoes	1500	2	1.3
Temperate cereals	2000	2	1.8
Rice	3000	2	2.6

(Mooney *et al.* 1978). The higher water use efficiency of C_4 species (see Chapter 10) that arises from their CO_2 concentration mechanism, and their consequent ability to sustain any given photosynthetic rate with more closed stomata than C_3 species, is probably a more important factor determining their common occurrence in many deserts and other subtropical regions (e.g. Gupta & Saxena 1971; Teeri & Stowe 1976). The dominance of C_3 species in cool or shady habitats probably results from their lower quantum requirement (and hence greater potential productivity) at temperatures below 30 °C (Ehleringer 1978). In spite of this, some C_4 *Atriplex* and *Spartina* species can operate effectively at low temperatures (Björkman *et al.* 1975; Long *et al.* 1975), while some C_4 trees (genus *Euphorbia*) occur in the Hawaiian rainforest understorey (Pearcy & Troughton 1975).

The very high water use efficiency and ability to survive long periods without rainfall of CAM plants enables them to survive in extremely arid regions, in spite of their low maximum productivity. In a multiple regression study of species distribution in North America in relation to climatic variables, Teeri *et al.* (1978) showed that CAM abundance was greatest in areas of low soil moisture (the dry Arizona deserts) while C_4 species occurrence was associated with high minimum temperature or high evaporative demand (Teeri & Stowe 1976).

In any one region it is common to find a balance between species having different photosynthetic pathways. For example in the East Thar Desert in northwest India, both C_3 and C_4 shrubs and annual species are common, although the grasses are almost exclusively C_4. The C_3 species are probably active mainly in the cooler winter months when they might be expected to be superior, and the C_4 species during the hotter months. Probably because of the high night temperatures in the summer, CAM species are rare, with only *Euphorbia caducifolia* being widespread in this environment.

It is clear from the relationships of species having the different photosynthetic pathways that the C_4 and CAM pathways each evolved independently on several occasions. There are many genera (e.g. *Atriplex* and *Panicum*) that include C_3 and C_4 species, while *Euphorbia*, for example, has C_3, C_4 and CAM species.

Many of the species that are known as CAM plants are in fact facultative CAM plants, using the C_3 pathway as seedlings, or when night temperatures are above 18 °C, or in conditions of adequate water supply, but switching to the CAM pathway under conditions of salinity or water stress (Kluge & Ting 1978). In these plants, the pattern of stomatal opening changes to increased night opening as the CAM pathway develops. The facultative CAM plants are found in the Aizoaceae and Portulacaceae while other families (e.g. Cactaceae) tends towards being obligate CAM plants. There are also suggestions that certain species may switch between C_3 and C_4 pathways as a function of environment (see Shomer-Ilan *et al.* 1979). Some species (e.g. *Panicum miloides*) are apparently intermediate between C_3 and C_4 in several characters, such as $\mathbf{P}_{max}$ and Γ, though the primary product of photosynthesis is the 3-carbon acid PGA (see Ray & Black 1979).

Because of the apparently greater photosynthetic potential in C_4 plants, there have been attempts to screen large numbers of cultivars or mutagen-treated plants of C_3 crop species for occurrence of C_4 photosynthesis (e.g. Menz *et al.* 1969). The technique has been to enclose the test material in a sealed, illuminated chamber together with a C_4 plant. C_4 plants can deplete the CO_2 in the chamber below Γ for C_3 plants so the C_3 plants lose CO_2 until they eventually die. No C_4 mutants have been detected among the several 100000 lines tested by this means.

In other studies, attempts have been made to cross C_3 and C_4 species from the same genus. In the work with *Atriplex* species (Nobs 1976), although the cross between *A. rosea* and *A. triangularis* has been taken beyond the first generation, no fully C_4 recombinants have been detected even though individuals with Kranz anatomy or high PEP carboxylase activity have been found.

In an interesting analysis, Körner *et al.* (1979) have shown a correlation between the maximum stomatal conductance and photosynthetic capacity of different species from different ecological groups. All C_3 species fell on one line, from plants such as conifers with low photosynthetic capacity to the more photosynthetically active herbs from open habitats and plants from aquatic habitats and swamps. C_4 plants fell on a different line, having a higher photosynthetic capacity for any stomatal conductance than C_3 plants.

Sample problems

7.1 For a leaf where the steady state level of fluorescence (all in arbitrary units) is 1.2, the maximum fluorescence achieved with a saturating flash (applied during steady state photosynthesis) is 3.2, F_m (1 h dark) is 3.7, F_o is 1.0 and F'_o is 0.9, calculate (i) F_v/F_m, (ii) q_O, (iii) q_P, (iv) q_N and (v) the quantum yield of electron transport through photosystem II.

7.2 Using a well-stirred open gas-exchange system where the volume rate into the cuvette is 5 cm^3 s^{-1}, leaf area in the chamber is 10 cm^2, $T_\ell = 23$ °C, $P =$ 100 kPa, $e_e = 0.5$ kPa, $e_o = 1.5$ kPa, $c'_e = 600$ mg m^{-3} and $c'_o = 450$ mg m^{-3}; calculate (i) u_e, (ii) u_o, (iii) x'_e, (iv) $\mathbf{P}^m$, (v) the molar gas-phase conductance to CO_2.

7.3 For the control photosynthesis response curve in Fig. 7.16, calculate the stomatal limitation to photosynthesis according to (i) the resistance analogue method (equation 7.25), (ii) Farquhar & Sharkey's method, or (iii) the sensitivity approach (equation 7.27).

7.4 Assuming that incident global radiation is 14.5 MJ m^{-2} day^{-1} in May and 17 MJ m^{-2} day^{-1} in July, calculate for the data in Table 7.7 the efficiency of (i) gross photosynthesis, (ii) net productivity in terms of incident radiation or incident PAR.

8

Light and plant development

The ability of plants to modify their patterns of development appropriately in response to changes in the aerial environment is a major factor in their adaptation to specific habitats. These morphogenetic responses are usually taken to include quantitative changes in growth (both cell division and cell expansion), as well as differentiation of cells and organs and even changes in metabolic pathways. Important examples include: the tendency for stem elongation to be greater in certain classes of shade plants, thus enabling them to outgrow competitors; the development of characteristic 'sun' and 'shade' leaves with appropriate biochemical and physiologial characteristics (see Chapter 7); the induction of flowering or other reproductive growth at an appropriate season and the induction of dormancy. For many of these developmental responses some feature of the light environment provides the main external signal, though other important signals can include temperature (see Chapter 9) and water availability. This chapter provides an introduction to the role of light in the control of plant morphogenesis, though the actual mechanisms involved in the tight control and coordination of developmental responses is outside the scope of this book, and the reader is referred to introductory plant physiology texts such as Salisbury & Ross (1985) or to more specialist works (e.g. Roberts & Hooley 1988).

In addition to affecting development through effects on photosynthesis (for example, rapid growth depends in part on high rates of photosynthesis), light can influence growth and development in a number of other ways. These photomorphogenetic responses include:

(*a*) Phototropism – those directional alterations in growth that occur in response to directional light stimuli.
(*b*) Photonasty – reversible light movements and related phenomena that occur in response to directional and non-directional light stimuli.
(*c*) Photoperiodism – non-directional developmental responses to non-directional but periodic light stimuli.

(*d*) Photomorphogenesis – other non-directional developmental responses to non-directional and non-periodic light stimuli.

A particularly useful review of photomorphogenesis is Kendrick & Kronenberg (1986), while additional information may be found in Smith (1975; 1981 *a*, *b*), Vince-Prue *et al.* (1984), and Shropshire & Mohr (1983).

Detection of the signal

Perception of any light stimulus must involve an appropriate receptor. The most important photoreceptors in plants are chlorophyll, a blue-light absorbing receptor sometimes called cryptochrome (probably a flavoprotein) and phytochrome (a family of red/far-red reversible chromoproteins). Different photomorphogenetic responses involve at least one, and often more, of these pigments.

Phytochrome. Phytochrome is apparently unique among higher plant photoreceptors in that it can exist in two photo-interconvertible forms: a red-light absorbing form (Pr) that has an absorption maximum in the red (660 nm) and a far-red absorbing form (Pfr) that has an absorption maximum in the far-red (730 nm). It is in this region of the spectrum that natural radiation shows the greatest variation so phytochrome is perhaps well fitted as a detector of the spectral quality of incident radiation. Absorption spectra of Pr and Pfr are illustrated in Fig. 2.4, along with the absorption spectra of riboflavin and chlorophyll.

The main interconversions of phytochrome are illustrated in Fig. 8.1: absorption of light by the Pr form converts it to Pfr, while light absorption by Pfr converts it back to Pr. Red light therefore tends to convert most of the phytochrome present to the Pfr form. These transformations, which each proceed through a number of intermediate structures, are driven by the light energy absorbed with quantum yields (ε_q) of between about 0.07 and 0.17 (Jordan *et al.* 1986).

Phytochrome is synthesised in the dark as Pr which accumulates rapidly, indicating a relatively slow rate of Pr destruction. Photoconversion to Pfr, a much more labile form (at least in dark-grown plants), results in rapid loss of phytochrome by destruction. The physiologically active form of phytochrome is probably Pfr, which is likely to have its effect through the differential regulation of gene expression, though as yet the mechanism involved is not known. The amount of Pfr present at any time depends both on processes that determine the relative amount of Pr and Pfr, and on processes that change the total amount of phytochrome (synthesis and destruction). The relative amounts of Pr and Pfr in a population of phytochrome molecules are determined both by phototransformation and,

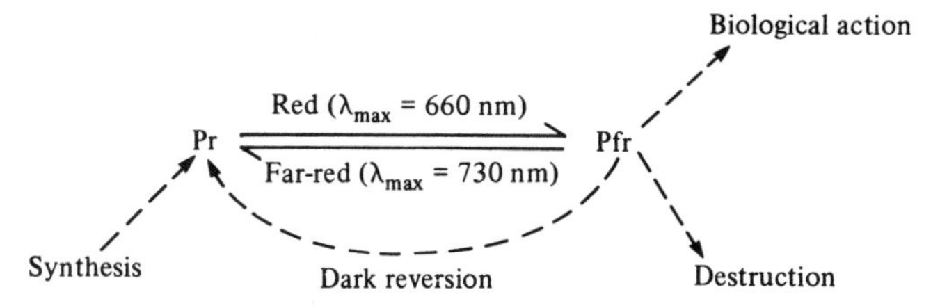

Fig. 8.1. Phytochrome interconversions.

at least in dicotyledons, by an additional dark reaction involving the thermal reversion of Pfr to Pr. This dark reversion appears to be absent in monocotyledons and in the Centrospermae, and though it is a moderately rapid process ($t_{\frac{1}{2}} \simeq 8$ min), this is so much slower than phototransformation that it is only likely to affect the ratio of Pfr to total phytochrome at very low irradiances (less than about 3 μmol m^{-2} s^{-1}). The significance of dark reversion is not clear.

Recent studies involving a comparison of results from spectrophotometric and immunological studies of phytochrome (the former providing evidence on the chromophore and the latter providing information on the protein moieties) have shown that the phytochrome system of higher plants is more complex than is indicated in Fig. 8.1 and comprises at least two populations of photoreversible chromoproteins with similar chromophores but different protein components. One population (Type I) builds up in the dark and is the form found in dark-grown seedlings; it is labile as Pfr, and it is the form which has been most studied. The other form of phytochrome (Type II) is relatively stable as Pfr, is present at all stages of plant development and is present at similar levels in light and dark. Type II is itself made up of at least two immunochemically distinct protein populations. Thus far, all kinetic information on phytochrome destruction has been obtained for Type I phytochrome. A further complication is that phytochrome normally exists as a dimer, with the possibility that the heterodimer (Pfr:Pr) is more unstable than either of the homodimers.

Phytochrome control of development. An enormous range of developmental responses apparently use phytochrome as the light receptor. These phytochrome responses are often divided into **Inductive** (or photoreversible) responses and **High Irradiance Responses** (HIR). Inductive responses, which include rapid (15–30 s) effects of light on leaflet closing, to slow (days to weeks) 'photoperiodic' effects on flowering, use the so-called low-energy phytochrome system and are typically saturated by radiant exposures of less than 1000 J m^{-2} (often 1–60 J m^{-2}), and are usually proportional to the logarithm of the incident energy up to this saturation point. This implies that a high irradiance for a short time or a low irradiance for a longer time are interchangeable in their effect. In addition these responses are usually red/far-red reversible, though in some cases reversibility is incomplete, either because the response is initiated extremely rapidly or because the amount of Pfr required to saturate the response is in excess of that established by the far-red (reversing) treatment.

High irradiance responses classically refer to phenomena such as stem extension and anthocyanin formation in dark-grown seedlings, and they characteristically show irradiance dependence and a requirement for continuous exposure to relatively high irradiances. They often involve an action spectrum with a peak in the blue (implying some involvement of cryptochrome) in addition to the peak in the far-red that would be expected for a phytochrome response. These responses have been most extensively studied in etiolated systems where it has been concluded that they depend on the total amount of Pfr present, while there are indications that the corresponding responses in green plants may be somewhat different. An important factor in all phytochrome responses to continuous irradiation is that in any constant radiation environment, a dynamic equilibrium is set up, such that the rate of conversion of Pr to Pfr exactly balances the rate of the reverse transformation of Pfr to Pr. Such an equilibrium is termed a photoequilibrium which can be expressed either in terms of the ratio Pfr/Pr (f) or more commonly as ϕ, the steady-state ratio of Pfr to total phytochrome, i.e.

$$\phi = \mathrm{Pfr}/(\mathrm{Pr}+\mathrm{Pfr}) = \mathrm{Pfr}/\mathrm{P}_{\mathrm{total}} = f/(1+f) \qquad (8.1)$$

The value of ϕ at equilibrium depends on the spectral distribution of the incident radiation, with ϕ decreasing as the proportion of far-red increases. Even monochromatic radiation gives rise to a photoequilibrium because both the Pr and Pfr forms each absorb some radiation over a wide range of wavelengths (Fig. 2.4).

As an example, for monochromatic light at 660 nm, the rate of the conversion Pr→Pfr is given by the product of the concentration of Pr multiplied by the amount of light absorbed and the quantum yield ($c_{\mathrm{Pr}} \times \mathbf{I}_{\mathrm{p660}} \times \alpha_{660(\mathrm{Pr})} \times \varepsilon_{\mathrm{q(Pr)}}$), at equilibrium this equals the rate of the reverse reaction Pfr→Pr (given by $c_{\mathrm{Pfr}} \times \mathbf{I}_{\mathrm{p660}} \times \alpha_{660(\mathrm{Pfr})} \times \varepsilon_{\mathrm{q(Pfr)}}$). Equating these two expressions and rearranging gives

$$\mathrm{Pfr}/\mathrm{Pr} = f = (\alpha_{660(\mathrm{Pr})}\,\varepsilon_{\mathrm{q(Pr)}})/(\alpha_{660(\mathrm{Pfr})}\,\varepsilon_{\mathrm{q(Pfr)}}) \qquad (8.2)$$

Substituting appropriate values has led to estimates of ϕ_{660} ranging between about 0.84 and 0.87 (see Jordan *et al.* 1986).

Unfortunately the amount of Pfr present at any time depends not only on the photoequilibrium, but also on the total amount of phytochrome present. For example, it is possible that continuous far-red light may result

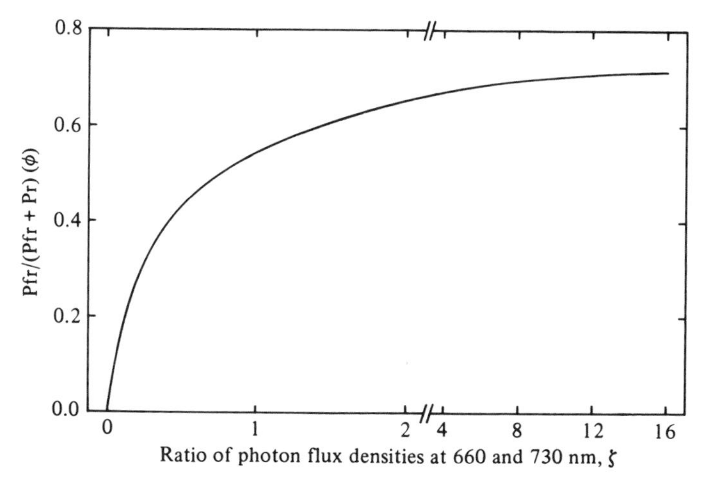

Fig. 8.2 Relation between phytochrome photoequilibrium (ϕ) and photon flux density ratio (ζ). (Data of Smith & Holmes 1977.)

in a higher total concentration of Pfr than does red light, even though it leads to a smaller ϕ. It has been proposed that such an effect could result from the greater susceptibility of Pfr than Pr to destruction, thus leading to a smaller amount of total phytochrome in red light. Other phenomena that may be involved in phytochrome responses in continuous light include 'photoprotection' (where it is postulated that phytochrome is increasingly protected from destruction as irradiance increases because high light causes rapid cycling between Pr and Pfr, so that a smaller proportion of time is spent in forms that are susceptible to destruction), and an irradiance dependent inactivation.

Another difficulty is that it is difficult to measure ϕ in an intact green plant because, with the usual photometric technique, absorption due to the small amount of phytochrome is swamped by that of the much larger amount of chlorophyll present. However, it is possible to determine ϕ spectrophotometrically in dark-grown etiolated tissues (which have no chlorophyll). This value of ϕ can be related to the ratio of the photon flux densities in the red (655–665 nm) and far-red (725–735 nm) portions of the spectrum (ζ), so that ζ is given by

$$\zeta = \mathbf{I}_{p(660)}/\mathbf{I}_{p(730)} = 0.904\,\mathbf{I}_{e(660)}/\mathbf{I}_{e(730)} \tag{8.3}$$

The factor 0.904 is required to convert radiant (energy) flux densities to photon flux densities as required by the definition of ζ. The relationship between ϕ and ζ is shown in Fig. 8.2, so that ζ can be used to estimate ϕ. Although not identical to the value of ϕ in green tissues in a corresponding

Table 8.1. *Approximate values of ζ and calculated photoequilibrium (ϕ) for the spectral distributions in Fig.* 8.3 (*data from Smith* 1975)

Time of day	No. of leaves	ζ	ϕ
Midday	0	1.00	0.50
	1	0.12	0.20
	2	< 0.01	0.06
Sunset or sunrise	0	0.63	0.35
	1	0.08	0.09
	2	< 0.01	0.02

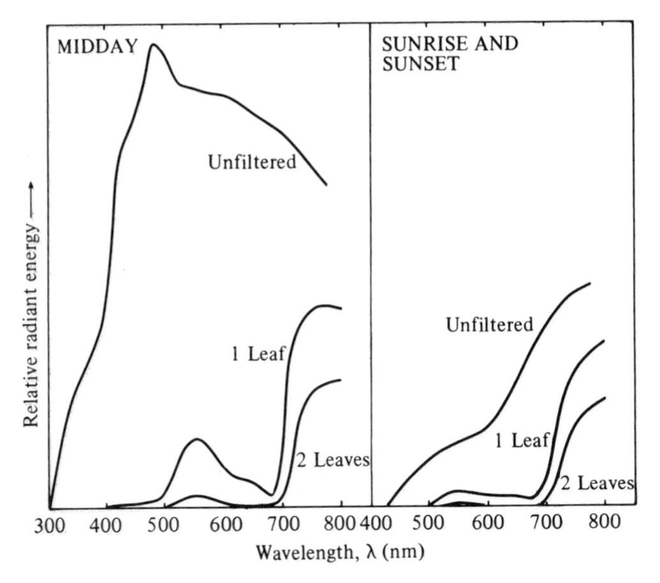

Fig. 8.3. Spectral distribution of relative radiant energy in direct sunlight at midday and at sunrise or sunset, together with the effects of filtering through one or two layers of sugar beet leaves. (After Smith 1973.)

radiation climate, the value obtained for etiolated tissue is often assumed to be a good estimate. Figure 8.2 shows that ϕ is most sensitive to ζ over the natural range of variation ζ. The maximum value of ζ in unfiltered sunlight is approximately 1.2, while the main factor that affects ζ is the filtering of light through leaves, though there is also a tendency for ζ to decrease with decreasing solar angle (Table 8.1 and Fig. 8.3) As was pointed out by Hughes *et al.* (1984), however, the variation in ζ as solar angle decreases is probably too variable and unreliable a signal to provide an effective photomorphogenetic signal. The values of ϕ and ζ corresponding to the spectra shown in Fig. 8.3 are summarised in Table 8.1. As described in Chapter 2, the relative enrichment in longer wavelengths when the sun is low in the sky, results from Rayleigh scattering removing most of the shorter wavelengths from the direct solar beam. Even further enrichment in the far-red component occurs as radiation passes through a canopy of leaves. This results from the much greater transparency of leaves to IR than to PAR (see Fig. 2.12, p. 30), so that ζ can fall to below 0.1 in the deep shade below several layers of leaves. Certain artificial light sources such as fluorescent tubes that are used in controlled environment chambers may give values of ζ from 2 to more than 9.5, with correspondingly high values of ϕ. This can be an important factor in the somewhat unnatural growth often experienced in growth chambers.

A more precise estimate of ϕ than can be obtained using measured photon irradiances at only two wavelengths can be achieved by taking into account the distribution of photon irradiance over the whole shortwave spectrum (see Smith 1981*b*), but any improvement in precision is likely to be small in most situations.

The detailed mechanisms of phytochrome action are incompletely understood and will not be discussed in detail, and though many responses apparently involve action at a membrane, phytochrome itself is a soluble cytosolic protein. It is worth noting, however, that the recent isolation and cloning of a number of different phytochrome genes opens up new opportunities for developing transformed plants which, by means of varying the promoters used, can synthesise differing amounts of phytochrome according to environmental conditions. Availability of this material will provide a valuable tool for improving our understanding of phytochrome action and unravelling the complexities of the interconversions and reactions of the different forms of phytochrome and of their relevance to the full range of different photomorphogenetic phenomena.

Phototropism

Phototropic responses are widely distributed in the plant kingdom, occurring in fern and moss tissues and the sporangiophores of some fungi, as well as in the growing regions of leaves and stems of higher plants. In a natural environment, phototropic alterations in the direction of growth (usually towards the light source for shoots) are of great importance in optimising the interception of available solar energy. Unfortunately most research has concentrated on the short-term phototropic responses of etiolated grass coleoptiles, but these are probably rather unrepresentative of normal green plant tissues. As with other tropic movements, phototropic curvature can occur only in tissues that still have some capacity for growth; fully differentiated stems do not bend.

Phototropic curvature of etiolated coleoptiles has been found to be related to the radiant exposure, that is the magnitude of the response is a function of the total radiation dose ($\int \mathbf{I} \mathrm{d}t$) rather than either flux density or time. This is often known as the *Bunsen–Roscoe Reciprocity Law* and is taken as evidence that only one photoreceptor system is operative. In fact the dose–response curve is rather complex with positive curvature (towards the light) increasing up to a maximum radiant exposure of about 0.1 J m^{-2}, but as the dosage increases further the response falls off and a slight negative curvature can occur. With still greater doses the curvature again becomes positive (the second positive curvature), and in this region the Reciprocity Law also breaks down. Although the directional responses are primarily sensitive to blue (< 500 nm) light (and hence do not involve phytochrome) the actual sensitivity observed is known to depend on the phytochrome system.

Details of the location of the sensor for directional radiation and the mechanism of the control of differential growth necessary to produce curvature are still controversial (see Firn & Digby 1980), though it was shown as long ago as 1880 by Charles Darwin that the tip region of coleoptiles appears to be the most sensitive to the phototropic stimulus. One difficulty in interpreting many experimental results is that plant tissues can transmit light to shaded regions perhaps by a light-piping mechanism similar to that occurring in optical fibres. The suggestion that the differential growth results from gradients of auxin concentration also remains to be proven. Although research has concentrated on the first two responses of coleoptiles, it is much more likely that something corresponding to the second-positive curvature is most relevant for green shoots growing in natural environments where they will rapidly receive a great deal more than 0.1 J m^{-2}!

The movements of leaves in relation to the sun (especially common in the Leguminosae), though sometimes termed helionastic, are probably better termed heliotropic in view of their directional nature. Some examples of heliotropic movements have been reviewed by Ehleringer & Forseth (1980). When well watered, the leaves of appropriate species tend to remain perpendicular to the solar beam throughout the day thus maximising light interception and photosynthesis (Fig. 2.19, p. 40): that is, they are diaheliotropic. When water-stressed, however, the leaves tend to become paraheliotropic and orientate parallel to the incident radiation, thus reducing irradiance at the leaf surface, conserving water and preventing overheating. Other organs, such as the flower heads in sunflower, also move in relation to the sun. The effects of heliotropic movements on light interception were discussed in Chapter 2, and ecological implications for temperature regulation and drought tolerance are discussed in Chapters 9 and 10.

Photonasty

Examples of photonastic phenomena include the opening or closing of flowers with a change in the general level of irradiance, and the nyctinastic 'sleep movements' of leaves that fold up at night. Although some nastic movements involve differential growth, the reversible photonastic movements of leaves of many Leguminosae (such as the lupin, *Lupinus albus*, and the sensitive plant, *Mimosa pudica*) are brought about by reversible turgor changes in the specialised hinge-cells or pulvini at the base of the leaflets. These changes in turgor apparently involve a similar mechanism to that of stomatal movement, including H^+ extrusion from the hinge cells with consequent uptake of K^+ as a major osmoticum. A similar mechanism is involved in many heliotropic movements.

The light sensor for the 'sleep movements' is not known, but they have a strong endogenous circadian rhythm, and movements can continue for several days even in continuous light.

Photoperiodism

In all parts of the world, except at the equator, the daylength changes with season (Fig. 8.4), it is constant throughout the year at the equator and the seasonal variation increases with latitude, changing by as much as 5.5 h between winter and summer at 40° (north or south) and by 24 h at the Arctic and Antarctic circles. These seasonal changes are utilised by many species as a reliable signal for phasing the conversion of the growing apices from a vegetative to a floral state as well as in the regulation of various other developmental processes. Using the photoperiod as the signal for floral induction ensures that flowering can occur at the optimum time for any particular species in relation to the climate at that location; for example, long enough before the average onset of frosts for seeds to ripen or before the onset of the dry season in more tropical climates.

The shortwave irradiance is sufficient to influence the photoperiodic detection system during the period of twilight when the sun is just below the horizon, so it is necessary to allow for this extension of daylight in calculation of the daylength. Although the exact threshold irradiance for photoperiodic effects is not known, and probably varies with species and climate, an appropriate, though arbitrary, definition of daylength for

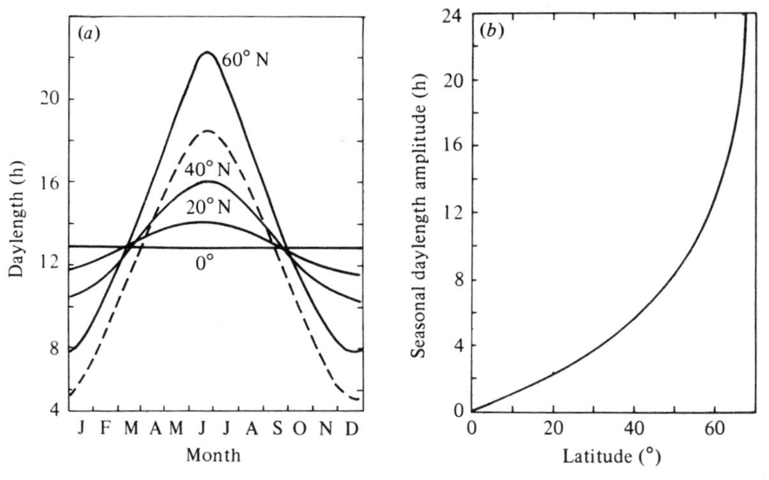

Fig. 8.4. (*a*) Seasonal variation of daylength at different latitudes. Solid lines include civil twilight, dashed line shows the daylight period when sun is above the horizon (at 60° N). (*b*) Amplitude of seasonal daylength changes at different latitudes (calculated according to Appendix 7).

photoperiodic studies is the period of daylight between sunrise and sunset together with the period of civil twilight (i.e. the period when the sun is less than 6° below the horizon). This period may be calculated by suitable rearrangement of equation A7.1 (p. 362), giving the curves in Fig. 8.4.

Since Garner & Allard's work in 1920, it has been known that angiosperms can be classified into at least three main groups on the basis of their flowering response to daylength. Some plants, such as tobacco and the weed *Chenopodium rubrum* are found to require a certain minimum number of days where the uninterrupted dark period exceeds a certain minimum duration for flower induction to occur. These are called short-day (SD) plants. In long-day (LD) plants, such as ryegrass, on the other hand, flowering only occurs when the photoperiod exceeds a certain value for a minimum number of cycles. Other species, particularly those of tropical origin, are day-neutral in their flowering response, being unaffected by daylength. This class includes, among others, many desert ephemerals.

In many species either the daylength requirement changes with age, or else they have a requirement for a particular sequence of environmental conditions such as a period of long-days followed by a period of short days (LSD plants), or short-days followed by long-days (SLD plants) in order to flower. Similarly some plants have an absolute or obligate photoperiodic requirement that must be satisfied before flowering can occur, while others may show hastened flowering in appropriate photoperiods, but will

Table 8.2. *Some examples of photoperiodic plants* (*from Wareing & Phillips* 1981)

Short-day plants	
(*a*) Obligate SD plants:	
Chenopodium rubrum	*Glycine max* (soybean)
Coffea arabica (coffee)	*Xanthium strumarium* (cockleburr)
Fragaria vesca (strawberry)[a]	
(*b*) Quantitative SD plants:	
Cannabis sativa (hemp)	*Oryza sativa* (rice)
Gossypium hirsutum (cotton)	*Saccharum officinarum* (sugar cane)
Long–short-day (LSD) plants	
Bryophyllum crenatum	
Long-day plants	
(*a*) Obligate LD plants:	
Avena sativa (oats)	*Raphanus sativus* (radish)
Lolium temulentum (ryegrass)	*Spinacia oleracea* (spinach)
(*b*) Quantitative LD plants:	
Beta vulgaris (beet)	*Poa pratensis* (Kentucky blue grass)
Hordeum vulgare (spring barley)	*Triticum aestivum* (spring wheat)
Short–long-day (SLD) plants	
Campanula medium (Canterbury bell)	
Trifolium repens (creeping clover)	

[a] Some varieties of strawberry are day-neutral.

ultimately flower even under unfavourable daylength (quantitative photoperiodic plants). Some examples of photoperiodic plants are listed in Table 8.2.

The SD plant group includes many plants that originated in low latitude regions, where daylength never exceeds about 14 h, including important crop species such as maize, millet, rice and sugar cane. There are, however, some SD plants in temperate regions including plants such as chrysanthemums that only flower late in the summer as the days shorten. LD plants are typically temperate region plants that flower during the long days of summer, and include many of the common temperate crop species.

Although a very precise photoperiodic response may be advantageous for a plant growing in a specific environment, it limits its adaptability to other regions. Wild species and crop plants that occur over a wide latitudinal range tend to be differentiated into ecotypes or local races with different daylength responses, thus restricting the area over which they can grow successfully. Many modern varieties of soybean, wheat and rice, however, are less rigorously controlled by daylength than are related wild species. In fact, breeding for independence of daylength has been an objective in the development of the new 'green-revolution' varieties of wheat in Mexico and rice in the Philippines, to aid trans-world adaptation. This selection for photoperiod insensitivity can, however, have some disadvantages. For example, some daylength sensitivity is useful in wheat grown at high latitudes to delay ear development until the risk of frost is past. Another example is in 'floating rice', where photoperiod sensitivity is used to delay flowering until the monsoon floods recede, or else the grain cannot be harvested.

Various other plant processes are known to be controlled or influenced by photoperiod. These include onset and breaking of bud dormancy in woody perennials, leaf abscission, rooting of cuttings, seed germination, bulb and tuber formation in herbaceous plants and the development of frost resistance. For example, the increasing length of the dark period in late summer is used as a signal by many woody plants for the initiation of dormant buds and for enhancing winter hardiness in anticipation of winter, even though climatic conditions may still be favourable for growth. This could provide an important ecological adaptation where the relatively slow process of developing winter hardiness is started before the first frosts occur, thus avoiding damage to the plant. Conversely the long-day induced formation of resting buds in the desert liverwort *Lunularia cruciata* provides an example of an adaptation reducing the effects of summer water stress.

Photoperiodic responses in higher plants involve complex interactions of phytochrome with an endogenous timekeeping system; these have been studied by a wide range of types of experiment including those that involve short periods of illumination during the dark period (night-breaks). These studies have shown that whether or not short-day plants flower in light/dark cycles is dependent primarily on the length of the dark period rather than on the duration of light. The fact that plants (such as some rice varieties) are sensitive to changes in daylength of as little as 15–20 min implies a precise endogenous time-measuring mechanism that must be combined with accurate detection of the light 'on' and light 'off' signals. There is also good evidence for a role for endogenous rhythms in these responses, and many involve some interaction with temperature. It is assumed that the end of the light period is detected by phytochrome action as a loss of the Pfr, though, paradoxically, there is a requirement for some Pfr to be present during darkness for induction to occur in short-day plants.

Photomorphogenesis

The term photomorphogenesis is given to the wide range of light-controlled developmental responses that were not conveniently discussed under any of the above headings. These include effects on seed germination, stem elongation, leaf expansion, the development of chloroplasts and the synthesis of chlorophyll and many secondary products.

Germination. Although many seeds are not affected by light during germination, there are some species that are strongly light-dependent. In some plants (e.g. varieties of lettuce, *Lactuca sativa*, and beech, *Fagus sylvatica*) germination is stimulated by white light, while in others (e.g. varieties of *Cucumis sativus*) white light is inhibitory. Interestingly, seeds of relatively few cultivated species are light-sensitive, perhaps as a result of artificial selection. Some weed species, however, are polymorphic for their light response. One can easily envisage the adaptive significance of light adaptation by seeds. Light inhibition would ensure germination only when the seeds were buried. On the other hand, light-stimulated seeds could remain dormant for long periods until exposed by soil disturbance. This behaviour could be advantageous in spreading out seed germination over many years. The light required for germination varies with species, some such as lettuce needing as little as one minute's irradiation by low light, while others may require repeated exposure for several hours per day. Studies by Bliss & Smith (1985) have shown that seeds can respond to the very low irradiances occurring within the surface layers of the soil. For example, *Digitalis pupurea* required light for germination, yet germinated well when covered by 10 mm soil. Transmission through soil attenuates the shorter wavelengths preferentially, thus decreasing the red/far-red ratio; this effect is greatest for dry sand.

The evidence suggests that both types of light response are mediated by the phytochrome system, with germination promoted by Pfr, since red light promotes germination and far-red inhibits. The complexity of the responses in seeds may result from the presence of several different forms of phytochrome, as in other parts of plants.

Plant morphology. Both quality and quantity of light have complex effects on plant morphology. Seedlings grown in complete darkness become etiolated: that is, they grow very long and pale. This provides a clear adaptation enabling a plant to extend above the soil surface into the light before expanding its leaves. The process of de-etiolation is known to require action of all three main photoreceptor systems.

The further development of plants once they have been exposed to light

Table 8.3. *Some developmental changes in* Tripleurospermum maritimum *and* Chenopodium album *grown under fluorescent* (F, $\phi = 0.71$) *or incandescent* (I, $\phi = 0.38$) *lights* (*equal photon flux densities in the PAR*) (*data from Holmes & Smith* 1975)

	T. maritimum		*C. album*	
	F	I	F	I
Height (cm)	29.7	59.4	15.0	28.4
Internode length (cm)	0.8	3.5	—	—
Leaf dry weight (g)	0.48	0.46	0.34	0.31
Stem dry weight (g)	0.33	0.58	0.10	0.20
Leaf area (cm^2)	—	—	107	78
Chlorophyll a + b (mg/g FW)	—	—	112	94

is also controlled by the phytochrome and blue light systems. For example there is extensive evidence that stem and leaf expansion are sensitive to light quality acting through the high-energy phytochrome system. Different light sources give rise to different photoequilibria, with different proportions of the phytochrome in the Pfr form. Much of the information on morphological responses has derived from research to determine appropriate lighting systems for controlled environments and glasshouses for producing 'normal' plants or for obtaining particularly short or tall specimens, or for maximising productivity.

For example, illumination under fluorescent or incandescent lamps arranged to provide equal photon flux densities in the PAR resulted in significant differences in growth (Table 8.3). In particular, the incandescent lamps gave greater total dry matter production and a much greater stem extension rate. These results are typical of what happens when the proportion of far-red light is increased: the incandescent lamps have a high proportion of radiation in the red and far-red ($\phi = 0.38$), while the fluorescent radiation is mainly in the blue ($\phi = 0.71$). In general, it is found that the logarithmic rate constant for stem extension tends to be inversely proportional to ϕ, but exponentially related to ζ (see Fig. 8.5).

Even brief exposure to supplemental far-red at the end of a light period can have large morphogenetic effects (Table 8.4). These include increased internode elongation, petiole extension and leaf expansion.

In a natural environment this sensitivity to light quality is an important factor in adaptation to shade. As shown in Fig. 8.3, a characteristic feature of shade light is a relative enrichment in the longer wavelengths with

Table 8.4. *Summary of the effect on growth of tobacco seedlings* (Nicotiana tabacum) *of* 0.5 *h illumination with fluorescent* (F) *or incandescent* (I) *lamps at the end of an* 8.5 *h photoperiod with* F+I *at* 670 $\mu mol\ m^{-2}\ s^{-1}$. *Average for* 5 *cultivars (calculated from Downs & Hellmers* 1975)

	F	I
Stem length (cm)	7.1	12.9
Leaf fresh weight (g)	11.4	14.5
Stem fresh weight (g)	2.0	4.1
Width of fifth leaf (cm)	16.4	19.0
Length of fifth leaf (cm)	9.3	9.7

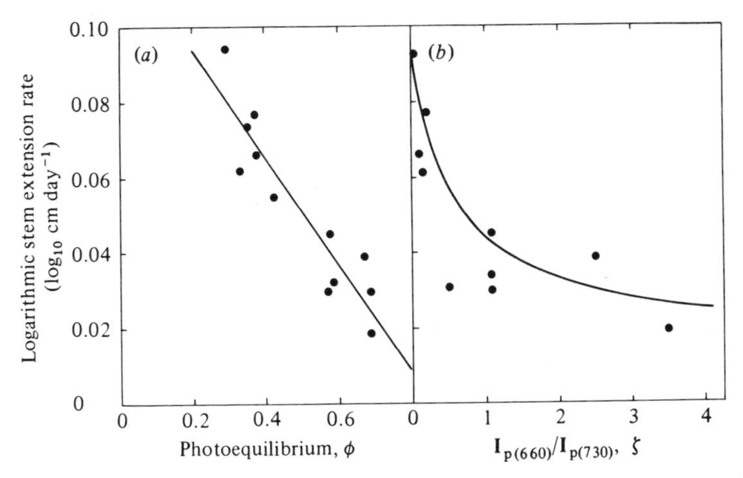

Fig. 8.5. Relation between stem growth rate and (*a*) photoequilibrium ϕ and (*b*) photon flux density ratio ζ for *Chenopodium album*. (After Smith 1981*b*.)

consequent lowering of the phytochrome photoequilibrium (ϕ). Many species, particularly arable weeds, are shade avoiders that show dramatic stem extension when shaded by other plants, thus enabling them to outgrow competitors. Shade-adapted woodland herbs, on the other hand, show much smaller responses to shade light. Although much of the response to shade light is a response to its *quality*, the characteristically low total flux density or *quantity* is also an important factor in many shade responses (see e.g. Evans 1972; Young 1975). Low irradiances tend to result in maximum leaf development at the expense of stem extension. This contrasts with the effect of low values of ζ which result in maximum stem extension at the expense of leaf expansion. Low irradiance also tends to increase the length of stem per unit weight and give larger but thinner leaves. In fact the morphological and biochemical differences between sun and shade leaves (see Chapter 7) involve interactions between quality and quantity perception.

It has been suggested that the red/far-red ratio is a particularly important signal for plant density. Even in sparse canopies the selective scattering of far-red radiation from neighbouring plants may be enough to signal their proximity, without the need to consider spectral modification of radiation by passage through leaves (Smith *et al.* 1990). The precise mechanism, however, whereby differences in plant spacing are distinguished from, for example, values of ζ that occur at different times of day is unclear. It is possible that the response may be weighted for the irradiance received and the differential properties of type I and type II phytochrome may be relevant.

Sample problem

8.1 Assuming that the incident energy in sunlight is 1.1 times greater at 660 nm than at 730 nm, (i) what is (*a*) ζ in unattenuated sunlight, (*b*) ζ in sunlight filtered through one layer of leaves ($\tau = 0.08$ at 660 nm, $\tau = 0.35$ at 730 nm)? (ii) Estimate the corresponding values of the phytochrome photoequilibrium (ϕ).

9

Temperature

Plants can survive the whole range of atmospheric temperatures (−89 to 58 °C: McFarlan 1990) that occur on the surface of the earth, as well as the even higher temperatures (up to about 70 °C) that occur in desert soils and in the surface tissues of slowly transpiring massive desert plants such as cacti (Nobel 1988). Seeds are particularly hardy, though other tissues of some species can also survive an extremely wide temperature range. Most plants can only grow, however, over a much more limited range of temperatures from somewhat above freezing to around 40 °C, while growth approaches the maximum over an even more restricted temperature range that depends on species, growth stage and previous environment. Several useful articles on plants and temperature may be found in Long & Woodward (1988).

In this chapter the physical principles underlying the control of plant temperatures are described and the physiological effects of high and low temperatures outlined. The final section considers the more ecological aspects of plant adaptation and acclimation to the thermal environment.

Physical basis of the control of tissue temperature

As outlined in Chapter 5, the temperature of plant tissue at any instant is determined by the energy balance. Neglecting any metabolic storage, the energy balance equation (5.1, p. 106) reduces to

$$\mathbf{\Phi}_n - \mathbf{C} - \lambda\mathbf{E} = \mathbf{S} \qquad (9.1)$$

Any imbalance in the energy fluxes goes into physical storage, thus altering tissue temperature. Although what follows refers mainly to leaf temperatures, the same principles apply to all above ground tissues. Further discussion of biophysical aspects of control of plant temperature may be found in texts by Campbell (1977), Gates (1980), Monteith (1981*b*) and Monteith & Unsworth (1990).

Steady state

In the steady state, when leaf temperature is constant, equation 9.1 reduces to

$$\mathbf{\Phi}_\mathrm{n} - \mathbf{C} - \lambda\mathbf{E} = 0 \tag{9.2}$$

This may be expanded by substituting the following versions of equations for the sensible heat (equation 3.29, p. 60) and latent heat (equation 5.17, p. 112) losses:

$$\mathbf{C} = \rho_\mathrm{a} c_p (T_\ell - T_\mathrm{a})/r_\mathrm{aH} \tag{9.3}$$

$$\begin{aligned}\lambda\mathbf{E} &= (0.622\rho_\mathrm{a}\lambda/P)(e_{s(T\ell)} - e_\mathrm{a})/(r_\mathrm{aW} + r_{\ell\mathrm{W}}) \\ &= (\rho_\mathrm{a} c_p/\gamma)(e_{s(T\ell)} - e_\mathrm{a})/(r_\mathrm{aW} + r_{\ell\mathrm{W}})\end{aligned} \tag{9.4}$$

Using the resulting expanded equation, it is possible to determine leaf temperature when values for absorbed radiation, air temperature, humidity and leaf and boundary layer resistances are known by means of an iterative computing procedure (Gates & Papain 1971). A rather more convenient analytical expression for leaf temperature can be obtained by means of a derivation similar to that used for the combination equation for evaporation (equation 5.23, p. 114). The procedure is to use equation 5.18 to replace the leaf–air vapour pressure difference in equation 9.4 by the humidity deficit of the ambient air (δe) and the leaf–air temperature difference. Subsequent combination of equations 9.2–9.4 yields the following:

$$T_\ell - T_\mathrm{a} = \frac{r_\mathrm{aH}(r_\mathrm{aW} + r_{\ell\mathrm{W}})\gamma\mathbf{\Phi}_\mathrm{n}}{\rho_\mathrm{a} c_p[\gamma(r_\mathrm{aW} + r_{\ell\mathrm{W}}) + s\,r_\mathrm{aH}]} - \frac{r_\mathrm{aH}\,\delta e}{[\gamma(r_\mathrm{aW} + r_{\ell\mathrm{W}}) + s\,r_\mathrm{aH}]} \tag{9.5}$$

This equation shows that the leaf temperature excess is given by the sum of two terms, one depending on net radiation and the other on the vapour pressure deficit of the air.

There are two important approximations involved in the derivation of equation 9.5. The first is the assumption that the rate of change of saturation vapour pressure with temperature (s) is constant between T_a and T_ℓ. This introduces negligible error for normal temperature differences. The other approximation is that net radiation is an environmental factor unaffected by leaf conditions, but $\mathbf{\Phi}_\mathrm{n}$ is actually a function of leaf temperature itself. It is possible to allow for this effect by using the concept of isothermal net radiation (Chapter 5). Replacing $\mathbf{\Phi}_\mathrm{n}$ in equation 9.5 by $\mathbf{\Phi}_\mathrm{ni}$ and replacing r_aH by r_HR gives

$$T_\ell - T_\mathrm{a} = \frac{r_\mathrm{HR}(r_\mathrm{aW} + r_{\ell\mathrm{W}})\gamma\mathbf{\Phi}_\mathrm{ni}}{\rho_\mathrm{a} c_p[\gamma(r_\mathrm{aW} + r_{\ell\mathrm{W}}) + s\,r_\mathrm{HR}]} - \frac{r_\mathrm{HR}\,\delta e}{\gamma(r_\mathrm{aW} + r_{\ell\mathrm{W}}) + s\,r_\mathrm{HR}} \tag{9.6}$$

Using equation 9.6 we can now investigate how the leaf–air temperature

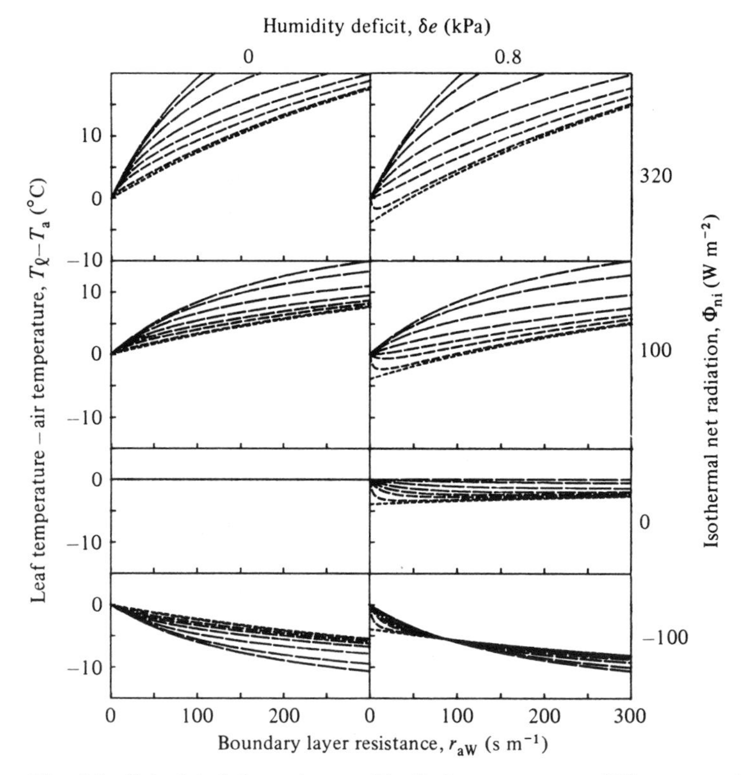

Fig. 9.1. Calculated dependence of leaf–air temperature difference on δe, $\mathbf{\Phi}_{ni}$, r_a and r_ℓ for $T_a = 20$ °C. The different lines represent r_ℓ ranging from 0 s m^{-1} with short dashes, through 10, 50, 100, 200, 500, 2000 to ∞ with the longest dashes.

difference depends on environmental and plant factors. Although in practice there are complex feedback effects on r_ℓ (see Chapter 6), a useful summary of how the various factors in equation 9.6 affect leaf temperature can be obtained by varying each independently. The results of such an approach are shown in Fig. 9.1 and summarised below:

(1) *Leaf resistance*. Where the surface is dry so that there is no latent heat term in the energy balance (this is equivalent to $r_\ell = \infty$), equation 9.6 reduces to

$$T_\ell - T_a = \mathbf{\Phi}_{ni} r_{HR}/\rho_a c_p \qquad (9.7)$$

In this case, $T_\ell - T_a$ is proportional to $\mathbf{\Phi}_{ni}$, with the leaf being warmer than air when $\mathbf{\Phi}_{ni}$ is positive (as it usually is during the day), and cooler than air when $\mathbf{\Phi}_{ni}$ is negative. Because r_{HR} includes both a radiative and a convective component, $T_\ell - T_a$ is not linearly related to r_{aH}.

When the surface is perfectly wet, as might occur when it is covered in dew, $r_\ell = 0$. In this case the latent heat cooling is maximal for any boundary layer resistance. As r_a tends to 0, the value of $T_a - T_\ell$ tends to $\delta e/(\gamma + s)$, the theoretical wet bulb depression.

When r_ℓ is finite, leaf temperature tends towards air temperature as the boundary layer resistance approaches zero. With normal values for the boundary layer resistance, the amount of transpirational cooling increases as r_ℓ decreases. Whether this transpirational cooling is adequate to cool the leaf below air temperature depends on other factors, particularly $\mathbf{\Phi}_{ni}$ and δe. Leaf temperature always tends to rise as r_ℓ increases (see Fig. 9.1). The anomalous behaviour in Fig. 9.1, when net radiation is negative, arises when condensation (i.e. dewfall) is occurring. In this case the leaf resistance is zero so the calculated curves for higher values of r_ℓ are physically unrealistic.

(2) *Vapour pressure deficit.* The effect of humidity deficit on T_ℓ depends on the total resistance to water vapour loss. Where the surface is dry (or where $r_\ell = \infty$) so no latent heat loss can occur, δe is irrelevant to leaf temperature (equation 9.7). In all other cases, any increase in δe lowers T_ℓ especially when r_ℓ is low.

(3) *Net radiation.* Increasing the radiative heat load on a leaf while maintaining other factors constant always tends to increase T_ℓ (Fig. 9.1). When $\mathbf{\Phi}_{ni}$ is negative (it is commonly as low as -100 W m^{-2} on a clear night) T_ℓ must be below T_a. The net radiation absorbed by a leaf is very dependent on the value of the reflection coefficient for solar radiation (ρ_S, see Chapter 2).

(4) *Boundary layer resistance.* The effect of r_a on leaf temperature is complex, since increasing r_a can increase or decrease T_ℓ depending on the environmental conditions and on r_ℓ. When T_ℓ is above T_a, increases in r_a always tend to increase T_ℓ. The value of r_a itself is dependent on windspeed and leaf size and shape as outlined in Chapter 3.

(5) *Air temperature* (T_a). The effect of ambient air temperature on leaf temperature is two-fold. First, it provides the reference temperature to which T_ℓ tends. Secondly, there are two major effects of T_a on the value of $T_\ell - T_a$: the value of s increases with temperature so that any leaf temperature excess decreases with increasing temperature (equation 9.6), and for any given value of relative or absolute humidity, δe increases with increasing temperature, therefore increasing latent heat loss and lowering T_ℓ with repect to T_a as shown in Fig. 9.2. These two latter effects lead to large positive values of $T_\ell - T_a$ at low temperatures.

(6) *Water deficit.* Because the leaf–air temperature differential is related to the leaf conductance, it has been suggested that T_ℓ or the leaf–air temperature differential can be used as measure of the degree of water stress to which a plant is subject. Both empirical and theoretical approaches have

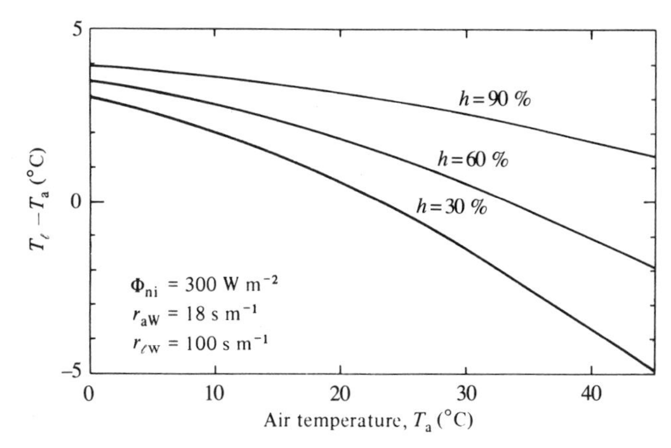

Fig. 9.2. Effect of air humidity (h, %) and air temperature on leaf–air temperature difference (equation 9.6).

been used to calculate a 'stress index', based on this property. The theoretical and practical basis of such a stress index is discussed in detail in Chapter 10.

Non-steady state

In a natural environment, irradiance and windspeed, particularly, are continually varying so that steady plant temperatures are rarely attained. When any component of the energy balance changes so that **S** is no longer zero, leaf temperature alters in the direction needed to return the net energy exchange to zero. For example, if $\mathbf{\Phi}_n$ increases, leaf temperature also increases until the increased sensible and latent heat losses again balance the new value of $\mathbf{\Phi}_n$.

The rate of change of leaf temperature depends on the heat capacity per unit area of the tissue ($\rho^* c_p^* \ell^*$)

$$\frac{\mathrm{d}T_\ell}{\mathrm{d}t} = \mathbf{S}/(\rho^* c_p^* \ell^*) = (\mathbf{\Phi}_n - \mathbf{C} - \lambda\mathbf{E})/(\rho^* c_p^* \ell^*) \tag{9.8}$$

where ρ^* and c_p^* are the density and specific heat capacity, respectively, of leaf tissue. ℓ^* is a volume to area ratio that equals the thickness for a flat leaf, $d/4$ for a cylinder or $d/6$ for a sphere (where d is the diameter). If the equilibrium temperature (T_e) for any environment is defined as that value of T_ℓ attained in the steady state, equation 9.8 (see Appendix 9) can be rewritten as

$$\frac{\mathrm{d}T_\ell}{\mathrm{d}t} = \frac{\rho_a c_p (T_e - T_\ell)}{\rho^* c_p^* \ell^*} \{(1/r_{HR}) + [s/\gamma(r_{aW} + r_{\ell W})]\} \tag{9.9}$$

Equation 9.9 is in the form of *Newton's Law of Cooling* which states that 'the rate of cooling of a body under given conditions is proportional to the temperature difference between the body and the surroundings'. This is a first-order differential equation (like equation 4.30, p. 100) that after substitution of appropriate boundary conditions (i.e. the leaf is initially at equilibrium at T_{e1} and the environment is altered instantaneously at time zero to give a new equilibrium T_{e2}) can be solved by standard techniques to give

$$T_\ell = T_{e2} - (T_{e2} - T_{e1}) \exp(-t/\tau) \tag{9.10}$$

where the time constant τ is given by

$$\tau = \frac{\rho^* c_p^* \ell^*}{\rho_a c_p \{(1/r_{HR}) + [s/\gamma(r_{aW} + r_{\ell W})]\}} \tag{9.11}$$

Thermal time constants for plant organs

The thermal time constant (τ) provides a measure of how closely tissue temperatures track T_e in a changing environment. The value of τ depends on the size and shape of the organ (thickness affects heat capacity per unit area, while size and shape affect r_a), on stomatal resistance and on air temperature (which affects the values of the constants, particularly s). Thermal properties of various materials are summarised in Appendix 5. If one assumes an average specific heat capacity of 3800 J kg^{-1} K^{-1} for leaves and fruits (this is close to the value for pure water (4180 J kg^{-1} K^{-1} at 20 °C) because usually between 80 and 90% of tissue fresh weight is water), and an average leaf density of 700 kg m^{-3}, this gives $\rho^* c_p^*$ as $\simeq$ 2.7 MJ m^{-3}.

Using this value for all plant tissues enables us to calculate approximate time constants for plant organs of different size and shape and for two windspeeds (Table 9.1). This table shows that τ is likely to be significantly less than a minute for all but the largest leaves. Stems and fruits have longer time constants than leaves because they have a larger mass per unit area, so that for the trunks of mature trees τ can be of the order of one day. The value of τ for leaves is significantly increased by stomatal closure.

Figure 9.3 illustrates just how rapidly the temperature of medium-sized (*c*. 10 cm^2) transpiring leaves can fluctuate in the field. Note that the air temperature itself also shows rapid changes that are not detected when using a measuring instrument with a long time constant such as a mercury-in-glass thermometer. Observed values of τ for leaves of different species are close to the values predicted in Table 9.1, ranging from about 0.15 to 0.45 min for species as diverse as vine, cotton, *Salix arctica* and *Pinus taeda*

Table 9.1. *Thermal time constants for leaves, stems and fruits treated as simple geometric shapes at 20 °C calculated using equation* 9.11. *Values for τ are for non-transpiring organs except those in brackets which assume an r_ℓ of* 50 s m^{-1} (*after Monteith* 1981*b*)

	Dimensions		Calculated time constant (τ, min)	
	d (cm)	ℓ^* (cm)	$u = 1$ m s^{-1}	$u = 4$ m s^{-1}
Leaves				
Grass	0.6	0.05	0.18 (0.13)	0.09 (0.08)
Beech	6	0.10	0.94 (0.52)	0.55 (0.36)
Alocasia	60	0.15	2.9 (1.34)	2.0 (1.01)
Stems				
Small	0.6	0.15	1.4	0.68
Medium	6	1.5	31	16
Large	60	15	540	330
Fruits				
Rowan	0.6	0.1	0.71	0.33
Crab apple	6	1	16	7.7
Jack fruit	60	10	300	170

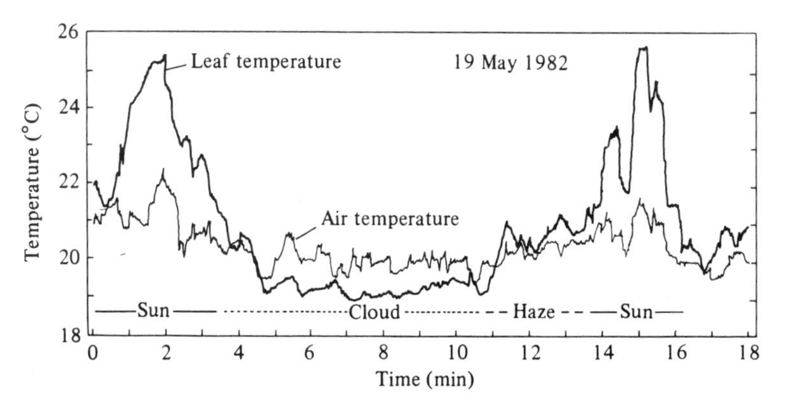

Fig. 9.3. Temperature fluctuations of an apple leaf (approx. 10 cm^2) in the field. Temperatures measured with 42-gauge copper–constantan thermocouples.

(Warren Wilson 1957; Ansari & Loomis 1959; Thames 1961; Linacre 1972) though reaching 7 min for the very thick leaves of *Graptopetalum* in a low wind (Ansari & Loomis 1959).

One approximation in the derivation of equation 9.9 is the assumption of a uniform surface temperature, a condition that is not usually satisfied in the field, though the error involved is usually small. A second problem is that with bulky tissue, the rate of heat conduction to the surface is important, because the thermal conductivity of plant tissues is quite low (being of the same order as water: Appendix 5). The lateral thermal conductivity of leaves, for example, ranges from about 0.24 to 0.50 (Vogel 1984) while the thermal conductivity of apple fruits is less than 0.9 W m^{-1} K^{-1} (Thorpe 1974). This means that the temperature at the centre of large organs lags behind that at the surface, and that the time constant at the centre of stems may be longer that that given by equation 9.11, which may in turn be longer than that at the surface. The low thermal conductivity of plant tissue also means that unequal radiation absorption on different sides of large organs can lead to large temperature gradients. For example, temperature differences as large as 10 °C between the two sides of an apple (Thorpe 1974) or a cactus (Nobel 1978) have been observed with high irradiance. Nobel (1980) has developed a model for calculating stem surface temperatures in different *Ferocactus* species that includes effects of plant size, apical pubescence and shading by spines. The results of the model were in close agreement with field observations and could be related to the natural distribution of the different species.

Particular cases of the time course of temperature changes

(*a*) *Step change*. Where the environmental temperature changes instantaneously from one value to another the time course of leaf temperature, for example, is given by equation 9.10 and is illustrated in Fig. 9.4*a*.

(*b*) *Ramp change*. Where the environment is changing at a steady rate, tissue temperature lags behind the equilibrium temperature, but when the time of the steady change exceeds about $3 \times \tau$, the rate of increase of tissue temperature equals the rate of increase of equilibrium temperature (Fig. 9.4*b*).

(*c*) *Harmonic change*. A particularly important situation in environmental studies is where the equilibrium temperature oscillates. Both the diurnal and the seasonal changes in temperature, for example, can be approximated by sine waves. The general effect is illustrated in Fig. 9.4*c* for the same value of the time constant as in Figs 9.4*a* and 9.4*b*. The effect of increasing τ is twofold: firstly it causes a damping of the amplitude of the oscillation, and secondly it increases the magnitude of the phase lag between the driving temperature and the sample temperature. The appropriate response functions are given in the caption to Fig. 9.4. This effect leads to the amplitude of diurnal temperature changes decreasing with increasing depth into massive organs such as tree trunks, or in the soil.

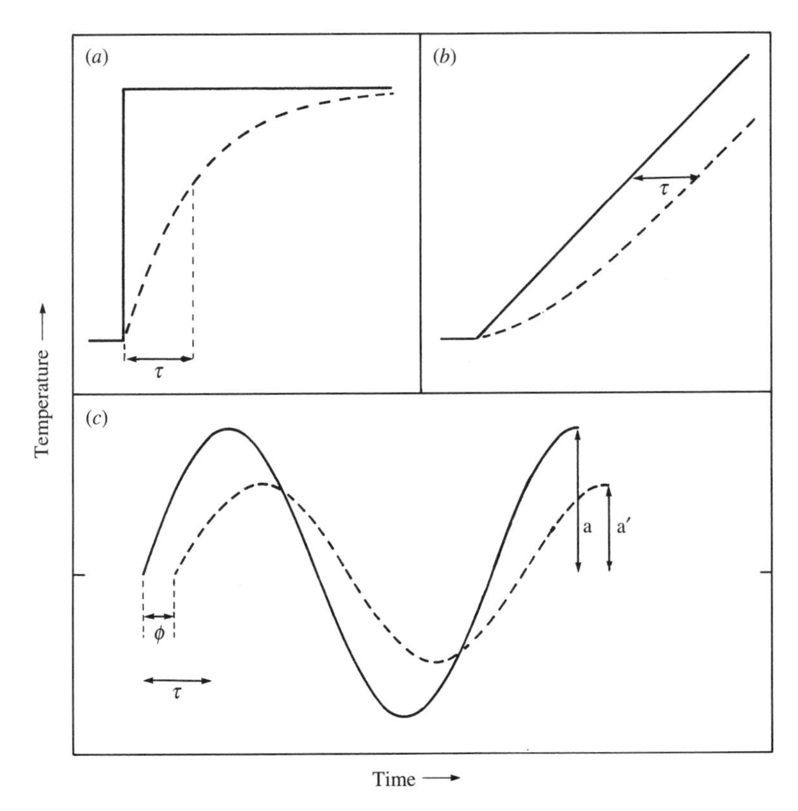

Fig. 9.4. The change in surface temperature (broken line) in response to changing environmental temperature (solid line) (modified from Monteith & Unsworth 1990). (*a*) Response to a step change, where τ is the t for a 63% change (see also Fig. 4.13); (*b*) response to a ramp change, where τ is the constant time lag eventually established; (*c*) response to a sinusoidally varying environmental temperature where $T_e = \bar{T}_e +$ a sin $(2\pi t/P)$, where a is the amplitude and P is the period of oscillation. The phase lag of the response (ϕ) is given by $\phi = \tan^{-1}(2\pi\tau/P)$, and the modified amplitude of the tissue temperature $a' = a \cos \phi$.

Physiological effects of temperature

Effect of temperature on metabolic processes

Although most metabolic reactions are strongly influenced by temperature, some physical processes such as light absorption are relatively insensitive, while the rate of diffusion is generally intermediate in sensitivity.

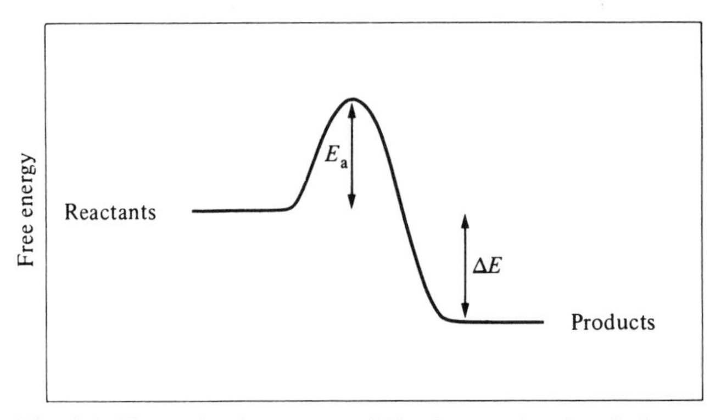

Fig. 9.5. The activation energy (E_a) of a reaction in relation to the net free energy change (ΔE).

Temperature dependence arises where the process requires that the molecules involved have a certain minimum energy (usually in the form of kinetic energy). In general, a high minimum energy requirement leads to greater temperature sensitivity.

The reason for this temperature effect may be discussed in relation to a simple chemical reaction where, before the reaction can take place, the molecule or molecules involved must be raised to a state of higher potential energy (Fig. 9.5). The energy involved in this reaction 'barrier' is called the activation energy (E_a). How the value of the activation energy affects the temperature response can be seen if one considers the distribution of energies between different molecules in a population of similar molecules at a given temperature. Although the mean kinetic energy increases with temperature, the number in the 'high-energy tail' of the frequency distribution increases more rapidly. The number ($n(E)$) that have an energy equal to or greater than E is given by the Boltzmann energy distribution, which can be expressed on a molar basis in the following form:

$$n(E) = n \exp(-E/\mathscr{R}T) \tag{9.12}$$

where n is the total number, $\mathscr{R}$ is the gas constant, and T is the absolute temperature. The rate of a reaction that has a particular activation energy would be expected to be proportional to the number of molecules with the appropriate energy so that the rate constant (k) is given by

$$k = A \exp(-E_a/\mathscr{R}T) \tag{9.13}$$

where A is approximately constant and depends on the type of process. If one takes logarithms of this equation, which is known as the *Arrhenius equation*, one obtains

$$\ln k = \ln A - E_a/\mathscr{R}T \tag{9.14}$$

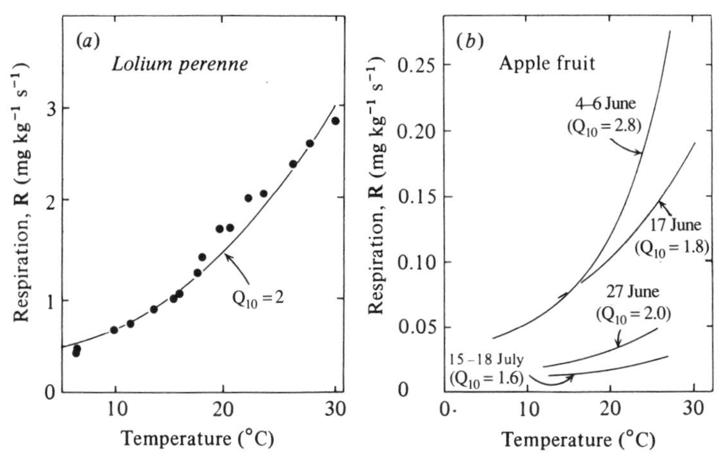

Fig. 9.6. Temperature response of respiration for (*a*) simulated sward of rye grass (*Lolium perenne*) with curve for $Q_{10} = 2$ (data from Robson 1981); (*b*) fitted exponential curves for apple fruits on different dates (data from Jones 1981*b*).

which predicts that the natural logarithm of the rate constant should be linearly related to $1/T$ with a slope of $-E_a/\mathscr{R}$.

Another way of describing the temperature sensitivity of biochemical processes is in terms of the temperature coefficient, Q_{10}, which is the ratio of the rate at one temperature to that at a temperature ten degrees lower. This coefficient is somewhat arbitrary and potentially misleading, especially when applied outside the rather limited range of conditions where the temperature response is exponential, but as long as its limitations are remembered it can be useful, and is widely used. The equations that follow can readily be reformulated for coefficients over temperature ranges other than 10 °C. Therefore from equation 9.13

$$Q_{10} = \frac{A\exp[-E_a/\mathscr{R}(T+10)]}{A\exp(-E_a/\mathscr{R}T)}$$

$$= \exp[10\,E_a/\mathscr{R}T(T+10)] \qquad (9.15)$$

From this it can be shown that, at 20 °C, a Q_{10} of 2 arises where the activation energy is 51 kJ mol^{-1} (i.e. $2 = \exp[(10 \times 51\,000)/(8.3 \times 293 \times 303)]$. The rate of respiration, for example, often has a Q_{10} of 2 at normal temperatures, though it varies with the state of the tissue and decreases at high temperatures (Fig. 9.6). Where the activation energy is lower, as for example for the diffusion of mannitol in water where $E_a = 21$ kJ mol^{-1}, the Q_{10} is lower: in this case 1.3.

In practice the Q_{10} may be obtained from the reaction rates k_1 and k_2 at any two temperatures T_1 and T_2 by using the approximation

$$Q_{10} \simeq (k_2/k_1)^{[10/(T_2-T_1)]} \tag{9.16}$$

Although the rates of simple chemical reactions increase exponentially with increasing temperature, most biological reactions show a clear optimum temperature, with reaction rates declining with any temperature increase above the optimum. There are several reasons why the rates of biological reactions do not continue to increase indefinitely with increasing temperature. One factor is that the rate limiting reactions for any process may change, as rates increase, from highly temperature-sensitive ones to those such as diffusion which have lower temperature coefficients. Another factor is that many processes are the net result of two opposing reactions with different temperature responses. The most important reason, however, is that most biological reactions are enzyme catalysed. Enzymes act to lower the activation energy, and hence to decrease temperature sensitivity and to increase the rate at any given temperature. However, as temperatures rise the catalytic properties of most enzymes are harmed and the total amount of enzyme present may fall as a result of increased rates of denaturation. These factors are discussed below in relation to net photosynthesis.

It is particularly notable that the rates of many plant processes, such as development, which integrate many individual components are frequently approximately *linearly* related to temperature over a wide range of normal temperatures, though an optimum with a subsequent decline is reached at high temperatures.

A convenient empirical equation to simulate many temperature responses, such as photosynthesis (see below) is (for $0 \leqslant k \leqslant 1$)

$$k = \frac{2(T+B)^2(T_{max}+B)^2-(T+B)^4}{(T_{max}+B)^4} \tag{9.17}$$

where T_{max} is the temperature at which the coefficient k reaches a maximum of 1.0 and B is a constant (Fig. 9.7).

Temperature response of net photosynthesis

Photosynthesis is one of the most temperature-sensitive aspects of growth. Some photosynthetic temperature responses for species from different thermal environments are shown in Fig. 9.8, illustrating the tendency for net photosynthesis of temperate zone plants to be maximal between about 20 and 30 °C, with species from hotter habitats having higher temperature optima. In addition many species show marked temperature acclimation when grown in different temperature regimes, as illustrated for *Eucalyptus* and *Larrea*.

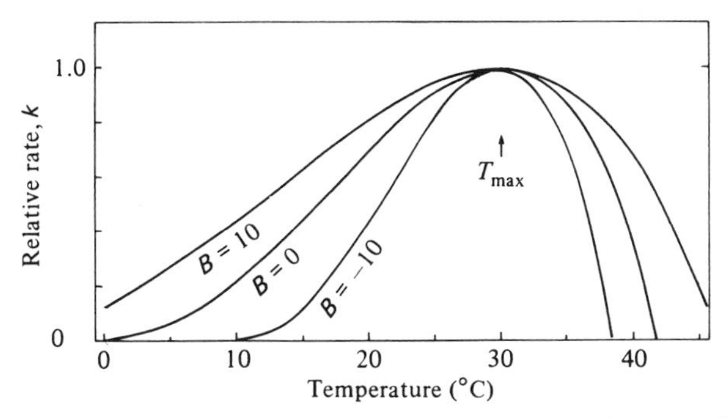

Fig. 9.7. Simulation of temperature responses using equation 9.17, for T_{max} = 30 °C, and for different values of *B*.

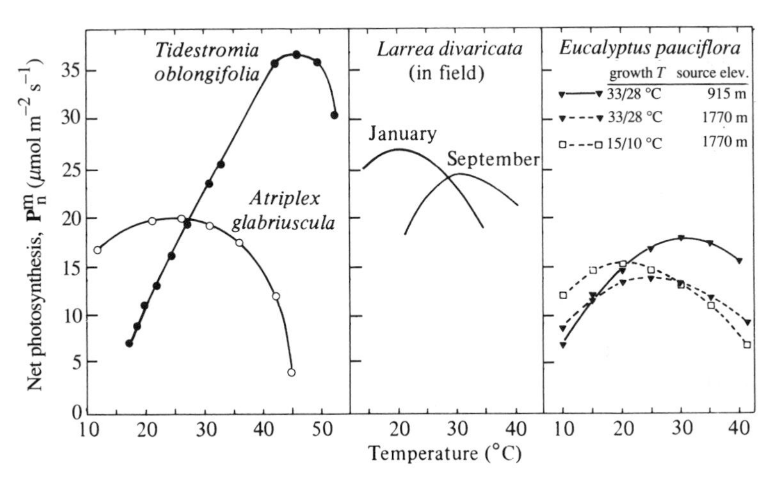

Fig. 9.8. Photosynthetic temperature responses for different species. (Data from Björkman *et al.* 1975 (*Tidestromia, Atriplex*); Mooney *et al.* 1978 (*Larrea*) and Slatyer 1977 (*Eucalyptus*).)

At higher temperatures the shapes of the temperature response curves depend on the duration of exposure to these temperatures, because thermal instability leads to time-dependent inactivation of the photosynthetic system (as was demonstrated by Blackman 1905). There are species differences in the temperature at which this inactivation occurs. For example, thermal inactivation occurs when leaf temperatures exceed about 42 °C in *Atriplex sabulosa* (a C_4 species from a cool coastal environment), but does not occur in the desert species *Tidestromia oblongifolia* until about 50 °C (Fig. 9.9).

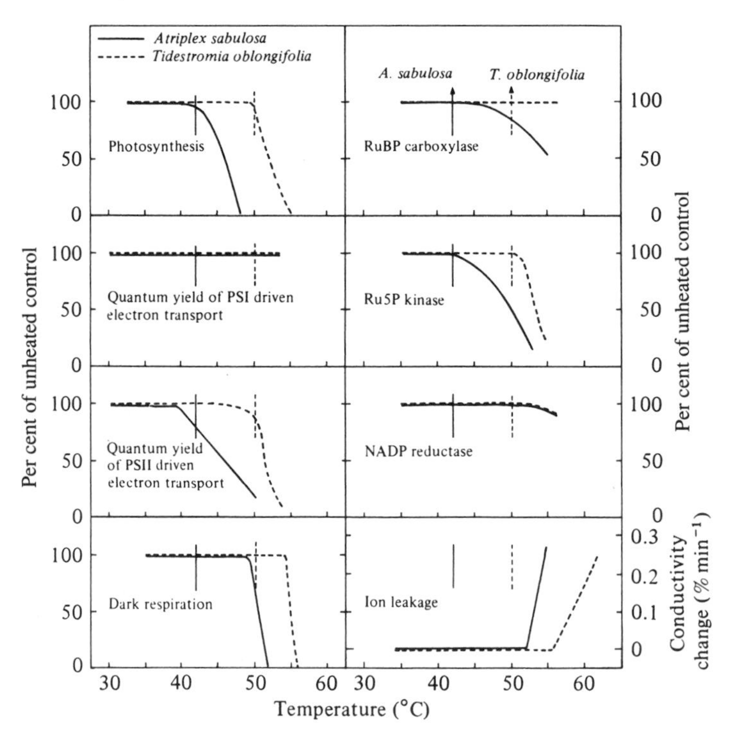

Fig. 9.9. Time-dependent high temperature inactivation of various photosynthetic components in *Atriplex sabulosa* and *Tidestromia oblongifolia* (see Björkman *et al.* 1980 for data sources). Rates were measured at 30 °C after 10 or 15 min pretreatment at the indicated temperature. Vertical lines indicate the temperatures at which time-dependent inactivation of photosynthesis sets in. PS I and PS II = photosystems I and II, respectively.

Some results from an extensive series of experiments investigating differences between *A. sabulosa* and *T. oblongifolia* in photosynthetic stability at high temperatures are presented in Fig. 9.9. Some processes, such as photosystem I driven electron transport and activity of enzymes such as NADP reductase, showed little sensitivity to short exposures to high temperatures. Other features, such as membrane permeability (meaured by ion leakage), dark respiration and carboxylase activities, were sensitive to high temperatures, with species differences being apparent. However, the inhibitory temperatures for these processes were significantly above those damaging photosynthesis. The temperature sensitivity of photosynthesis was most closely related to that of the quantum yield of photosystem II driven electron transport and some photosynthetic enzymes (such as ribulose 5-phosphate kinase). Other evidence indicated that species

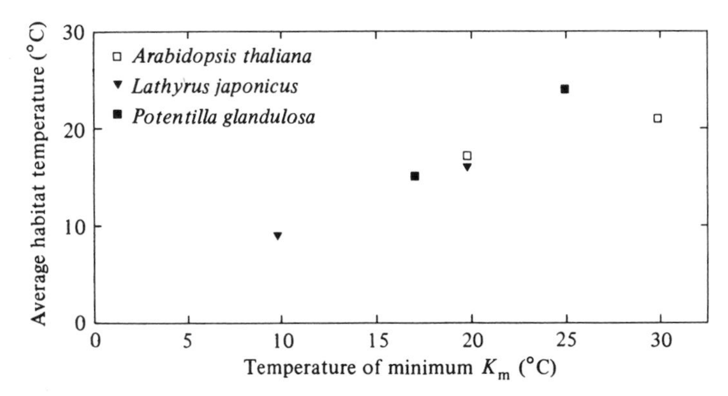

Fig. 9.10. The relationship between the assay temperature giving the minimum K_m and the average habitat temperatures for two populations of each of three species. The enzymes studied were glucose 6-phosphate dehydrogenase for *Arabidopsis* and malate dehydrogenase for the other species. (Data from Teeri 1980.)

differences did not result from differences in stomatal behaviour. Overall, the available evidence supports the conclusion that much of the difference in thermal stability between species results from differences in thermal stability of the chloroplast membranes, and in particular the integrity of photosystem II. The involvement of primary damage at the PS II system is particularly clearly demonstrated by the sharp increase of basal level of chlorophyll fluorescence (F_0) above a threshold temperature that corresponds to the onset of leaf necrosis (Bilger *et al.* 1984).

In contrast, there is evidence that differences between species in their photosynthetic rate at low temperatures is strongly correlated with the capacity of specific rate-limiting enzymes such as RuBP carboxylase and fructose bis-phosphate phosphatase. There is also evidence, at least for some enzymes, that their kinetic properties may be adapted to the normal environmental temperature of that ecotype. The assay temperature giving the minimum Michaelis constant (K_m) can be closely related to the habitat temperature (Fig. 9.10). Small changes in K_m of about two-fold over the normal temperature range may provide a mechanism for maintaining the rate of the catalysed reaction relatively insensitive to temperature fluctuation. At least for malate dehydrogenase in *Nicotiana*, acclimation of K_m has been related to alterations in the proteins (isozymes) synthesised (de Jong 1973).

Effects of temperature on plant development

The rates of many plant developmental processes, and hence the timing of phenological stages, are strongly temperature dependent. If under a particular set of conditions a particular developmental stage takes t days, the corresponding rate of development (k_d) is $1/t$; this implies that the time taken to complete the developmental stage is inversely proportional to k_d. This rate of development is usually a strong function of temperature, so that in a constant environment

$$1/t = k_d = f(T) \tag{9.18}$$

We can denote the state of plant development at any time t (measured from a suitable starting date such as sowing) as $S(t)$, where S is a fractional state of development on a zero to one scale. In a fluctuating environment where T is a function of time (written as $T(t)$), $S(t)$ is given by

$$S(t) = \int_{t=0}^{t} k_d \, dt = \int_{t=0}^{t} T(t) \, dt \tag{9.19}$$

In practice it is often found that, between a threshold temperature for development to occur (T_t) and an optimum temperature (T_o), the rate of development is approximately linearly related to temperature so that between these temperatures

$$k_d = a(T - T_t); \quad \text{for } T_t \leqslant T \leqslant T_o \tag{9.20}$$

where a is a constant. Thus if temperature remains constant over time one can substitute into equation 9.19 to get

$$\begin{aligned} S(t) &= a(T - T_t) \int_0^t dt \\ &= a(T - T_t)\, t \end{aligned} \tag{9.21}$$

It follows that the value of the temperature sum $(T - T_t)\,t$ that is required to complete the developmental stage being considered (i.e. $S(t) = 1$) is equal to $1/a$. Under fluctuating temperature conditions we can write

$$S(t) = a \int_0^t (T(t) - T_t) \, dt \tag{9.22}$$

Calculation of this integral requires a knowledge of the relationship between temperature and time. For convenience the temperature sum (given the symbol D and measured in day-degrees) is often obtained by summing for each day the daily mean temperatures (T_m) according to

$$D = \sum_{d=0}^{n} (T_m - T_t); \quad \text{for } T_t \leqslant T_m \leqslant T_o \tag{9.23}$$

Completion of the developmental stage ($S = 1$) requires that this temperature sum, D, often incorrectly referred to as a 'heat-sum', equals or exceeds $1/a$. The term $1/a$ is known as the thermal time or accumulated temperature required for completion of the developmental stage being considered.

Estimates of the appropriate threshold or base temperature can be obtained in two main ways. Where studies can be made in controlled environments at constant temperature, it is possible to plot rate of development against temperature and T_t is given by the intercept on the x-axis. Where temperatures are fluctuating, as in the field, it is necessary to calculate D with different thresholds and determine which gives the best linear fit to equation 9.21.

Extensions of thermal time. The concept of thermal time as a replacement of chronological time has been in use for phenological studies for over 200 years (Wang 1960). Although the simple formula is adequate for most purposes, non-linearities in the temperature responses and interaction with other environmental factors have been incorporated in a number of more sophisticated equations. For example, polynomial or other functions of T can be substituted for the linear form used in equations 9.21 and 9.22.

For seed germination studies it is common to assume that germination rate increases linearly with temperature between T_t and T_o, and then decreases linearly above T_o (see Fig. 9.11), so that above the optimum

$$k_d = (T_c - T)/D_2; \quad \text{for } T_o \leqslant T \leqslant T_c \tag{9.24}$$

where T_c is the upper limit for germination and D_2 is an appropriate temperature sum. The behaviour at supraoptimal temperatures, however, is complicated by the fact that there is a time $\times$ temperature dependency. Although high temperatures may damage the seed there can still be an underlying tendency for germination to be speeded up by higher temperatures. This is illustrated for the cowpea data in Fig. 9.11 where the earliest germinating seeds emerged before high temperature damage became apparent.

Effective day-degrees. Inclusion of other environmental variables enables one to describe development where temperature is not the only environmental variable affecting the process. A particularly useful approach is the definition of what have been called 'effective day-degrees' (D_{eff}, Scaife *et al.* 1987) to include effects of radiation and temperature. In this approach D is replaced by D_{eff}, defined by

$$D_{eff}^{-1} = D^{-1} + b \cdot \mathbf{I}^{-1} \tag{9.25}$$

where b is a constant describing the relative importance of irradiance (**I**)

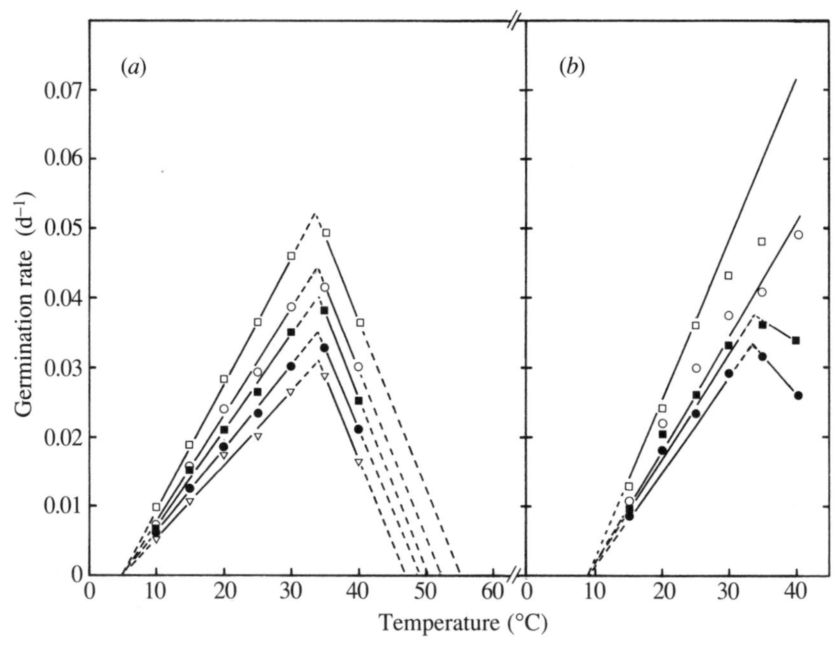

Fig. 9.11. The relation between temperature and rate of progress to germination (at constant temperature) for (*a*) soyabean and (*b*) cowpea. Broken lines represent extrapolation beyond experimental values, and the different symbols represent various percentage germination – 10% (□), 30% (○), 50% (■), 70% (●) and 90% (▽). (Data from Covell *et al.* 1986.)

and temperature. Where b is zero, D_{eff} becomes equal to conventional day-degrees.

Vernalisation. Daylength is not the only factor showing annual variation: temperature also shows marked seasonal fluctuation, particularly at higher latitudes, though there may be considerable short-term variability. These seasonal changes are a major factor in the control of flowering, often interacting with photoperiodic control. For example 'winter' varieties of wheat must be exposed to several months of low temperatures if they are to flower. If sown in the spring (rather than the autumn), they remain vegetative as a consequence of failure to satisfy their vernalisation requirement. Spring varieties, on the other hand, have only a minimal or no vernalisation requirement. Such a requirement for a period of low temperatures can ensure an adequate period of growth before flowering, otherwise, in natural situations, seeds that germinate in late summer might flower in the same year and fail to mature. Many perennial plants also require a period of cold for optimal flower development.

As with photoperiodic responses, vernalisation requirements vary from obligate to quantitative. The actual rate of vernalisation is a function of temperature. In wheat it is very slow at 0 °C or below, reaches a maximum at about 3 °C, with an upper limit of about 11 °C (Evans *et al.* 1975). Wheat can be vernalised as seeds, though other species require exposure as leafy plants to low temperatures.

An interesting feature of vernalisation is the very low temperature optimum. A hypothetical scheme that could give rise to such a response would be a two-stage reaction:

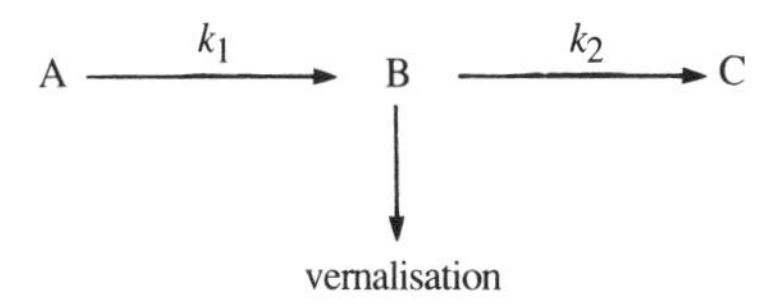

where the intermediate B is required for the vernalisation process. The concentration of B would increase as temperature falls if the rate constant k_1 has a lower temperature coefficient than k_2. Combining this with a final reaction where the rate of utilisation of B increases with temperature could give the observed temperature response.

Many perennial plants, such as temperate fruit trees, also have an annual requirement for a period of low temperatures. This is necessary for optimal flowering in the spring. In some cases, however, it is possible, at least partly, to substitute the cold requirement by drought or by defoliation; for example Jones (1987) was able to initiate in apple trees a second phase of flowering within one year by drought treatments that were severe enough to cause defoliation on re-watering. The mechanism of low winter temperature requirement for effective flowering is complex as there is also evidence from weather–yield correlations that abnormally high late winter temperatures can be disadvantageous for apple yields (Jackson & Hamer 1980), even with adequate total winter chilling.

Many seeds also have a requirement for an often extensive period of low temperature before they can germinate. These low temperatures are most effective at overcoming seed dormancy if given when the seeds are moist. The process of vernalisation of seeds is known as stratification.

Dormancy and leaf abscission. Many perennial species show a well-marked period of dormancy, usually during the winter in temperate species, but often in the summer in mediterranean species. Temperature also plays an important role in the induction of dormancy and in leaf abscission, in conjunction with light effects.

High temperature injury

A common response to high temperatures in many organisms is a marked change in the pattern of protein synthesis (for reviews see Craig 1986; Ho & Sachs 1989); as the temperature rises above a critical value (often around 40 °C in plants) normal protein synthesis stops and is replaced by the rapid and coordinated synthesis of a characteristic set of heat shock proteins (HSPs). Many of these HSPs show a very high degree of homology in all organisms that have been studied (e.g. HSP70), though higher plants also synthesise a unique group of small (15–18 kDa) proteins. The rapid synthesis of HSPs is related to increased transcription and the appropriate *m*RNAs can increase within 3–5 min of high temperature stress. HSP levels tend to decline fairly rapidly, even if the high temperatures continue. Although the functions of many HSPs are not known, they have been associated with the induction of thermotolerance by exposure to brief periods at high temperature. It is likely that HSPs act to maintain cellular structure and function at high temperatures, though the fact that they are not constitutively synthesised suggests that their continued presence can be disadvantageous. A common HSP, ubiqitin, is thought to be involved in tagging thermally denatured proteins for subsequent proteolysis by a special protease. Interestingly, some of the plant HSPs can be induced by other factors such as abscisic acid, heavy metals, osmotic stress, arsenite or anaerobiosis, though the physiological significance is not clear. In addition many of these stresses and other stresses such as pathogenesis also give rise to specific stress proteins (Ho & Sachs 1989).

High temperature damage to cells and tissues normally involves loss of membrane integrity with consequent ion leakage. Cell death can readily be assessed by means of the ability of cells to take up a vital stain such as neutral red. On the tissue scale, high temperature damage can be seen as tissue necrosis.

High temperature tolerance

As has already been indicated (e.g. Fig. 9.9) many plants have a great capacity to adapt to temperature extremes. This ability to acclimate is widespread: for example, Nobel (1988) has collated results from a number of experiments on 33 species of agaves and cacti where it was shown that the temperature leading to 50% apparent cell death increased by between 1.6 and 15.8 °C for different species when the growth temperature was raised from 30°/20° (day/night temperature) to 50°/40°. For nearly all species the temperature tolerated under the higher temperature growth regime was in excess of 60 °C. Nobel (1988) calculates that a number of cacti including *Opuntia ficus-indica* (prickly pear) can tolerate temperatures in excess of 70 °C for one hour, a common duration for the most extreme temperatures in the field. Such a high temperature appears to be lethal to other vascular plants that have been studied.

The underlying basis of acclimation to high temperatures is not well understood, and although there is likely to be a role for heat shock proteins and for changes in saturation of membrane lipids, the details are not clear.

Low temperature injury

There are two main types of low temperature injury. The first, which is common in plants of tropical or subtropical origins (such as beans, maize, rice and tomatoes) is called chilling injury. It is usually manifest as wilting or as inhibited growth, germination, or reproduction or even complete tissue death, and occurs in sensitive species when tissue temperatures are lowered below about 10 °C (though this varies with species and degree of acclimation and may occur as high as 15 °C). As long as exposure to chilling temperatures is of short duration, the damage is usually reversible. The other major type of low temperature injury is freezing injury, which occurs when some of the tissue water freezes. All growing tissues are sensitive in some degree to frost though sensitivity varies. In many plants which have not been given a chance to acclimate, tissues are killed by freezing to only −1 to −3 °C. After acclimation the range of temperatures that can be survived is very wide, depending on species and tissue. Many seeds, for example, can withstand liquid nitrogen temperatures (−196 °C), while many tissues can survive −40 °C or below.

Mechanisms of damage – chilling

Current evidence suggests that an early, if not primary, effect of chilling involves damage to the cell membranes. For example, chilling-damaged tissues tend to lose electrolytes rapidly because of increased permeability of the plasma membrane, while there is also evidence for damage to chloroplast and mitochondrial membranes. There have been many studies that suggest that membrane properties in chilling-sensitive plants undergo a sudden change at about the temperature where chilling injury occurs, while chilling-resistant plants show no such abrupt change. It has been suggested, on the basis of changes in the slopes of Arrhenius plots (ln (rate) against $1/T$) at this critical temperature, that this results from a phase change in the membranes from a relatively fluid form to a more solid gel structure, so that normal physiological activity can only occur above the critical temperature. Recent evidence, however, suggests that this phase change does not occur in the bulk membrane though it may be localised in small 'domains' within the membrane. There is some evidence that the temperature of any phase change is correlated with the fatty acid composition of the lipids, with a high proportion of saturated fatty acids occurring in the membranes of chilling-sensitive species, but the correlations are not always very good. The existence of any phase change and the effects of altered fatty acid saturation are still very controversial, and other factors, such as membrane protein composition, may be involved. The evidence for these various possibilities is discussed by Lyons *et al.* (1979) and Levitt (1980), though little is known for certain beyond the fact that chilling affects membranes.

Another suggestion has been that chilling injury may be related to the breakdown of cytoplasmic microtubules; this is supported by evidence that microtubule-disrupting agents enhance chilling injury, while abscisic acid tends to increase chill resistance and also retards disruption of microtubules (Rikin *et al.* 1983).

Mechanisms of damage – freezing

With freezing injury the damage relies on the formation of ice crystals within the tissue, though freezing-tolerant species can apparently withstand some ice formation. A precise value for frost sensitivity in different plants is difficult to determine because the actual damage depends on the rate of thawing, as well as on the lowest temperature reached. Membrane damage is a universal result of freezing damage, though it is still not certain whether it is the primary effect. As temperatures are lowered, ice starts to form in the extracellular water (e.g. in the cell walls). Because ice has a lower vapour pressure (and chemical potential) than liquid water at the same temperature, extracellular freezing causes water to be removed from within the cells to the sites of extracellular freezing. This leads to rapid dehydration of the cell (see Levitt 1980).

The amount of water that must be lost from a cell before equilibrium is reached between intracellular water and extracellular ice depends on the temperature and on the osmotic properties of the cell. Water continues to be lost until the reduction in cell volume causes the cell water potential to balance the extracellular water potential.

The cell water potential that is in equilibrium with pure extracellular ice at the same temperature can be obtained, using equation 5.11 (p. 110), from

$$\Psi = (\mathscr{R}T/\bar{V}_W)\ln(e_{ice}/e_s) \tag{9.26}$$

where e_{ice} is the vapour pressure over pure ice and e_s is the vapour pressure over pure water at the temperature T (see Appendix 4). The value of Ψ

given by equation 9.26 decreases by approximately 1.2 MPa $°C^{-1}$ below 0 °C. The equivalent increase in solute concentration is given by the van't Hoff relation (equation 4.8, p. 77) as approximately 530 osmol m^{-3} per °C lowering of temperature (i.e. $-\Psi/\mathcal{R}T = 1.2 \times 10^6/2270$).

Since solute concentration, $c_s \propto 1/V$, it follows that relatively large absolute changes in cell volume are required to maintain equilibrium for a 1 °C drop in temperature near the freezing point compared with lower temperatures. That is the relationship between cell volume and temperature at equilibrium should be hyperbolic. As expected from this it is found that the amount of liquid water present in tissue at any temperature, expressed as a fraction (f) of the liquid water in unfrozen tissue, decreases hyperbolically as temperature falls below that at which freezing first occurs (Gusta *et al.* 1975) according to

$$f = (\Delta T_f/T) + b \tag{9.27}$$

where ΔT_f is the freezing point depression, T is the temperature (°C) and b is the intercept on a plot of f against $1/T$, representing the amount of 'bound' water.

The main physiological effects of freezing seem to be related to the dehydration and consequent concentration of cell solutes with possible solute precipitation, protein denaturation or alteration in membrane properties. Levitt's sulphydryl–disulphite hypothesis (see Levitt 1980, and also Chapter 10) provides an example of one such possible mechanism. In this case it is proposed that protein denaturation and subsequent aggregation by the formation of disulphide bridges may occur as intermolecular distances decrease with removal of cell water.

A difficulty with simple dehydration hypotheses of freezing damage is that most water freezes at relatively high temperatures (above about −10 °C), so that differences in tolerance of lower temperatures would require sensitivity to relatively small changes in hydration. For example, for a typical cell sap concentration, ΔT_f is 1.5 °C so it follows from equation 9.27 that less than 20% of the original water would remain in the cell at −10 °C.

In some sensitive species, damage by freezing may occur as a result of intracellular freezing or of direct damage by the extracellular ice crystals.

Hardening and mechanisms of frost tolerance

Many plants show some degree of acclimation to chilling or freezing temperatures such that the damage temperature is lowered after exposure to a period of low temperatures. Figure 9.12 shows the changes in minimum

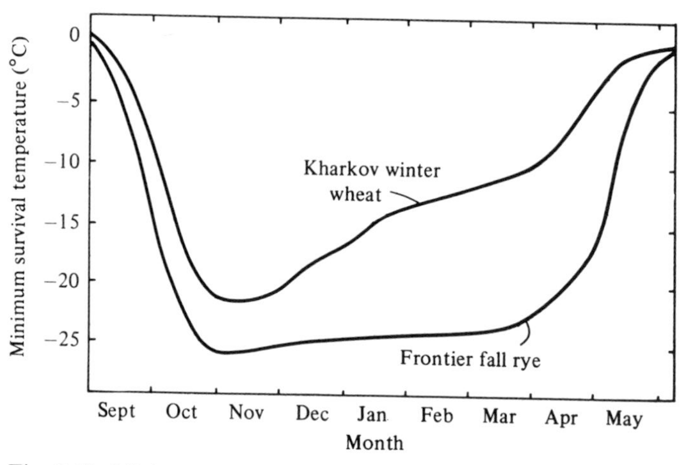

Fig. 9.12. Minimum survival temperatures (LD_{50}) for a winter wheat and a winter rye, showing development of cold tolerance during the winter. (After Gusta & Fowler 1979.)

survival temperatures that occur in two winter cereals over the winter period. Some degree of winter hardiness can also be induced by exposure to drought or salinity, while conversely, frost hardening can induce a degree of drought or salinity tolerance. At least in insects cold hardening can be rapid, occurring within two hours, where it has been associated with glycerol accumulation (Lee *et al.* 1987).

Various plant factors have been associated with freezing tolerance in different species, including lower osmotic potentials, increased levels of soluble carbohydrates (especially the so-called compatible solutes such as glycine-betaine), lipid composition and small cell size (e.g. Levitt 1980). The plant growth regulator abscisic acid (ABA) has also been implicated in the development of freezing resistance. It may be that there are several alternative mechanisms that can increase freezing tolerance. It is interesting to note that although agaves and cacti also show the ability to increase their frost hardiness when grown at low temperatures, the alteration in damage temperature is much smaller than the degree of hardening observed at high temperatures.

Another mechanism that may be important in the frost tolerance of some species is supercooling. The equilibrium state is for ice to form when the temperature falls below the freezing point depression (ΔT_f) appropriate for the solute concentration. As ΔT_f is approximately 1 °C for every 1.2 MPa, typical cell osmotic potentials of down to −3 MPa would lower the freezing point by less than 3 °C. In fact the cell contents rarely freeze. This is partly because of the absence of suitable ice nucleation sites within the cells, allowing the water to remain in the unstable supercooled condition. Similarly, the cell wall water may remain liquid far below the theoretical freezing point, though in most plant tissues only a few degrees of supercooling can be achieved. Ice nucleation usually starts in the extracellular water, either because of the lower solute concentration there or because of the presence of ice-nucleating bacteria. Some hardwoods, including various oak, elm, maple and dogwood species, however, can supercool to the homogeneous nucleation temperature (Burke & Stushnoff 1979). This is the temperature at which ice forms spontaneously without a requirement for nucleation sites, being between about −41 and −47 °C for plant tissues.

Certain tissues and cells in these hardwood species, such as ray parenchyma cells and flower buds, show supercooling. This mechanism for avoiding freezing damage cannot work below the deep supercooling temperature of about −41 °C, so that species that rely on deep supercooling cannot survive in areas where lower temperatures occur. Those species that occur in regions where lower temperatures are likely (e.g. pines, willows, etc.) survive by tolerating extracellular freezing.

Frost protection

Because of the economic importance of frost damage in temperate climates, much work has been done to develop techniques for protecting high value crops from the damaging effects of frost (see Rosenberg *et al.* 1983). Plant and air temperatures may fall below freezing, either by advection of a cold air mass (e.g. from polar regions), or as a result of the net heat loss by longwave radiation that can occur on calm clear nights. Such a radiation frost causes the build up of a stable inversion layer where the air near the ground is cooler than that at higher levels.

Various methods of frost protection include:

(i) *Water sprinkling*. As water freezes it liberates a large amount of heat (the latent heat of fusion is 334 J g^{-1}). By spraying water onto sensitive tissues during a frost, this heat liberation prevents tissue temperatures falling below zero, as long as enough water is supplied to maintain liquid on the surface of the tissue.

(ii) *Direct heating*. Various types of orchard heater have been used to keep air temperatures above freezing, though their main effect is to help break up the inversion layer and to replace the cold air near the surface by warmer air from higher levels.

(iii) *Air mixing*. With radiation frosts it is also possible to break up the inversion layer near the ground by using fans or propellers.

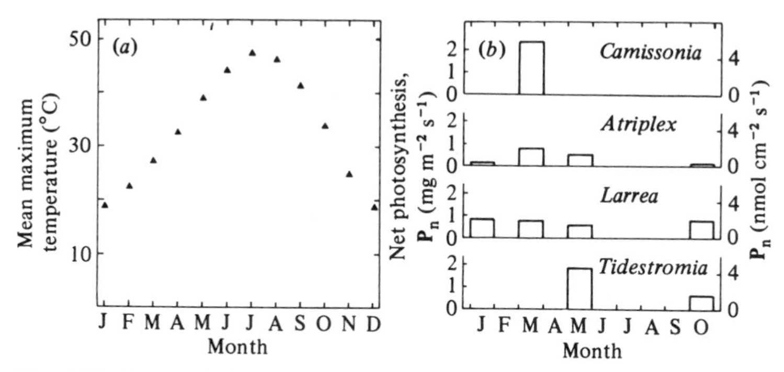

Fig. 9.13. Seasonal changes of (*a*) monthly mean maximum daily temperature for Furnace Creek, Death Valley, and (*b*) maximum photosynthetic capacity of four natives species: *Camissonia claviformis, Atriplex hymenelytra, Larrea divaricata* and *Tidestromia oblongifolia* (after Mooney *et al.* 1976).

(iv) *Radiation interception.* Another technique of use in radiation frosts is to prevent the loss of longwave radiation from the ground by covering the crop or by using smoke.

(v) *Choice of site.* Because cold air is dense, it flows to low-lying valleys and hollows. For this reason sensitive crops should not be planted in such situations.

(vi) *Insulation.* Covering the crop with an insulating mulch can also be effective for small plants.

Ecological aspects

Different plants exhibit a wide range of adaptations that enable them to live in different thermal climates. These include biochemical tolerance mechanisms as well as seasonality and morphological or physiological adaptations that lead to avoidance of temperature extremes. Not only is the success of any species dependent on its tolerance or avoidance of extreme temperatures, but also it depends on its capacity to grow and compete at more normal temperatures. An ability to tolerate the short-term extreme temperatures that occur during bush fires is an important factor in environments where fires are common.

Some species are widely adaptable, being able to grow over a wide temperature range, while others are more specialised. Examples of these two types of plant are found in the flora of Death Valley, California, USA, where the mean maximum temperature ranges from less than 20 °C in January to more than 45 °C in July (Fig. 9.13). Some species, such as the evergreen perennials *Larrea divaricata* and *Atriplex hymenelytra* grow all

Table 9.2. *Total dry matter yields (final dry weight/initial dry weight) for four species over a 22 day growth period in two contrasting thermal regimes (after Björkman* et al. 1980)

Species	16 °C day/ 11 °C night	45 °C day/ 31 °C night
Atriplex glabriuscula (coastal C_3)	24.4	0.1 (died)
Atriplex sabulosa (coastal C_4)	18.2	0.3 (died)
Larrea divaricata (desert C_3)	5.4	3.2
Tidestromia oblongifolia (desert C_4)	< 2.5	88.6

the year, acclimating to the changing temperatures. Others, for example the perennial *Tidestromia oblongifolia*, are summer active and cannot grow at normal winter temperatures, while winter annuals such as *Camissonia claviformis* grow only at low temperatures. These species differences are reflected in seasonal changes of maximum photosynthetic capacity (Fig. 9.13) and in their ability to grow in, and acclimate to, different temperatures in terms of dry matter yields (Table 9.2). Table 9.2 also illustrates the general point that the more widely adaptable species (e.g. *L. divaricata*) tend not to have as high growth rates in any environment as the appropriate 'specialists'.

Avoidance mechanisms

There are many possible avoidance mechanisms that can result in particular species not being subject to unfavourable tissue temperatures. (The term avoidance, and ones such as strategy used below, are used in a strictly non-teleological sense in that they do not imply that the plant can be rational or have a sense of purpose. Rather, they provide simple descriptions of particular physical or biological responses.)

Perhaps the most important avoidance mechanism is seasonality (see e.g. Fig. 9.13). Annual species may avoid periods of potentially damaging temperatures by completing their life cycle entirely within the period of favourable temperatures. Perennial species, on the other hand, often have the capability of going into a dormant state where the tissues are less sensitive to temperature extremes.

In other plants the sensitive tissues may avoid being subject to the extremes of daily temperature range as a result of the damping (long thermal time constant) that results from a high thermal capacity of the tissues or of their immediate surroundings or from a high degree of insulation. This is particularly important in the grasses, where the meristematic tissues (at least in the vegetative phase) are at or below ground level and thus not subject to the temperature fluctuations found in the aerial environment. Similarly, the long time constants for large stems or fruits (Table 9.1) can lead to a significant reduction in the amplitude of daily temperature fluctuations.

Because the thermal time constants for most other aerial plant tissues are relatively short (Table 9.1), the steady-state energy balance can be used to obtain a good indication of their temperature control mechanisms. These may be discussed in relation to the different components of the steady-state energy balance ($\mathbf{\Phi}_n$, $\mathbf{C}$ and $\lambda\mathbf{E}$), with radiation absorption, and leaf and boundary layer resistance being of prime importance. The principles described above (equation 9.6, and Figs 9.1 and 9.2) will now be discussed in relation to some examples.

One general factor in the avoidance of extreme tissue temperatures is a consequence of the temperature dependence of s and δe (Fig. 9.2). At high air temperatures (above about 30–35 °C), T_ℓ tends to be below T_a, but the converse occurs at low temperatures. The exact changeover temperature is a function of plant and environmental factors.

Hot environments. Leaf temperature may be kept low by decreasing net radiation absorbed or by increasing latent heat loss. Whether or not sensible heat transfer needs to be minimised or maximised depends on whether the leaf is below or above air temperature.

For the case where water is freely available, the most productive strategy is to maximise latent heat loss ($\lambda\mathbf{E}$) by having a low leaf resistance, but to minimise any sensible heat gain from the air by having large leaves with a large boundary layer resistance. This combination is common in drought-evading desert annuals and in some perennials with access to ample water. As an example, *Phragmites communis*, although a temperate species, can grow in summer in a wet part of Death Valley. It survives the high temperatures there because leaf temperature is maintained as much as 10 °C below air temperature by a large latent heat loss (Pearcy *et al.* 1972). A similar mechanism probably explains why *Caliotropis procera*, a shrub with leaves 10 cm or more wide, is common in areas of the Thar Desert of northwest India. It is worth noting that transpirational cooling to below air temperature is most effective for large leaves, as sensible heat transfer (which tends to negate the effect) is minimal.

Where water is limited, however, as it often is in hot environments, large leaves are likely to be disadvantageous. In this case, small leaves with their low boundary resistance and efficient sensible heat exchange can avoid heating up much above air temperature, but equally they cannot cool much below air temperature even when transpiring rapidly. For example, *Tidestromia oblongifolia* has small leaves that closely track air temperature (e.g. Pearcy *et al.* 1971). Many summer-active desert perennials have small leaves.

Either strategy for latent and sensible heat exchange can be combined with minimising radiation absorption by means of vertical leaf orientation (as in the Eucalypts of Australia), wilting, leaf rolling, heliotropic movements keeping leaves edge-on to the solar beam, or high leaf reflection coefficient. In a rather different fashion, shading by other tissues (as, for example, by the spines on cacti (Nobel 1978) can also protect sensitive tissues from an excessive heat load. The effects of leaf angle and reflectivity are additive and can be evaluated by means of equation 9.6. As an example, it has been calculated (Mooney *et al.* 1977) that an alteration of leaf angle from horizontal to 70° could be expected to lower leaf temperature of *Atriplex hymenelytra* in Death Valley by 2–3 °C, while halving the total absorptance for shortwave radiation can lower leaf temperature by a further 4–5 °C. A factor in determining the low leaf temperatures in *Phragmites* is that the leaves are normally fairly erect.

Leaf reflectance depends on a variety of characters (see Chapter 2) including leaf water content and the presence of crystalline surface salts, pubescence and the amount and structure of surface waxes. Of particular ecological interest is the observation that the spectral properties of leaves tend to alter with environmental aridity. The environmental and seasonal changes in reflectivity with variation in leaf water content in *Atriplex* (Fig. 2.13, p. 31) and with pubescence in *Encelia* (Fig. 2.14, p. 33) both help to minimise leaf temperatures in summer. The increased radiation absorption in winter may help to increase leaf photosynthesis both by raising leaf temperature when air temperatures are suboptimal and by increasing total absorbed PAR.

Cold environments. Many arctic and alpine species have 'cushion' or 'rosette' habits where the plant forms a dense canopy within a few centimetres of the ground surface. This gives rise to a high boundary layer resistance as air movement is inhibited within the canopy and the whole plant is within the layer of markedly reduced windspeed. Coupled with efficient radiation absorption, this enables the temperatures of leaves and flowers to be 10 °C or more above air temperature (see e.g. Warren Wilson 1957; Geiger 1965). It is even possible that the dense pubescence on the lower surfaces of the leaves of many alpine plants (e.g. *Alchemilla alpina*) may increase radiation absorption by reflecting transmitted light back into the leaf (Eller 1977). The metabolic component of tissue energy balance can also significantly affect tissue temperature; spadix temperatures of certain

Araceae can even be raised between 15 and 35 °C above air temperature by thermogenic respiration in the early spring (see p. 172: Knutsen 1974).

In other species, particularly the large rosette plants of genera such as *Senecio* and *Lobelia*, that grow at high altitudes in the tropics, the meristematic tissues are protected from frost damage by the formation of 'night buds' where the adult outer leaves fold inwards by a nyctinastic movement (see Beck *et al.* 1982). The insulation provided by the adult outer leaves (that can tolerate freezing) is sufficient, especially when combined with a degree of supercooling, to prevent the more sensitive tissues from freezing.

Temperature avoidance or efficient water use?

It is questionable whether the primary effect of many of the high temperature avoidance mechanisms described above is to avoid direct thermal damage or to minimise transpirational water loss. Any mechanism lowering T_ℓ lowers the water vapour pressure inside the leaf and tends to reduce water loss and conserve moisture as well as tending to increase the ratio of photosynthesis to water loss (particularly at supraoptimal temperatures). The efficiency of water use is discussed in more detail in Chapter 10. It is worth noting that water conservation tends to have higher priority than minimising leaf temperature (at least by means of evaporative cooling).

Thermal climate and plant response

Different features of the temperature regime are critical at different times of year. In winter, the minimum temperatures may determine what species survive, while the occurrence of blossom-damaging frosts in spring can be crucial for other species. Similarly, the sensitivity of different species to variation in growing-season temperature depends on how close the environment is to their natural optimum. For example, for a hypothetical yield response of the form of that in Fig. 9.6 (for $B = 0$), a 5 °C rise in temperature would double yield at 9 °C, have no affect at 27 °C and cause complete failure at 38 °C.

Where the complete information required for crop or ecosystem modelling is not available, two simple methods, in addition to mean temperatures, are often used to characterise the temperature regime. The first is to use the concept of thermal time (see above) and to calculate the total number of growing degree-days typically available during the season. Examples of the thermal time required for maturation of different crops range from 1500 °C d for spring barley to about 4000 °C d for rice. The

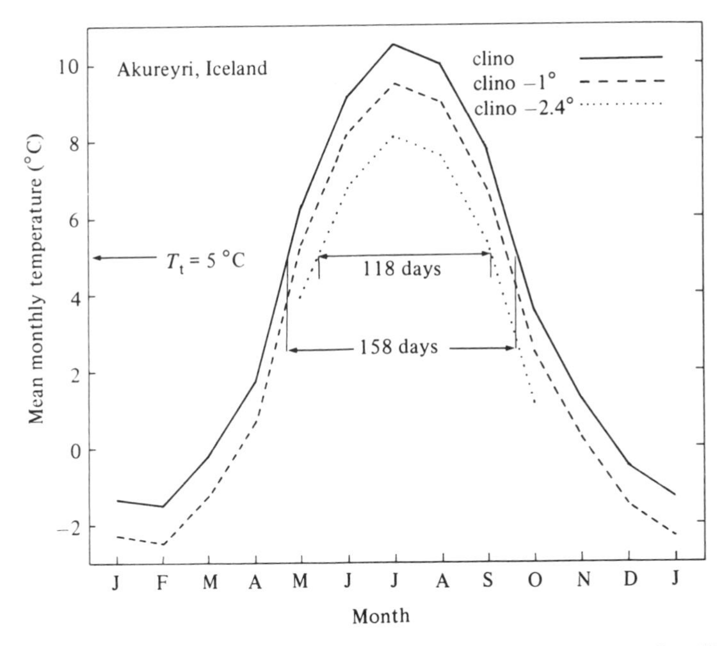

Fig. 9.14. Effect of reduction of the annual mean temperature for Akureyri, Iceland, by 1 °C or 2.4 °C below the climatic normal (clino). (After Bryson 1974.)

other approximate approach is to determine the length of the growing season, assuming an appropriate minimum temperature for growth. A minimum of 5 or 6 °C is often assumed for temperate crops and about 10 °C for crops such as maize. Although these thresholds are not necessarily exact for any species, the length of period during which mean temperatures remain consistently above the appropriate threshold, called the 'length of the growing season', provides a useful simple measure of climate. The potential growing period is particularly important for grain and fruit crops that need a minimum period to complete a reproductive cycle. Growing-season duration is most frequently a limiting factor in northern latitudes or at high altitude.

An example of temperature sensitivity. Both growing-season length and growing degree-days are particularly sensitive to small climatic shifts when mean temperatures are close to T_t. This is illustrated for Akureyri in Iceland in Fig. 9.14. Here a small shift in temperature has a large effect on growing-season length and an even more dramatic effect on growing degree-days. For example, a 2.4 °C reduction in temperature from the long-term mean

Table 9.3. *Effects of changes in mean temperature at Akureyri, Iceland, on growing season and growing degree-days (D) above 5 °C. 'Clino' is the climatic normal temperature (data from Bryson* 1974)

	Mean warm season temperature °C	Growing season		Growing degree-days (*D*)	
		days	%	°C-days	%
Clino	7.47	158	100	597	100
Clino −1 °C	6.47	144	91	443	73
Clino −2.4 °C	5.07	118	75	276	46

decreases the growing season to 75% of normal and decreases D to only 46% of normal (Table 9.3). Changes of this magnitude would be disastrous if crop yields were affected proportionately.

Although it is difficult to quantify the effect of temperature on crop yield using observational data alone, hay yields in Iceland for a period of 25 years are available and can be related to seasonal temperature using multiple regression techniques. As a first approach, Bryson (1974) noted that mean yields averaged 4.33 tonne ha^{-1} in the late 1950s, but only 3.22 tonne ha^{-1}, or 75%, in 1966 and 1967. In the same periods, corresponding mean warm-season temperatures were 7.65 °C and 6.83 °C. This yield reduction occurred even though more fertiliser was applied in the 1960s, so he attributed the effect to the lower temperature, the actual reduction corresponding approximately to that expected on the basis of the reduction in degree-days.

Unfortunately this clear result is an oversimplification as other factors are also important so that the magnitude of the temperature effect is not as great if other combinations of years are selected. In fact more objective regression studies (Table 9.4) showed that mean cool-season temperature (which averaged −0.08 °C, well below the assumed growth threshold) was a better yield predictor (perhaps because it measured winter kill of the grass). Although the effect of warm-season temperature remained significant, the average predicted effect of a 1 °C reduction was only to lower yield by 0.3 tonne ha^{-1} (less than 10%). This is much less than the prediction from the reduction in degree-days, but is similar to the sort of result obtained in other, less extreme climates (see e.g. Hooker 1905; Jones 1979*a*).

Table 9.4. *Multiple regressions for hay yields* (Y) *in Iceland* (1943–1967) *on mean warm season temperature* (W), *mean cool season temperature* (C), *nitrogenous fertiliser applied* (N), *and year* (A = *actual date* – 1900). *Year was included in an attempt to eliminate effects of advances in technology over the period.* (*From Dyer & Gillooly* 1977, *and H. G. Jones unpublished*)

	Variance accounted for %	Significance
$Y = -0.36 + 0.030\,A + 0.34\,W$	21	*
$Y = -15.2 + 0.592\,A - 0.0051\,A^2 + 0.30\,W$	49	***
$Y = 2.65 + 0.022\,A + 0.30\,C$	44	***
$Y = 1.54 + 0.28\,C + 0.15\,N + 0.26\,W$	61	***

Sample problems

9.1 The net radiation absorbed by a leaf is 400 W m^{-2}. What are (i) the leaf temperature if $T_a = 25$ °C, $h = 0.4$, $r_{aH} = 40$ s m^{-1}, $r_{\ell W} = 200$ s m^{-1}, (ii) the thermal time constant for this leaf if $\rho^* c_p^* = 2.7$ MJ m^{-3} and $\ell^* = 1$ mm? (iii) the time constant when the stomata are closed, (iv) the leaf temperature if the whole leaf is wet, (v) the corresponding thermal time constant?

9.2 Respiration rate of a fruit is 0.1 mg CO_2 m^{-2} s^{-1} at 13 °C and 0.19 mg CO_2 m^{-2} s^{-1} at 19 °C. What are (i) the Q_{10}, (ii) the activation energy?

9.3 The thermal time between flowering and maturity for a certain plant is constant at 600 degree-days above 6 °C, while the rate of net photosynthesis (= 50 g m^{-2} day^{-1} for $k = 1$) is given by equation 9.16 with $B = -5$ and $T_{max} = 30$ °C. Assuming a constant temperature over the period, plot the relationship between the net photosynthesis and temperature. What is the optimum temperature?

10

Drought and drought tolerance

Over large areas of the earth's surface, lack of water is the major factor limiting plant productivity. The average primary productivity of deserts is less than 0.1 tonne ha^{-1} yr^{-1}, which is at least two orders of magnitude less than the productivity achieved when water is non-limiting. Even in relatively moist climates, such as that of southern England, drought lowers the yields of crops such as barley by an average of 10–15 % each year, while the yields of more sensitive crops such as salads and potatoes may be reduced even further if unirrigated.

There are many possible definitions of drought that range from meteorological droughts defined in terms of the length of the rainless period, through definitions that allow for the water storage capacity of the soil and the evaporative demand of the atmosphere, to those that include some aspect of plant performance. In the following, drought is used to refer to any combination of restricted water supply (e.g. as a result of low rainfall or poor soil water storage) and/or enhanced rate of water loss (resulting from high evaporative demand) that tends to reduce plant productivity.

The first part of this chapter outlines the effects of water deficits on individual plant processes and on yield, while the remainder considers the various mechanisms that enable plants to grow or survive in dry environments. Further aspects related to the breeding of crop plants tolerant of drought are discussed in Chapter 12.

Plant water deficits and physiological processes

The effects of water deficits on processes such as stomatal behaviour and photosynthesis have already been discussed in some detail in Chapters 4, 6 and 7. Some useful reviews that cover these and other effects of water deficits on physiological processes in plants include the series of books edited by Kozlowski (e.g. 1968, 1972, 1974, 1981), and volumes by Levitt (1980), Turner & Kramer (1980), Blum (1988), Jones *et al.* (1989*a*) and Alscher & Cumming (1990).

Effects of water deficits

A simple description of the relative sensitivity to water deficits of different processes is difficult, not only because of the large species differences and the marked capacity for acclimation to stress, as has already been seen for stomata and photosynthesis, but also because of the uncertainty concerning the way in which water deficits have their effect. In particular, we have already seen that the effect of soil drying on processes such as shoot growth and photosynthesis may be mediated by signalling (possibly involving ABA) from the root to the shoot (see pp. 87–8 and 149–52), indicating that the water status of the responding organ is not all that has to be considered.

The most obvious effect of even mild stress is to reduce growth, with cell enlargement being particularly sensitive to water deficits (Hsiao 1973). As was outlined in Chapter 4, turgor pressure (rather than total water potential) in the growing cells provides the driving force for cell expansion, but the actual rate of extension is controlled by variation in the yield threshold (Y) and the extensibility (ϕ) (equation 4.16). It is probable that, at least in those cases where changes in shoot growth have been observed in the absence of detectable differences in shoot water status (e.g. Gowing *et al.* 1990), modulation of these parameters may involve some form of signalling from the root to the shoot (Davies & Jeffcoat 1990). Some synthesis of cell wall materials may continue during mild stresses that inhibit expansion growth. This can be manifest as 'stored growth' so that most of the growth lost during a short stress may be recovered after re-watering (Acevedo *et al.* 1971). Cell division, though affected by water stress, is normally less sensitive than cell expansion.

In addition to simple growth inhibition, water deficits can greatly modify plant development and morphology (Table 10.1). For example, the differential sensitivity of roots and shoots (with root growth being less sensitive to water deficits) leads to large increases in the root to shoot ratio in drought (Sharp & Davies 1985). Other effects on vegetative development include the reduction of tillering in grasses and the early termination of extension growth in perennials with the formation of dormant buds. Water deficits also increase the abscission of leaves and fruits, particularly after relief of stress. Not only does water stress decrease the size of leaves as a result of reduced cell expansion and cell division, but, at least in wheat, it leads to a reduction in the proportion of epidermal cells that form stomata and an increase in the number of epidermal trichomes (Quarrie & Jones 1977).

Water deficits also affect reproductive development, with a period of water stress being required to stimulate floral initiation in some species (e.g.

Table 10.1. *Evidence for the involvement of abscisic acid (ABA) in plant responses to water stress based on the similarity of responses to water stress and to exogenous ABA application (collated from Jones* 1981*a*; *Addicott* 1983 *and Davies & Jones* 1991)

Response	Water stress	ABA	
Short term			
Stomatal conductance	Decrease	Decrease	+++
Photosynthesis	Decrease	Decrease (Primarily a stomatal effect)	+++
Membrane permeability	Increase/decrease	Increase/decrease	+
Ion transport	Increase/decrease	Increase/decrease	+
Long term: biochemical and physiological			
Specific *m*RNA and protein synthesis[1]	Increase	Increase	++
Proline & betaine accumulation	Increase	Increase	++
Osmotic adaptation[2]	Yes	Yes	+
Photosynthetic enzyme activity	Decrease	Decrease	+
Desiccation tolerance[3]	Increase	Increase	+
Salinity and cold tolerance	Induces	Induces	++
Wax production[4]	Increase	Increase	+
Long term: growth			
General growth inhibition	Yes	Yes	+++
Cell division	Decrease	Decrease	+++
Cell expansion	Decrease	Decrease	+++
Germination	Inhibits	Inhibits	++
Root growth	Increase/decrease	Increase/decrease	++
Root/shoot ratio	Increase	Increase	++
Long term: morphology			
Production of trichomes	Increase	Increase	++
Stomatal index	Decrease	Decrease	++
Tillering in grasses[5]	Decrease	Decrease/increase	+
Conversion from aquatic to aerial leaf type	Yes	Yes	++
Induction of dormancy, terminal buds or perennation organs	Yes	Yes	++
Long term: reproductive			
Flowering in annuals	Often advanced	Often advanced	+
Flower induction in perennials	Inhibited	Inhibited	+

Table 10.1 (*cont.*)

Response	Water stress	ABA	
Flower abscission	Increased	Increased	+
Pollen viability	Decreased	Decreased	+
Seed set	Decreased	Decreased	+

Extra references: 1. Heikkila *et al.* (1984); Mundy & Chua (1988); Cohen & Bray (1990); 2. Henson (1985); 3. Gaff (1980); Bartels *et al.* (1990); 4. Baker (1974); 5. Hall & McWha (1981).

The strength of correlation is indicated as ranging from weak (+) to strong (+ + +).

Litchi: Menzel 1983). In other cases severe water stress can cause the emergence of ready-differentiated floral buds (Jones 1987). Water deficits tend to advance flowering in annuals and to delay flowering in perennials. In wheat, for example, mild deficits can advance flowering by up to a week, though with corresponding decreases in the number of spikelets and in pollen fertility and grain set (e.g. Angus & Moncur 1977; Morgan 1980).

Just about every aspect of cellular metabolism and fine structure has been reported to be affected by water deficits. Particularly characteristic changes include: increases in rates of degradative as compared with synthetic reactions; decreased protein synthesis; increases in the concentrations of free amino acids (particularly proline, which may increase to as much as 1% of leaf dry matter in some species), glycine-betaine, di- and poly-amines (with osmotic stress especially giving rise to increases in putrescine: Smith 1985) and sugars; all with corresponding changes in relevant enzyme activites. Many of these changes may be considered adaptive, but it is often difficult to distinguish changes that are a manifestation of cell or tissue damage from those that represent acclimation or responses to counter damage. For example, as with other types of environmental stress, water deficits tend to shift the cell redox potential to a more oxidised state and to increase concentrations of free radicals (both presumably damaging changes), but levels of reducing agents such as glutathione, and free radical scavenging systems such as superoxide dismutase which can counter the damage, both tend to increase (see Alscher & Cumming 1990).

All the effects described above contribute to the general decrease in dry matter production and seed yield that is characteristic of drought. Although the effects on carbon assimilation per unit leaf area (or on other physiological processes) may be important, the dominant factor contributing to reduced productivity in drought is generally the reduction in leaf area.

In addition, seed yield can be particularly sensitive to stress at critical periods of development (e.g. microsporogenesis: see Salter & Goode 1967).

Mechanisms and the role of plant growth regulators

How even small water deficits can have such major metabolic and developmental consequences is not clear. As has been pointed out by Hsiao (1973), it is difficult to see how the possible effects of mild stress on water activity, macromolecular structure or the concentration of molecules in the cytoplasm can be primary stress sensors. The best evidence is for a sensor(s) responding to turgor pressure or cell size (Chapter 4). At least in growing cells, small changes in turgor may reduce cell expansion with the consequent build-up of unused cell wall materials or other metabolites then affecting metabolism. There is also evidence that turgor can directly affect ion transport through the involvement of membrane stretch sensors or other turgor-dependent systems (Gutknecht 1968; Zimmermann *et al.* 1977; Tomos 1987). Most of the observed effects of water deficits are almost certainly secondary and result from the operation of plant regulatory responses.

There is good evidence that plant growth regulators are involved in the integration of the various responses (see Davies & Jeffcoat 1990). Although water stress may affect levels of gibberellins and auxins, there is little evidence for these growth regulators having a major role in regulating stress responses. Ethylene production is stimulated by many stresses including water deficits, and more importantly, by flooding; this stimulation has been implicated in a number of observed responses including abscission of leaves and fruits, leaf epinasty and also stomatal closure and decreased assimilation. Cytokinins have also been implicated in some drought-induced responses, such as leaf senescence and stomatal closure; data of Itai *et al.* (1968) and Blackman & Davies (1985) suggest that these may be a consequence of the reduced supply of cytokinins in the transpiration stream that occurs in drought.

Although these other plant hormones may be involved, the clearest evidence is for abscisic acid (ABA) performing the major role in integration of plant responses both to water stress, and to a wide range of other environmental stresses including salinity and high temperatures (see reviews by Jones 1981*a*; Addicott 1983; Davies & Jones 1991). Evidence for the involvement of ABA in stomatal closure in response to drought has been outlined in Chapter 6; other particularly important observations include the fact that ABA concentrations rise rapidly in stressed plants (this rise tends to be a function of Ψ_p rather than total water potential), and the very close correspondence between the responses to water deficits and to exogenously supplied ABA of a wide range of short- and long-term plant responses (Table 10.1). These observations, when combined with information obtained from the study of ABA deficient mutants (Chapter 6, p. 137), provide compelling evidence that ABA does indeed have a general role as an endogenous plant growth regulator involved in plant adaptation to water deficits and other stresses.

Drought tolerance

The term drought resistance has long been used to refer to the ability of plants to survive drought, but I prefer 'drought tolerance' to describe all mechanisms that tend to maintain plant survival or productivity in drought conditions. In an agricultural or horticultural context, a more drought-tolerant cultivar is one that has a higher yield of marketable product in drought conditions than does a less tolerant one. Many farmers and breeders also look for some degree of yield stability from year to year as a criterion of drought tolerance (see e.g. Fischer & Turner 1978). In natural ecosystems, however, a drought-tolerant species is one that has the ability to survive and reproduce in a relatively dry environment. In this case, drought tolerance does not necessarily rely on a high productivity. It follows, therefore, that the mechanisms favouring drought tolerance in typical agricultural monocultures may be distinct from those that have evolved in natural ecosystems.

Plants that can exist in dry environments are called xerophytes (cf. hygrophytes in wet habitats and mesophytes in intermediate habitats). Although xerophytes occur naturally only in dry situations, this is not necessarily because they are truly xerophilous (drought-liking) but rather because their competitive ability is greater than other species only in dry places. Under careful husbandry, when competition from other species is eliminated, growth and production by these species is improved by increasing the availability of water.

There are many different ways in which plants may be adapted to dry conditions so that there is no simple set of morphological or physiological criteria that can be used to distinguish xerophytes. Many of the methods that have been proposed for classifying xerophytes (see e.g. Kearney & Shantz 1911; Maximov 1929; Levitt 1980) have been in terms of the different ecological niches occupied, using descriptions such as drought escaping, drought evading and drought enduring. Partly because of difficulties in attributing plants to one or other group, I think that the most useful approach is to concentrate on the mechanisms that contribute to drought tolerance, recognising that one plant may have several. These

Table 10.2. *Drought tolerance mechanisms*

1. *Avoidance of plant water deficits*
   (*a*) *Drought escape* – short growth cycle, dormant period.
   (*b*) *Water conservation* – small leaves, limited leaf area, stomatal closure, high cuticular resistance, limited radiation absorption.
   (*c*) *Effective water uptake* – extensive, deep or dense root systems.
2. *Tolerance of plant water deficits*
   (*a*) *Turgor maintenance* – osmotic adaptation, low elastic modulus.
   (*b*) *Protective solutes, desiccation tolerant enzymes, etc.*
3. *Efficiency mechanisms*
   (*a*) *Efficient use of available water.*
   (*b*) *Maximal harvest index.*

mechanisms may conveniently be classified into three main types: (1) Stress avoidance – that is, those mechanisms that minimise the occurrence of damaging water deficits. (2) Stress tolerance – that is, those physiological adaptations that enable plants to continue functioning in spite of plant water deficits. (3) Efficiency mechanisms – that is, those that optimise the utilisation of resources, especially water. A useful subdivision of these groups is presented in Table 10.2, while some details are discussed below.

Avoidance of plant water deficits

(*a*) *Drought escape.* A plant that rapidly completes its life cycle, or at least its reproductive cycle, can escape periods of drought and grow during periods of favourable soil moisture. This mechanism is typical of the desert ephemerals that can complete their life cycles from germination to seed maturation in as little as four to six weeks. A similar, though less extreme adaptation is found in many crop plants, where the most drought-tolerant cultivars, at least in environments with a marked dry season, are frequently those that flower and mature the earliest, thus avoiding the worst of the dry season. Many annual plants even show a dynamic response of this type, flowering earlier than usual if they are subjected to water stress, as would occur as the soil dries out in a dry year. In general, this group does not rely on any other physiological mechanisms for surviving in dry climates.

(*b*) *Water conservation.* Plant adaptations that limit the rate of water loss can prevent the development of detrimental plant water deficits in two ways. They can either conserve soil water for an extended period, thus maintaining soil (and plant) water potential suitably high over a sufficient period for seed ripening, or else the reduced transpirational flux can reduce the depression of Ψ_ℓ that results from the frictional resistances in the transpiration pathway (see Chapter 4). Plants having mechanisms for restricting water loss have been termed water-savers (see Levitt 1980), and include many of the plants that are commonly thought of as xerophytes.

Particular adaptations that minimise transpiration include the following: a thick cuticle with a correspondingly low cuticular conductance (often less than 2 mmol m^{-2} s^{-1} (or 0.05 mm s^{-1}) in these plants); small leaves with a small total transpiring surface per plant (leaves may be reduced or even absent with some semi-desert plants, such as chuparosa, *Beloperone californica*, in the western USA which is summer deciduous but maintains some photosynthesis in the green stem tissue); a high leaf reflectivity and other adaptations that minimise radiation absorption (see Chapters 2 and 9); a low stomatal conductance as achieved by stomatal closure or by very small, sunken or sparsely distributed stomata. It is often suggested that a tomentum, or thick layer of leaf hairs, can potentially help to conserve water, particularly if the hairs are dead (see Fahn 1986 for morphological information on leaf trichomes), but in practice the effect on the total leaf diffusive resistance is usually rather small (for example, even a tomentum 1 mm thick has a diffusive resistance (see equation 3.21) equal to $\ell/\mathbf{D} = 1/24.2 = 0.041$ s mm^{-1} (or 1.025 m^2 s mol^{-1}) which is small in comparison with the stomatal resistance which can be about two orders of magnitude larger when the stomata are closed). Leaf hairs probably have greater significance for radiation balance and for the balance between water loss and assimilation. The role of these and other morphological characters that might be expected to favour water conservation will be discussed in more detail in the section on efficiency mechanisms.

CAM plants, with their reversed stomatal cycle and thick cuticles, are particularly effective at limiting water loss under stress conditions. For example, a specimen of *Echinocactus* was reported to lose less than 30 % of its weight in six years when maintained without water (see Maximov 1929).

In addition to these characters that have obvious implications for water conservation, many of these 'xeromorphic' plants have an extreme development of structural tissues such as schlerenchyma and collenchyma and the presence of spines and other protective structures. The latter have probably evolved not as direct tolerance mechanisms, but as characters reducing grazing damage in environments where little other vegetation may be present.

Although some of these characters that reduce water loss are fairly constant irrespective of environmental conditions (such as thick cuticle or stomatal morphology), most respond to drought in some degree. For example, wax development, cuticle thickness, hair development and leaf reflectivity have all been reported to increase with increasing drought.

(*c*) *Effective water uptake.* Many plants that are successful in dry habitats have no specific adaptations for controlling water loss but rely on the development of a very deep and extensive root system that can obtain water from a large volume of soil or from a deep water table. Many desert shrubs (e.g. mesquite, *Prosopis juliflora*) have deep root systems. Polunin (1960) even quotes Rubner as saying that roots of tamarisk were detected as deep as 50 m during excavation of the Suez Canal. Root development is very plastic, with more of a plant's resources (e.g. carbohydrate) being put into root growth relative to the shoot in dry conditions. In some cases the absolute root growth may be enhanced. Plants that have an effective water supply system often behave as water-spenders (Levitt 1980) and do not limit the rate of transpiration.

Tolerance of plant water deficits

There are several ways in which plants maintain physiological activity as water content or Ψ falls. These include:

(*a*) *Turgor maintenance.* Turgor maintenance by means of increases in cell solute concentration (lowering Ψ_π – see equation 4.7, p. 76) in response to water stress is widespread, occurring in leaves, roots and reproductive organs of many species (see Turner & Jones 1980; Morgan 1984). This process, often called osmotic adjustment or osmoregulation, is probably the most important mechanism for maintaining physiological activity as Ψ falls. Full turgor maintenance occurs if the decrease in Ψ_π equals any decrease in total Ψ, so turgor remains constant (see Fig. 10.1). Even partial turgor maintenance, where $(d\Psi_\pi/d\Psi) < 1.0$, can be advantageous. Many species show at least partial turgor maintenance, particularly where drought is imposed slowly, while full turgor maintenance is often observed over a limited range of Ψ. Some examples of full turgor maintenance over a limited range of Ψ and of partial turgor maintenance are shown in Fig. 10.2.

Both inorganic ions (especially K^+ and Cl^-) and organic solutes can be involved in osmotic adjustment of cells, though changes in organic solutes (particularly those known as compatible solutes – see below) tend to be much more important. In saline environments, especially, Na^+ and Cl^- can be exploited as vacuolar solutes, but it is necessary that they are maintained at low levels in the cytoplasm if cell metabolism is not to be damaged. In general, diurnal changes in Ψ_π are rather small, with osmotic adjustment usually taking days or even months. Although the capacity for osmotic adjustment has been positively related to yield in some crops, there are cases where species that have a large capacity for osmoregulation are actually more sensitive to drought than are non-osmoregulating species (Quisenbery

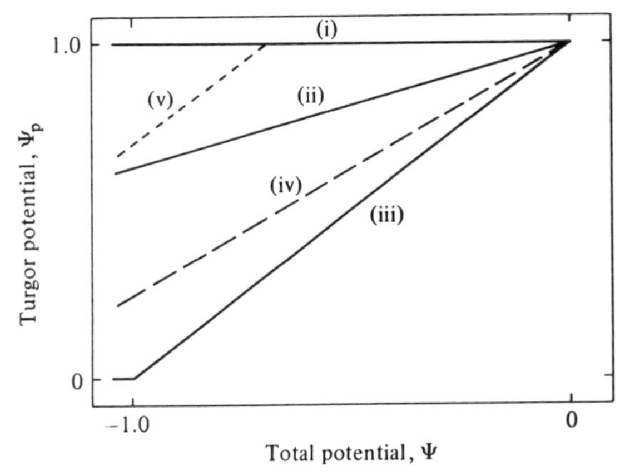

Fig. 10.1. Schematic representation of the relationship between turgor potential (normalised to 1.0 when $\Psi = 0$) and total water potential for (i) full turgor maintenance ($d\Psi_\pi/d\Psi = 1$); (ii) partial turgor maintenance ($0 < (d\Psi_\pi/d\Psi) < 1$); (iii) no turgor maintenance with constant solute concentration (extremely rigid walls); (iv) passive turgor maintenance with solutes concentrating as a result of cell shrinkage with elastic cell walls; (v) an example where turgor maintenance is only achieved over a limited range of Ψ.

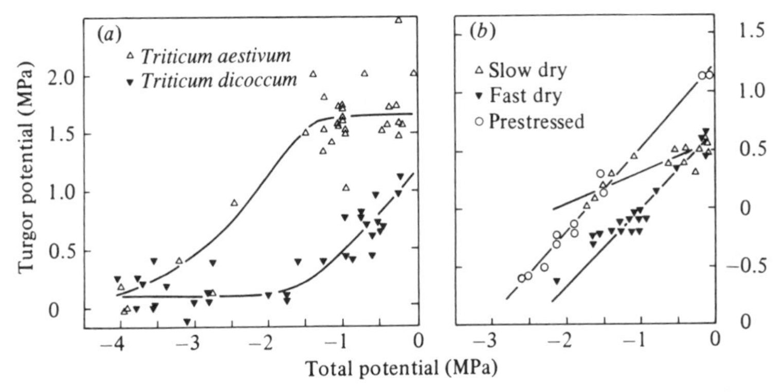

Fig. 10.2. Examples of relationships between turgor potential and leaf water potential for (*a*) two wheat genotypes and (*b*) sorghum plants: ▼ previously well-watered and dried rapidly; △ previously well-watered and dried slowly; ○ previously stressed to −1.6 MPa and dried rapidly. (Data from Turner & Jones 1980.)

et al. 1984); this suggests there may be a 'cost' associated with osmoregulation.

In addition to active adjustment of total cell solutes, there is also a passive concentration effect as cell water decreases. This effect is greatest for a given fall in Ψ for those plants such as succulents with relatively elastic walls (i.e. ε_B is small – see Chapter 4) as shown in Fig. 10.3. The possible contribution of this passive adjustment to turgor maintenance is illustrated in Fig. 10.1 for cells with differing ε_B. It has also been suggested that a decrease in cell size, as commonly occurs with drought, can also help in maintaining turgor (Cutler *et al.* 1977).

The extreme development of structural tissues in many xeromorphic plants that was noted above results in inextensible cells with a large modulus of elasticity (ε_B). This is an additional factor often associated with the ability to survive low water potentials as it allows the cells to withstand high turgor pressures (Fig. 10.3), thus permitting high osmotic concentrations with the consequent ability to maintain turgor down to very low values of Ψ. The associated constancy of cell volume may also be important in maintenance of physiological activity over a wide range of Ψ without a need for osmoregulation or compatible solutes.

The opposite extreme is found in water storage tissues of succulent species. In this case ε_B is very small. As a result large changes in cell volume occur for relatively small changes in Ψ (Fig. 10.3): that is, there is a large tissue capacitance. This large capacitance acts to increase the time constant for water exchange (cf. equation 4.32), and may help buffer against rapid environmental changes. It has been estimated that decreasing ε_B from a high value of 40 MPa to a low one of 4 MPa would increase the half-time for water exchange from 0.6 to 6 min (Tyree & Karamanos 1981). It is, however, difficult to envisage how changes over this time scale might be ecologically significant. Equally, the amount of water stored is generally small in relation to potential evaporation rates but, where the stomata are closed so that rates of water loss are slow, storage may be adequate to maintain turgor for quite long periods.

(*b*) *Protective solutes, desiccation tolerant enzymes, etc.* Survival is as important as the ability to continue functioning at low water potential. This survival ability, or desiccation tolerance, has often been measured in terms of the environmental water potential or tissue water content that causes 50% cell death (see Gaff 1980; Levitt 1980). Usually tissue is 'equilibrated' with air at a given humidity and Ψ (calculated according to equation 5.11, p. 110), but, in many published examples where short exposure periods were used, vapour equilibrium may not have been attained. It is, however, clear that an extreme capacity to survive severe desiccation is common in lower plants such as algae, lichens and bryophytes. Although many seeds and pollen grains of higher plants can tolerate extreme desiccation, there are relatively few 'resurrection' species in which the vegetative tissues can tolerate severe desiccation and recover rapidly on re-watering (see Gaff 1980). An example is *Myrothamnus flabellifolia*, whose air-dry leaves start

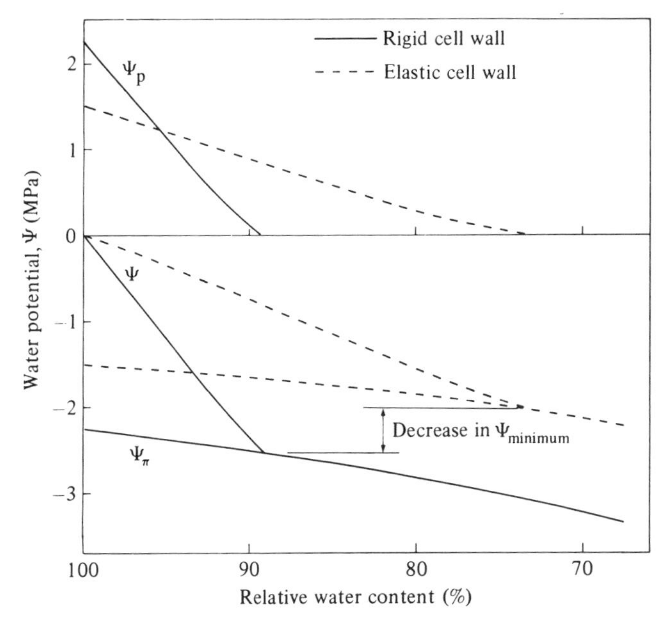

Fig. 10.3. Höfler diagrams for cells with elastic or rigid cell walls, illustrating the potentially lower minimum Ψ for positive turgor with rigid cell walls and the smaller volume changes for a given change of Ψ.

to respire within half an hour of re-watering and show the first signs of photosynthesis within 6.5 h. Lower plants, such as the moss *Tortula ruralis*, may recover even more rapidly. The remarkable desiccation tolerance of these plants can involve either an ability to maintain subcellular structure and physiological integrity (especially in seeds, mosses and in plants such as *Myrothamnus*), or an ability to repair damage rapidly on rehydration (as appears to be the case in the resurrection plant *Borya nitida*). In either case recovery is usually only possible where dehydration occurs slowly.

A number of types of solute, known as 'compatible solutes' are found to be particularly effective at protecting cytoplasmic proteins and cell membranes from desiccation, even when they increase to high concentrations during osmotic adjustment. They are particularly effective when compartmented in the cytoplasm, and include a number of sugars (e.g. trehalose), sugar alcohols (e.g. sorbitol and mannitol), amino acids (especially proline) and betaines. Although proline has been widely studied as a drought adaptation mechanism of this type, and though it may be important as a protective solute or osmoticum in lower plants and bacteria (Hellebust 1976), it does not appear to be quantitatively so significant in higher plants (Stewart & Hanson 1980). Similar mechanisms are also involved in freezing tolerance, where dehydration is also a cause of injury (see Chapter 9).

(*c*) *Antioxidants*. The cellular injuries arising from desiccation are probably a combination of physical damage to cellular components following water loss, consequences of increased solute concentrations and chemical damage (see Stewart 1989 for a review). There is increasing evidence that chemical injury in the form of free-radical damage (especially that caused by oxygen radicals) is an important factor in desiccation injury (see review in Winston 1990). The superoxide radical (O_2^-) and other oxygen radicals can be produced by a number of reactions in cells including autoxidation of a number of reduced compounds, and the 'Mehler' reaction in chloroplasts where O_2 rather than CO_2 becomes the ultimate acceptor for electron transport (as can occur when assimilation is blocked by water stress; this may also be involved in photoinhibitory damage: see Chapter 7). Once formed, O_2^- undergoes further reduction to form the very damaging hydroxyl radical (•OH), which can cause lipid peroxidation, and hydrogen peroxide. Plants contain a number of antioxidant mechanisms that protect against the production of oxygen radicals including: (a) water soluble reductants such as thiol-containing compounds (e.g. glutathione) and ascorbate; (b) fat soluble vitamins such as α-tocopherol and β-carotene; and (c) enzymic antioxidants such as catalase and superoxide dismutase. Athough there are a number of reports that differential expression of these systems may underlie differences in drought tolerance, further studies are required before clear conclusions can be drawn.

Efficiency mechanisms

Any mechanism enhancing survival in drought conditions tends to decrease the potential dry matter productivity. For example, total photosynthesis would be decreased by stomatal closure, by leaf rolling, by decreases in leaf area or even by reductions in growing season length. The 'ideal' plant for any environment involves a compromise between water conservation and productivity mechanisms with the *optimum* balance depending on the aridity of the environment. In addition to such optimal expression of stress-avoidance characters, drought tolerance is favoured by any mechanisms that improve either the efficiency with which a limited supply of water is used for photosynthesis or the efficiency of subsequent conversion of photosynthate into reproductive structures (or yield in the case of crops). This section concentrates on the efficiency with which water is used in dry matter production.

In many situations, efficiency alone is a poor competitive strategy because greater efficiency is often associated with a slow rate of water use.

Such 'efficient' plants might then lose some of their potentially available water to faster-growing competitors. This is, however, less of a problem in typical agricultural or horticultural monocultures. A related point that is often forgotten in discussion of efficiency mechanisms is that it is no use just having a high efficiency (e.g. of water use) if water supply is non-limiting. Other things being equal, productivity in dry environments is likely to be greater for a plant whose behaviour tends to maximise assimilation in relation to the amount of water available than for a plant that simply has a high ratio of assimilation to water evaporated.

Water use efficiency. Following common usage, the ratio of net assimilation to water loss will be given the general term water use efficiency (WUE), though it is, of course, not a true efficiency as it does not have a maximum value of unity. The precise definitions of assimilation or water use differ between authors: for example, water loss may be in mass units or molar units and assimilation may be expressed in terms of net CO_2 exchange ($\mathbf{P}_n$ in either molar or mass units), dry matter growth or yield (Y_d) or economic yield (Y). Total net daily CO_2 uptake may be converted into dry matter using a value of 0.61–0.68 g dry matter per g CO_2 (see Chapter 7), though for photosynthetic data for single leaves it is also necessary to allow for respiratory losses at night and from other tissues. Economic yield may be obtained from Y_d by multiplication by the harvest index (see Chapter 12). Unfortunately, many published estimates of Y_d have neglected root dry matter, which may vary from less than 20% of plant dry weight when ample water is available, to more than 50% in dry conditions.

Similarly water loss may be expressed in terms of either water transpired ($\mathbf{E}_\ell$) or total evaporation ($\mathbf{E}$). The yield per unit transpiration (transpiration efficiency = $Y/\mathbf{E}_\ell$) is a better measure of plant performance than yield per unit evaporation ($Y/\mathbf{E}$), though the latter is ecologically important because a significant proportion of total evaporation occurs directly from the soil.

Because of the sensitivity of evaporation to the environment (particularly δe), WUE varies markedly from place to place and from year to year. Typical of many results are those of N. M. Tulaikov with 'Beloturka' wheat in Russia (see Maximov 1929): WUE (in terms of dry matter) in one year (1917) ranged fom 2.13×10^{-3} at Kostychev to 4.22×10^{-3} in the relatively cool and humid climate at Leningrad, while at one site values ranged from 1.74×10^{-3} to 3.31×10^{-3} over the years 1911–1917.

In contrast, there is extensive evidence that, for any one species, the seasonal WUE in one environment can be surprisingly constant over a wide range of treatments (sowing date, planting density, water supply, nutrition, etc.). Some examples of the effects of different irrigation regimes (and total water use) on grain yields of wheat and barley in southern England are

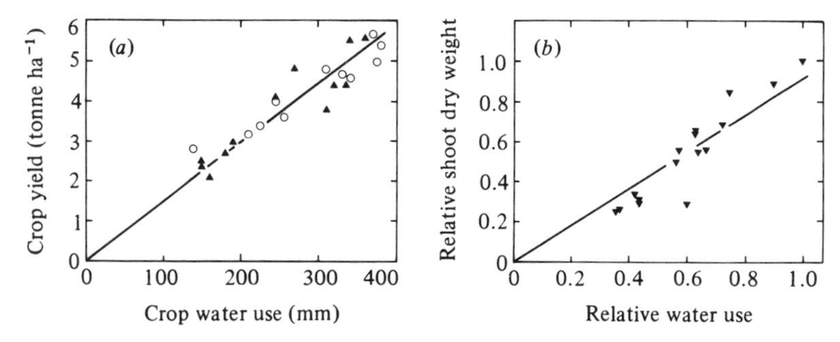

Fig. 10.4. (*a*) Grain yield as a function of water used under a range of irrigation treatments for barley in 1976 (○) and wheat in 1979 (▲) in south-eastern England (data from Day *et al.* 1978 and Innes & Blackwell 1981). (*b*) Relative shoot dry weight as a function of relative water use for cowpea in California in 1976 (after Turk & Hall 1980). The water use efficiency (WUE) for any point is given by the slope of a line joining that point to the origin, so that the lines join points of equal WUE.

shown in Fig. 10.4*a*. The close similarity between the years is somewhat fortuitous, but all treatments lie close to a straight line through the origin, implying nearly constant WUE. Similar constancy of WUE in terms of dry matter production by cowpea (*Vigna unguiculata*) is shown in Fig. 10.4*b*. In many cases where WUE is not constant over a range of treatments, deviations can be attributed to differing amounts of soil evaporation.

In an early analytical attempt to explain site-to-site variability of WUE, de Wit (1958) recognised that evaporation was a dominant feature and showed that for dry climates

$$Y/\mathbf{E}_\ell = k/\mathbf{E}_0 \tag{10.1}$$

where $\mathbf{E}_0$ was the mean daily free water evaporation and k is a parameter that was relatively constant for any species. This relationship broke down only if growth was seriously 'nutrition-limited', if the water available was excessive, or in humid regions (where $Y/\mathbf{E}_0$ was approximately constant).

There remained, however, large differences between species, as is clearly illustrated by the well-known results in Table 10.3. Although absolute values of WUE are variable (depending on weather), differences between species are very consistent and are clearly related to photosynthetic pathway. CAM plants have the highest dry matter efficiencies (e.g. 20×10^{-3} to 35×10^{-3} for *Agave* and pineapple: Joshi *et al.* 1965; Neales *et al.* 1968) followed by C_4 plants which are approximately twice as efficient as C_3 plants. Table 10.3 illustrates that differences between species with one photosynthetic pathway are no larger than differences between cultivars of

Table 10.3. *Values of water use efficiency* ($Y_d/\mathbf{E}_\ell$) *for some potted plants grown at Akron, Colorado (rearranged from Maximov* 1929, *after data of Shantz & Piemeisel* 1927)

	$Y_d/\mathbf{E}_\ell$ ($\times 10^3$)
C_4 plants	
Cereals	2.63–3.88
millet cvs.	2.72–3.88
sorghum cvs.	2.63–3.65
maize cvs.	2.67–3.34
Other C_4 Gramineae	2.96–3.38
Other C_4	2.41–3.85
Range for C_4 plants	2.41–3.88
C_3 plants	
Cereals	1.47–2.20
wheat cvs.	1.93–2.20
oat cvs.	1.66–1.89
rice cvs.	1.47
Other C_3 Gramineae	0.97–1.58
Other C_3 crops	1.09–2.65
alfalfa cvs.	1.09–1.60
pulses	1.33–1.76
sugar beet	2.65
Native plants	0.88–1.73
Range for C_3 plants	0.88–2.65

one species. Interestingly, there is no relationship between WUE and aridity of habitat.

Further analysis of water use efficiency

Simple models based on leaf gas exchange

A more complete analysis based on the approach of Bierhuizen & Slatyer (1965) can be used to explain variation in WUE. Using equations described earlier for leaf gas exchange, and using molar units (though expressing driving forces in terms of partial pressure, rather than mole fraction as used previously), one can write for transpiration

$$\mathbf{E}_\ell^m = \frac{e_\ell - e_a}{P(r_a + r_\ell)} \tag{10.2}$$

wher e_ℓ is shorthand for the saturation vapour pressure at leaf temperature, and for assimilation

$$\mathbf{P}_n^m = \frac{p'_a - p'_i}{P(r'_a + r'_\ell)} = \frac{p'_a - p'_x}{P(r'_a + r'_\ell + r'_m)} \tag{10.3}$$

where p'_x is an 'internal CO_2 concentration' usually taken equal to Γ (see Chapter 7). Combining equation 10.2 and 10.3 gives

$$\frac{\mathbf{P}_n^m}{\mathbf{E}_\ell^m} = \frac{(p'_a - p'_x)}{(e_\ell - e_a)} \frac{(r_a + r_\ell)}{(r'_a + r'_\ell + r'_m)} \tag{10.4}$$

Since p'_x can be assumed constant, WUE for a given environment is proportional to $(r_a + r_\ell)/(r'_a + r'_\ell + r'_m)$. This ratio is larger for C_4 than for C_3 plants because of the larger r'_m (and often smaller r_ℓ) in C_3 plants and explains the observed differences between plants with different photosynthetic pathways (Table 10.3).

This resistance approach is not altogether satisfactory because of non-linearity in the CO_2 response curve, but a useful simplification can be made using the observation that the ratio of gas-phase to liquid-phase resistances is often nearly constant over a range of conditions (nutrition, age, irradiance, etc.) so that p'_i/p'_a is approximately constant at near 0.3 for C_4 plants and 0.7 for C_3 plants (see Chapter 7). For example as irradiance decreases, both r'_m and r'_ℓ increase. Combining equation 10.2 and 10.3 gives

$$\frac{\mathbf{P}_n^m}{\mathbf{E}_\ell^m} = \frac{(p'_a - p'_i)}{(e_\ell - e_a)} \frac{(r_a + r_\ell)}{(r'_a + r'_\ell)} \tag{10.5}$$

from which the resistances can be eliminated (because the CO_2 resistances $(r'_a + r'_\ell) \simeq 1.6\ (r_a + r_\ell)$ – see Chapter 3), to give

$$\frac{\mathbf{P}_n^m}{\mathbf{E}_\ell^m} = \frac{p'_a[1 - (p'_i/p'_a)]}{1.6(e_\ell - e_a)} \tag{10.6}$$

where p'_i/p'_a is constant, this can be written

$$\mathbf{P}_n^m/\mathbf{E}_\ell^m = k^m/(e_\ell - e_a) \tag{10.7}$$

where k^m is a constant that depends on species. In agronomic studies, where one is concerned with dry matter yield, it is more usual to express equation 10.7 in terms of mass fluxes of CO_2 so that

$$\mathbf{P}_n/\mathbf{E}_\ell = (M_C/M_W)(\mathbf{P}_n^m/\mathbf{E}_\ell^m) = 2.44(\mathbf{P}_n^m/\mathbf{E}_\ell^m) \tag{10.8}$$

Equation 10.7 emphasises the role of leaf–air vapour pressure difference in determining WUE in different climates. The importance of this can be illustrated by studies of the effect of irrigation on productivity using small plots: in a monsoon climate where rainfall affects δe, the effect of rain on production may be larger than the effect of an equivalent amount of irrigation; similarly the effects on yield when water supply is controlled using rainshelters will not be the same as equivalent amounts of rain falling in the dry season.

When extrapolating this approach (based on leaf gas exchange) to the crop or community level it is necessary to take account of a number of additional factors such as evaporation from the soil and diurnal and seasonal variation, as discussed below.

Extrapolation to long term

Integration of equation 10.7 over daily, or longer, periods requires several additional assumptions. Most important is estimation of the proportion of $\mathbf{P}_n$ lost in respiration. The simple conversion from CO_2 to dry weight given above is appropriate if net CO_2 exchange is available for the whole plant for the full 24 h. Where only daytime $\mathbf{P}_n$ is available, one needs to allow for growth respiration (with a conversion efficiency $\simeq$ 0.53 g dry weight per g CO_2) and maintenance respiration (about 15–30 % of $\mathbf{P}_n$) giving about 0.37 to 0.45 g dry weight per g CO_2 overall (see Chapter 7 and Tanner 1981).

A second problem is estimation of $(e_\ell - e_a)$. First, leaf temperature is not usually known, so e_ℓ is usually approximated by the saturation vapour pressure at air temperature. Use of a model that allows for leaf–air temperature differences is described in the next section. Secondly, it is difficult to estimate an appropriate daily mean for $(e_\ell - e_a)$. For example, Tanner (1981) showed that an integrated daytime saturation deficit was 1.45 times the mean of the saturation deficit calculated at minimum and maximum temperatures.

Assuming that appropriate mean values for the leaf–air vapour pressure difference are available, equation 10.6 can be integrated to give a total WUE over the life of the plant by including a term (ϕ_C) to represent the losses of CO_2 not associated with assimilation through the stomata (respiration losses at night and from the roots), and a term (ϕ_W) for the losses of water vapour from the soil or through the cuticle:

$$\mathrm{WUE} = \frac{p'_a(1 - p'_i/p'_a)(1 - \phi_C)}{1.6(e_\ell - e_a)(1 + \phi_W)} \tag{10.9}$$

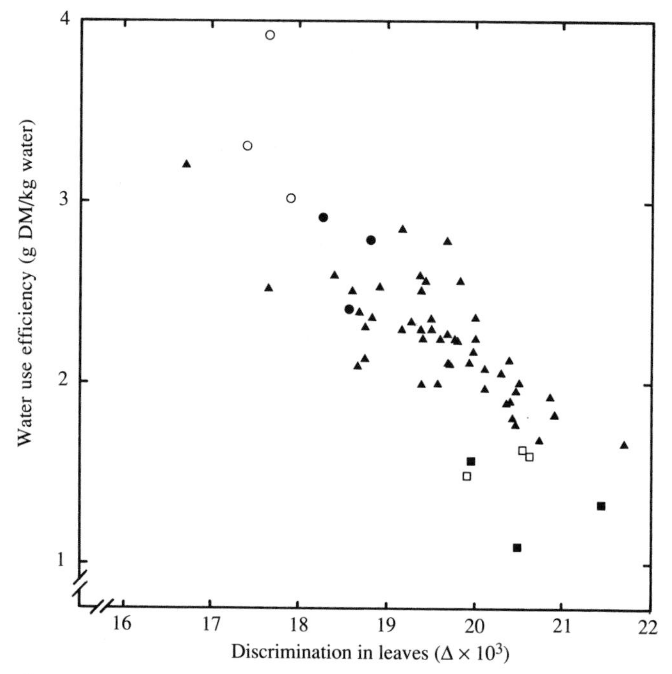

Fig. 10.5. Water use efficiency of whole peanut plants plotted against carbon isotope discrimination measured in dried leaf material. Solid symbols: well-watered plants; open symbols: water-stressed plants; Chico (□ ■); Tifton-8 (○ ●); F2 progeny of Chico × Tifton-8 (▲). (Data from Hubick *et al.* 1988.)

Carbon isotope discrimination

The basis of the discrimination (Δ) between the two stable isotopes of carbon (^{13}C and ^{12}C) was described in Chapter 7. It was shown there (equation 7.31) that Δ is a function of p'_i/p'_a in C_3 plants. Rearranging equation 7.31 to give p'_i/p'_a in terms of Δ and substituting into 10.9 gives

$$\text{WUE} = \frac{p'_a(0.030-\Delta)(1-\phi_C)}{1.6(0.0256)(e_\ell-e_a)(1+\phi_w)} \tag{10.10}$$

This equation implies that WUE should decrease linearly with increasing Δ, so that measurement of Δ might provide an indicator of differences in WUE for plants growing in a particular environment. A particular advantage of the technique should be that measurement of the carbon isotope composition of dry matter theoretically integrates $\mathbf{P}_n/\mathbf{E}_\ell$ over time.

Negative relationships between WUE and Δ of the plant dry matter as

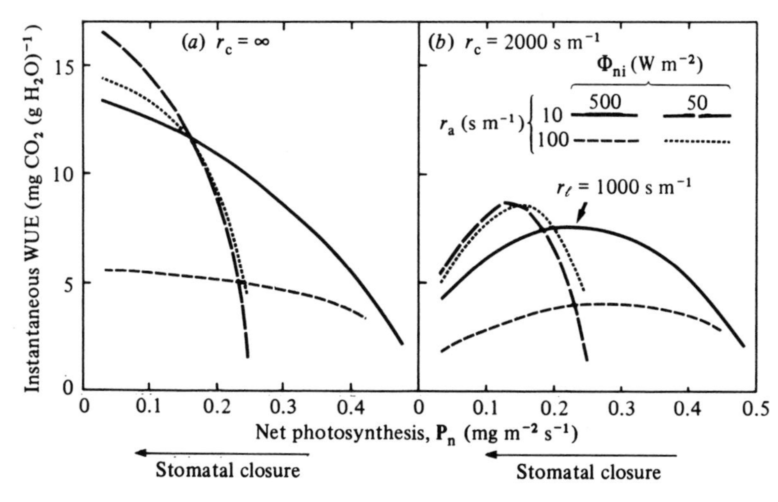

Fig. 10.6. Variation of instantaneous water use efficiency (WUE) with photosynthesis rate (which increases as stomata open) for C_3 leaves (*a*) asssuming an infinite cuticular resistance (r_c); (*b*) assuming an r_c of 2000 s m^{-1} (after Jones 1976). The right-hand end of each curve corresponds to a leaf resistance of 50 s m^{-1}, and the left-hand end to 6400 s m^{-1}. The value of r'_ℓ giving maximum WUE for $\mathbf{\Phi}_{ni} = 500$ W m^{-2} and $r'_a = 10$ s m^{-1} is shown by the arrow.

predicted by equation 10.10 have been found for a number of species both in pots and in the field (e.g. Fig. 10.5), though field results tend to be more variable. These results suggest that measurements of Δ may provide a useful selection method for WUE, at least in C_3 species, especially since reasonably high heritabilities have been found for both water use efficiency and Δ (Hubick *et al.* 1988).

More sophisticated leaf models

Unfortunately the predictions from the simple models (e.g. equations 10.4 or 10.9) are not always accurate because they do not allow for differences between leaf and air temperature or for the precise shape of the photosynthesis response curves. Models that include a more complete description of leaf photosynthesis and of leaf energy balance are available (e.g. Jones 1976; Cowan 1977; von Caemmerer & Farquhar 1981).

Some results obtained with a more detailed model are shown in Figs 10.6 and 10.7. As expected from equation 10.4, stomatal closure tends to increase the instantaneous WUE at the expense of absolute production (Fig. 10.6). The magnitude of the cuticular resistance, however, is crucial.

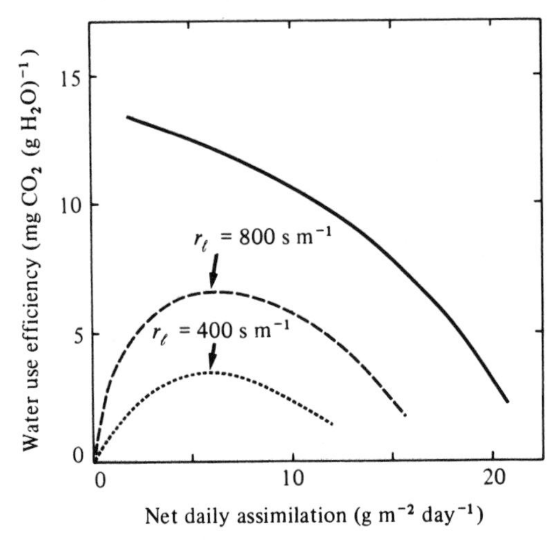

Fig. 10.7. Variation of daily water use efficiency (WUE) with photosynthetic rate for C_3 leaves with $\mathbf{\Phi}_{ni} = 500$ W m^{-2} and $r'_a = 10$ s m^{-1} for different amounts of respiration at night (**R**, mg m^{-2} s^{-1}). —— $\mathbf{R} = 0$; ---- $\mathbf{R} = 0.14\,\mathbf{P}_n + 0.05$; ------ $\mathbf{R} = 0.14\,\mathbf{P}_n + 0.15$. The arrows show values of r'_ℓ giving maximum WUE.

As r_c decreases, the advantage of stomatal closure disappears so that an optimum stomatal resistance becomes apparent (Fig. 10.6*b*). This effect occurs because the long liquid-phase component in this pathway to the chloroplast means that CO_2 uptake through the cuticle is negligible even where water loss may be significant. Although WUE usually tends to increase as stomata close, it is possible in some cases when the ratio of r'_m/r'_a is less than a critical value (e.g. in C_4 plants with large leaves) for WUE to *increase* with increasing stomatal aperture. This prediction has been confirmed experimentally for a crop (Baldocchi *et al.* 1985).

Figure 10.6 also illustrates the complexity of effects of alterations in leaf boundary layer resistance (e.g. resulting from leaf size, hairiness or exposure). Although increasing r_a, as one might get with larger leaves, normally leads to decreased WUE, particularly at high light or when stomata are open, the reverse can be true when stomata are closed. There are also interactions with the photosynthetic pathway in terms of the optimum combination of stomatal and boundary layer characters for any particular environment.

Although it can be instructive to calculate the stomatal resistance that gives the maximum instantaneous WUE (as in Fig. 10.6), of more relevance to plant growth is the resistance giving maximum WUE over daily or longer periods. This resistance is smaller because of the need to make up for respiration at night (see above). Figure 10.7 shows how this optimum resistance changes as respiration increases. Further details of this model and its extension to canopies are discussed by Jones (1976).

It is worth reiterating at this point the earlier comment that maximal production for the amount of water available is likely to be more generally advantageous than maximal WUE (i.e. it is no use maximising WUE if some water is left unused). In such cases, the optimum stomatal aperture may be between fully open and that giving maximum WUE. Calculation of this optimal behaviour is discussed in the next two sections (similar calculations for processes such as leaf area development are also possible).

One application of these studies is to provide information on the relative importance of different drought tolerance mechanisms, by comparing these predictions with the behaviour of real plants. A very close similarity is unlikely as evolution is determined by a plant's overall fitness, which is the summation of many factors over the whole life cycle. An alternative application of these studies is in plant breeding (see Chapter 12). The 'shorthand' terminology used in these sections and elsewhere should be interpreted in a non-teleological sense: for example, an 'optimistic plant' (see below) simply describes a pattern of response without implying purpose.

Optimum stomatal behaviour

Although at any one time there may be an optimum stomatal aperture giving maximum instantaneous WUE, optimum use of water over a period involves optimal distribution of stomatal opening as the environment changes. Clearly it is more efficient to restrict periods of open stomata and rapid photosynthesis to those times when potential evaporation is low, particularly in the morning. Therefore, stomatal closure during midday and in the afternoon, as commonly observed in water-stressed plants, will tend to improve WUE.

Cowan (1977) and Cowan & Farquhar (1977) have developed a model to quantify the optimal behaviour for stomata in a changing environment. Starting from the premise that the optimal behaviour is that in which the *average* evaporation rate is minimal for a given average rate of assimilation, they present a theorem that defines the optimal behaviour. The criterion used is equivalent to that suggested above of maximising the total assimilation for a given total water use and may also be framed in these terms.

For the majority of plants and environments, $\mathbf{E}_\ell$ is relatively more sensitive to changes in stomatal conductance than is $\mathbf{P}_\mathrm{n}$ (for a constant

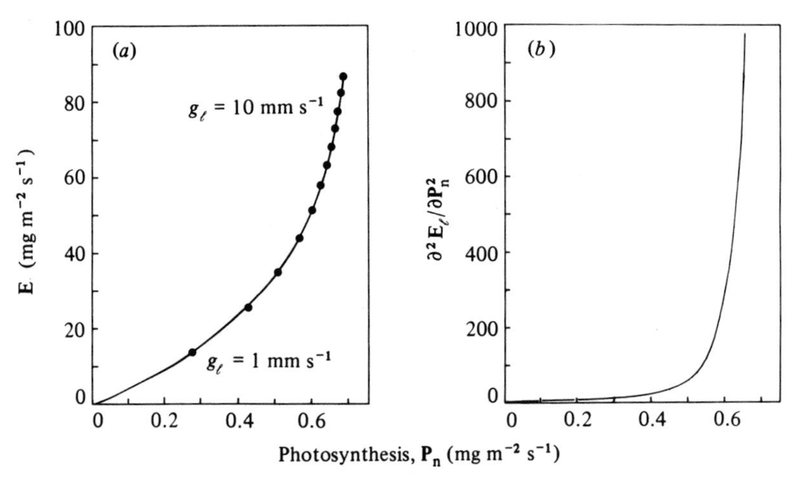

Fig. 10.8 (*a*) A typical relationship between $\mathbf{E}_\ell$ and $\mathbf{P}_n$ for a C_3 plant in a constant environment calculated using the model of Jones (1976). The relative increase in $\mathbf{E}_\ell$ (i.e. $\Delta\mathbf{E}_\ell/\mathbf{E}_\ell$) for any change in g_ℓ is always greater than the corresponding relative increase in $\mathbf{P}_n$ (i.e. $\Delta\mathbf{P}_n/\mathbf{P}_n$). (*b*) The second derivative $\partial^2\mathbf{E}_\ell/\partial\mathbf{P}_n^2$.

environment), as shown in Fig. 10.8. Mathematically this can be stated as: $\partial^2\mathbf{E}_\ell/\partial\mathbf{P}_n^2$ is everywhere > 0. For this particular case, the optimal stomatal behaviour is that which maintains

$$\frac{(\partial\mathbf{E}_\ell/g_\ell)}{(\partial\mathbf{P}_n/g_l)} = \frac{\partial\mathbf{E}_\ell}{\partial\mathbf{P}_n} = \lambda \qquad (10.11)$$

where λ is a constant depending on the average rate of assimilation required (or water available). If the amount of water available is small, λ is small. Using the terminology of economics, the optimal conductance is that which maintains the marginal cost of a change in conductance equal to the marginal *benefit*, so the ratio λ is small if water is 'expensive'. In practice this means that, as the environment changes during a day, the stomata adjust to keep the ratio of the sensitivities of evaporation and assimilation to infinitesimal changes in g_ℓ at a constant value. Putting this another way, it means that the stomata operate in such a way as to keep the assimilation rate at that position on a curve relating $\mathbf{P}_n$ and $\mathbf{E}_\ell$ where the slope is a constant value, irrespective of changes in the shape of this curve as the environmental conditions change.

Figure 10.9 shows for typical diurnal variation in climatic variables, daily trends in $\mathbf{E}_\ell$ and $\mathbf{P}_n$ for different constant values of λ. As the amount of available water, and therefore λ, decreases there is an increasing tendency

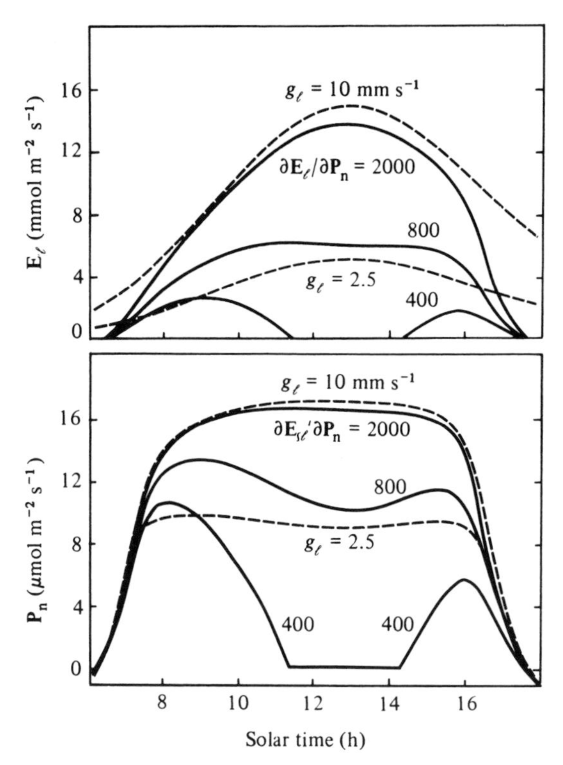

Fig. 10.9. Optimal time courses for transpiration ($\mathbf{E}_\ell$) and assimilation ($\mathbf{P}_n$) in a typical day for various magnitudes of $\partial\mathbf{E}_\ell/\partial\mathbf{P}_n$ (solid lines). These curves show that midday stomatal closure is an optimal behaviour when water is limiting and $\partial\mathbf{E}_\ell/\partial\mathbf{P}_n$ is small. Also shown are $\mathbf{E}_\ell$ and $\mathbf{P}_n$ for various constant values of leaf conductance (g_ℓ) (dashed lines) (after Cowan & Farquhar 1977).

towards midday stomatal closure. Midday stomatal closure is, in fact, a well-known phenomenon in water-stressed plants, and this analysis suggests that this behaviour may have evolved as a way of maximising WUE in drought conditions. Although the behaviour that maintains λ constant gives a higher value of WUE over a day than any constant stomatal conductance, surprisingly, perhaps, the difference is not large particularly at high rates of water use.

The conclusion that $\partial\mathbf{E}_\ell/\partial\mathbf{P}_n$ must remain constant over time for the optimal average WUE of a single leaf can be extended to a complete canopy, where it can be shown that optimal WUE is obtained if the stomata operate so that $\partial\mathbf{E}_\ell/\partial\mathbf{P}_n$ is constant over all the leaves in the canopy. Evidence that $\partial\mathbf{E}_\ell/\partial\mathbf{P}_n$ is in fact maintained approximately constant, as required by this hypothesis, has been provided for plants in controlled environments (Farquhar *et al.* 1980*a*) and in the field (Field *et al.* 1982).

The strategy of maintaining $\partial \mathbf{E}_\ell / \partial \mathbf{P}_n$ constant can be shown to be not the most efficient, but the most inefficient stomatal behaviour if $\partial^2 \mathbf{E}_\ell / \partial \mathbf{P}_n^2 < 0$. This relationship between $\mathbf{P}_n$ and $\mathbf{E}_\ell$ may occur in C_4 plants (particularly when r_a is large as occurs with large leaves or low windspeeds) but is rare for C_3 plants. In this case the most efficient behaviour is one where the stomata are either fully closed or fully open, with the period open being appropriate to use all the available water.

Optimal patterns of water use in an unpredictable environment

Even for a plant growing in a particular location with a known climate, it is difficult to define the most efficient pattern for water use during the season. This is because the various weather components (rainfall, temperature, etc.) usually vary from year to year both in amount and in the way in which they are distributed through the season. Although it is possible, at least in principle, to use techniques such as those outlined above to determine the best patterns of stomatal behaviour or leaf area development if the exact patterns of rainfall and other weather factors are known for the whole season, this information is never available to the plant (or to the farmer) at the beginning of the season. Therefore any optimal pattern for water use during the season must take account of the probability of future rainfall. Cowan (1982, 1986) has extended his earlier model for stomatal behaviour to take account of future rainfall probabilities and concluded that one would expect the stomatal conductance (and hence the assimilation rate) to decrease with increasing soil water deficit because this behaviour reduces the probability of plants completely drying the soil profile and then dying. Another related conclusion is that the optimal stomatal conductance declines during the season for an annual crop, even if the environment remains constant. The reasons for these results can be understood by means of the following, rather simplistic, analysis of the consequences of uncertainty in the pattern of future rainfall.

The pattern of water use that is adopted by any plant is inevitably a compromise that depends on the climate. This is because a plant that is over 'optimistic', for example, and produces a large transpiring leaf area on the assumption that more rain will fall, may not be able to produce any seed at all in the odd dry year. On the other hand, a relatively 'pessimistic' plant that ensures the production of some seed, even though no extra rain falls (e.g. by initiating reproductive growth early), may not be able to respond adequately in the wetter than average year. Clearly the more variable the climate, the more difficulty there will be in finding a type of behaviour that

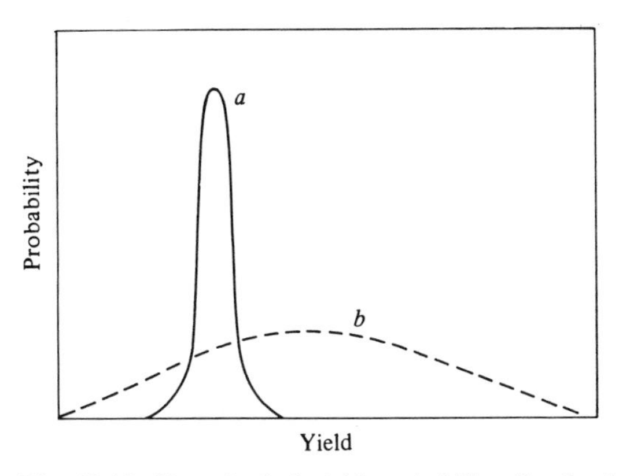

Fig. 10.10. Hypothetical yield probability distributions over many years at one site for two contrasting crop varieties: (*a*) stable but low yield; (*b*) more variable variety with higher mean yield.

maintains the productivity close to the retrospectively determined 'ideal' behaviour in most years.

Desert ephemerals tend to be relatively pessimistic, while perennials can be more optimistic since they have reserves to survive the odd bad year. In an agricultural situation, different farmers may have differing definitions of their ideal crops. A peasant farmer, for example, may be most concerned to avoid starvation in any year and might therefore prefer a crop variety that reliably produces some yield every year (e.g. *a* in Fig. 10.10), even though it has a low yield potential in the good years. A farmer with greater resources, on the other hand, may prefer a cultivar that has a greater long-term average yield, even though yield may be very low in dry years (*b* in Fig. 10.10).

The effect of climate and climatic variability on the productivity of plants with different types of water use behaviour (optimistic, pessimistic, and responsive or non-responsive to changes in soil moisture – Fig. 10.11) has been investigated by the use of a Monte Carlo modelling technique (Jones 1981*c*). Figure 10.12 illustrates the effects of these stomatal behaviours on the probability distributions of total assimilation over a period for a particular climate with moderately variable rainfall. It is apparent from Fig. 10.12 that, although relatively pessimistic and optimistic behaviours may produce the same average yield over many years, the more pessimistic behaviour is likely to lead to very low yields much less frequently.

Although desert ephemerals may be regarded as pessimistic in terms of their overall water use pattern, as they have a very short life cycle whatever the water supply (and hence a low potential yield), their stomatal behaviour

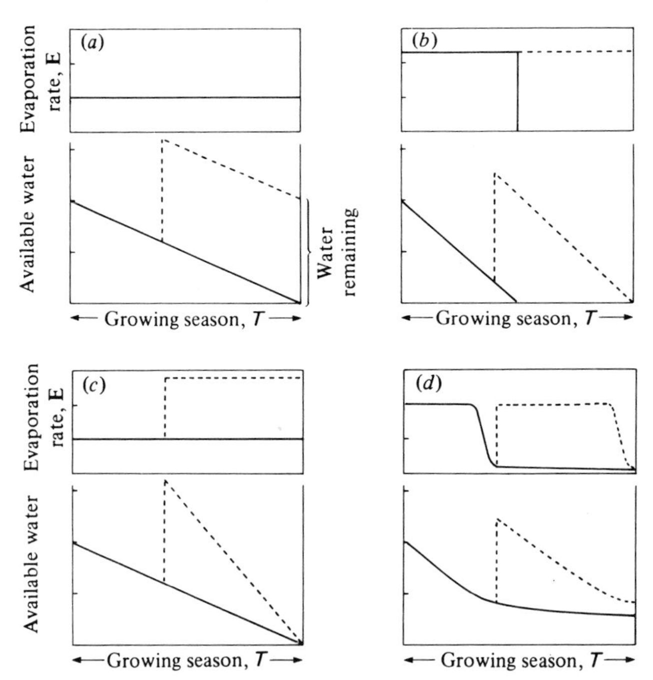

Fig. 10.11. Four classes of stomatal behaviour and their effects on water use. The solid lines represent the time course of evaporation rate (upper half of each figure) or water remaining in the soil (lower half) if no rain falls. The dotted lines illustrate the consequences of rain part-way through the season.

(*a*) Stomata fixed to use initial soil water by end of season (pessimistic, non-responsive); (*b*) stomata set to use water faster than justified on initial water in soil (optimistic non-responsive); (*c*) stomata respond to keep rate of water use at rate required to use all currently available water by the end of the season (pessimistic, responsive); (*d*) stomata initially optimistic, but with ability to close preventing complete desiccation (after Jones 1980).

may be relatively optimistic. This illustrates that the optimal stomatal behaviour depends on all aspects of the environment and of the plant life cycle.

Figure 10.12 also shows the effects of one type of stomatal behaviour where the plants respond to the amount of water available. Interestingly, such responsive behaviour is very little better, on average, than the best 'constant' behaviour shown in Fig. 10.12*a*, but it is more adaptable to a wide range of climates. Stomatal closure in response to increasing water deficit (Fig. 10.11*d*) as found in most plants, provides a good general compromise. It is an optimistic behaviour that permits achievement of high productivity in a wet year, while preventing serious damage in short

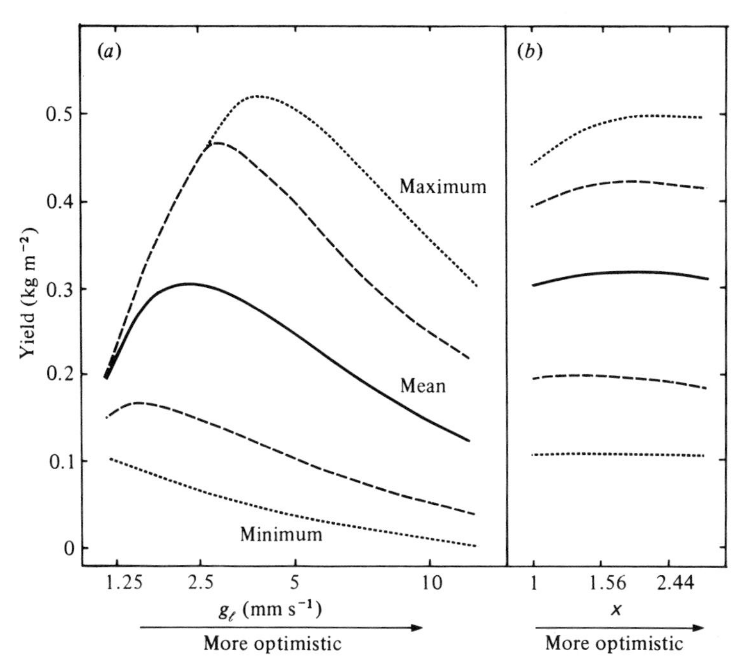

Fig. 10.12. Examples of how the yield probability distribution for a particular climate changes depending on whether stomatal conductance is optimistic or pessimistic. The mean yield over a large number of simulations (500) is given by the solid line, the dotted lines give the highest and lowest yields observed, and the dashed lines give the 5% tails. (*a*) The yield distribution with a range of fixed stomatal apertures; (*b*) the results if the stomata can respond to the amount of water available in such a way that the stomatal aperture is set at x times the value required to finish all currently available water by the end of the season (after Jones 1981*c*).

drought periods. Only in a long drought would a more conservative or pessimistic behaviour be better, and then only for annual species.

Optimisation of stomatal behaviour is only one of the plant responses involved in maximising assimilation with limited water availability. In a particularly interesting extension of Cowan's analysis to include the optimal allocation of carbon to the synthesis of leaves and roots, Givnish (1986) demonstrated that the following summary relationship is optimal:

$$r'_m/r'_\ell = f/(1-f) \tag{10.12}$$

where f is the proportion of carbon allocated to leaves as compared with roots. This remarkable relationship suggests that the ratio of r'_m (important for determining assimilation) to r'_ℓ (important for determining water loss)

Table 10.4. *Permeabilities of various materials to water vapour, carbon dioxide and oxygen at 25 °C unless stated otherwise (data from Brandrup & Immergut* 1975)

Material	Permeability ratio (H_2O/CO_2)	Permeability ($\times 10^{14}$) ($cm^2\ s^{-1}\ Pa^{-1}$)		
		H_2O	CO_2	O_2
Polychloroprene (Neoprene G)	3.5	6825	194	30
Polypropylene	5.8	383	66.2[a]	17.3[a]
Polyethylene (density 0.914)	7.1	675	94.5	21.6
Polystyrene	11.4	9000	78.8	19.7
Gutta Percha	14.4	3825	266	46.2
Natural rubber	14.9	17180	1148	175
Polyvinylidene chloride (Saran)	16.3	3.75	0.23[a]	0.040[a]
Butyl rubber	21.3	825[b]	38.7	9.75
Polyvinyl chloride	1750	2060	1.18	0.34
Nylon 6	1770	1330	0.75[a]	0.29[a]
Cellulose (Cellophane)	404000	14300	0.035	0.016

[a] = 30 °C; [b] = 37.5 °C.

should be equal to the ratio between the allocation of carbon for synthesis of new leaves and the allocation to new roots.

Antitranspirants

As mentioned above (e.g. equation 10.4), there is a tendency for WUE to improve as stomata close. This has provided the impetus for many attempts to improve crop WUE by the application of antitranspirants (see reviews by Gale & Hagan 1966; Solárová *et al.* 1981). The main types of antitranspirant are:

(*a*) *Compounds that close stomata*, such as ABA, phenyl mercuric acetate and decenylsuccinic acid. Unfortunately these compounds are often ineffective, or too expensive, or else toxic. Even when they do close stomata, improvements in WUE are often not observed in the field (see above).

(*b*) *Film-forming compounds*, such as silicone emulsions or plastic films, are also rarely very effective or long lasting. Unfortunately these compounds are less permeable to CO_2 than to H_2O (see Table 10.4), so application tends to reduce WUE. Equally important perhaps is that they are all even less permeable to O_2. The fact that no 'ideal' antitranspirant materials are known that are more permeable to CO_2 than to H_2O is hardly surprising in view of the greater molecular weight of CO_2 and the similar polar nature of the two molecules.

(*c*) *Reflecting materials.* Increasing leaf reflectivity by application of materials such as kaolinite can help to lower leaf temperature and hence decrease $\mathbf{E}_\ell$. Although reflection in the PAR is increased more than in the infra-red, at high irradiances (above those saturating $\mathbf{P}_n$), this may not be too much of a disadvantage, so that overall improvements in WUE can be obtained.

Although antitranspirants have proved to be of value for minimising water loss and improving plant survival, especially of high value perennials during transplanting, in only a very few cases have they proved beneficial for improving WUE. The lack of success in field crops is not surprising when one considers the degree of coupling of the crop to the atmosphere (Chapter 5): although a single plant is closely coupled to the environment, and **E** (and hence WUE) is determined by stomatal conductance, as the area of crop grown increases, the effective boundary layer conductance decreases so that **E** becomes increasingly decoupled and hence insensitive to antitranspirants.

Why is WUE so conservative?

Although the instantaneous ratio $\mathbf{P}_n/\mathbf{E}_\ell$ can vary over a wide range, it remains to be explained why WUE over a whole season (after allowance is made for the humidity deficit) is so remarkably constant for one species (e.g. Fig. 10.4). The major factor is almost certainly that the long-term mean is primarily determined when the stomata are fully open, because transpiration and photosynthesis when the stomata are closed, or nearly so, are small proportions of the total. Cowan & Farquhar (1977) have pointed out that, particularly when stomata are fairly wide open, the advantage for WUE of the optimal pattern of stomatal behaviour over that obtained by a constant aperture throughout the day can be relatively small (of the order of 10%). The other factor is the tendency of p'_i to be relatively stable in the long term so that equation 10.7 is nearly true (i.e. r'_ℓ adjusts as r'_m changes maintaining $(r_a+r_\ell)/(r'_a+r'_\ell+r'_m)$ more nearly constant).

Crop water stress index

There has been much interest in developing short-cut methods for assessing the degree of 'water stress' to which a crop is subject. Methods for measuring water status were outlined in Chapter 4, where it was pointed out there that measures based on the *response* of the plants to the stress can be useful. Stomatal closure, as one of the most sensitive plant responses to

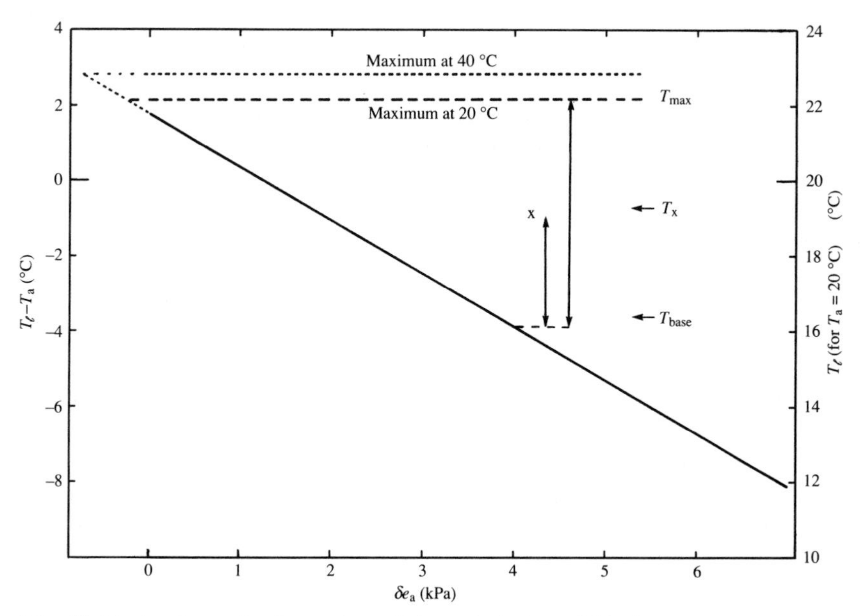

Fig. 10.13. A schematic illustration of the calculation of Idso's crop water stress index, showing the decline in $(T_\ell - T_a)$ with increasing δe_a for well-watered crops (the 'non-water-stressed baseline'), the upper limit of $(T_\ell - T_a)$ for two temperatures, and the calculation of the index for a sample point x (see text).

water deficits, has been suggested as one such indirect measure of stress, particularly because variation in g_ℓ can readily be detected by remote sensing techniques as variation in canopy temperature. Unfortunately it rapidly became apparent that the leaf–air temperature differential $(T_\ell - T_a)$ was a function, not only of stomatal aperture (and hence crop water status), but also of environmental conditions, especially air vapour pressure deficit (δe_a). Idso *et al.* (1981) and Jackson *et al.* (1981) proposed a 'crop water stress index' that normalised the value of $(T_\ell - T_a)$ for environment. In this approach it is assumed (see Fig. 10.13) that a plot of $(T_\ell - T_a)$ against δe_a for well-watered crops is linear with $(T_\ell - T_a)$ decreasing solely as the humidity deficit of the air increases; this line is termed a 'non-water-stressed baseline'. The value of $(T_\ell - T_a)$ for stressed crops (with the stomata only partly closed) at any δe_a is intermediate between the non-water-stressed baseline value and a potential maximum (assumed constant at all δe_a) obtained when stomata are completely closed (see sample point x in Fig. 10.13). An index in the range 0–1 is then calculated as $(T_x - T_{base})/(T_{max} - T_{base})$ as illustrated, with larger values being indicative of increasing water deficits.

The theoretical basis of the approach (see also Jackson 1982) and its limitations are readily apparent from Chapter 5 and especially from the analysis leading up to equation 5.35. In particular it is easy to show that rather than directly estimating stomatal conductance, the water stress index is better related to crop transpiration, and is really a measure of $(\mathbf{E}_p - \mathbf{E})/\mathbf{E}_p = (1 - \mathbf{E}/\mathbf{E}_p)$ where $\mathbf{E}_p$ is the potential transpiration from the well-watered crop.

Although this approach appears to work well in some hot dry climates (such as in Arizona, USA) where it was developed, there have been difficulties in its application more generally. There are a number of reasons for this, including the fact that in addition to the dependence on δe_a and g_ℓ, there is a strong dependence of $(T_\ell - T_a)$ on radiation and an effect of varying windspeed and crop surface roughness (see equations 9.5 and 9.6). Although differences between crops and even between growth stages can be eliminated by the use of empirically determined non-water-stressed baselines, other factors such as plot size and environmental coupling can be important. In applying the index it is important to recognise that, in spite of its name, it is not a measure of stress, rather it is a measure of crop response.

11

Wind, altitude, carbon dioxide and atmospheric pollutants

This chapter considers a number of related aspects of the aerial environment that have not been adequately treated elsewhere in this book – these include wind and the effects of altitude, the 'Greenhouse effect' and the implications for plant growth of climate change and increased atmospheric CO_2 concentration, and the effects of atmospheric pollutants. All these areas bring together principles that have been introduced earlier. Further details of these and other features of the microenvironment of plants are discussed by Geiger (1965), Grace *et al.* (1981), Oke (1987) and Monteith & Unsworth (1990).

Wind

Not only is wind directly involved in heat and mass transfer by forced convection (see Chapter 3), but it is important to plants in many other ways including the dispersal of pollen and seeds and other propagules and in shaping vegetation, either directly or, particularly at coastal sites, by means of transported sand or salt (Grace 1977; Nobel 1982).

Measurement and variability

Wind is very variable both in direction and velocity. In general windspeeds tend to be greater during the day than at night (Fig. 11.1), largely as a result of the convection processes set up by solar heating of the earth's surface during the day. Standard meteorological estimates of windspeed are obtained at 10 m above the surface where windspeeds only rarely fall below 1 m s^{-1}. For instance, mean annual windspeed is greater than 4.5 m s^{-1} over most of Britain with the highest values occurring on mountains and near coasts. However, as seen in Chapter 3, the windspeed decreases rapidly as one approaches a surface (e.g. Fig. 3.7, p. 67), so that windspeeds near and within vegetation tend to be much slower than at 10 m. At agro-

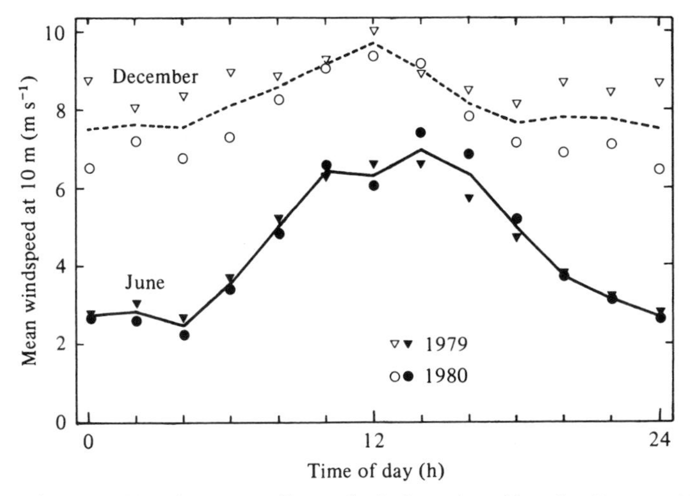

Fig. 11.1. Hourly mean values of windspeed at 10 m for June and December for two years at East Malling, Kent. The mean standard deviation of hourly values is 2.8 m s^{-1} for June and 4.6 m s^{-1} for December.

meteorological sites windspeeds are commonly recorded at 2 m above the ground.

The various methods available for measuring wind are described by Grace (1977) and Fritschen & Gay (1979). The rotating-cup anemometer is widely used, often in conjunction with a directional vane. Sensitive types can be used within crop canopies to measure windspeeds as low as 0.15 m s^{-1}. For studies of smaller-scale, more rapid velocity fluctuations as occur in eddies, hot-wire anemometers are commonly used (Fig. 5.8). Other types of anemometer include sonic anemometers and Pitot pressure tubes.

Although mean windspeeds are commonly measured, some processes such as damage are more dependent on maximum gust speeds than on average speeds.

Wind and evaporation

Increases in windspeed decrease the boundary layer resistance (see Chapter 3); this generally causes evaporation rate to increase. If, however, leaf temperature is significantly above air temperature (for example, in conditions of high irradiance with moderate stomatal closure: see Fig. 9.1, p. 233), increasing windspeed can decrease evaporation (Fig. 11.2). This is because the increased heat loss lowers leaf temperature and therefore the water vapour pressure in the leaf. The resulting reduction in driving force for evaporation can be greater than the reduction in the transfer resistance.

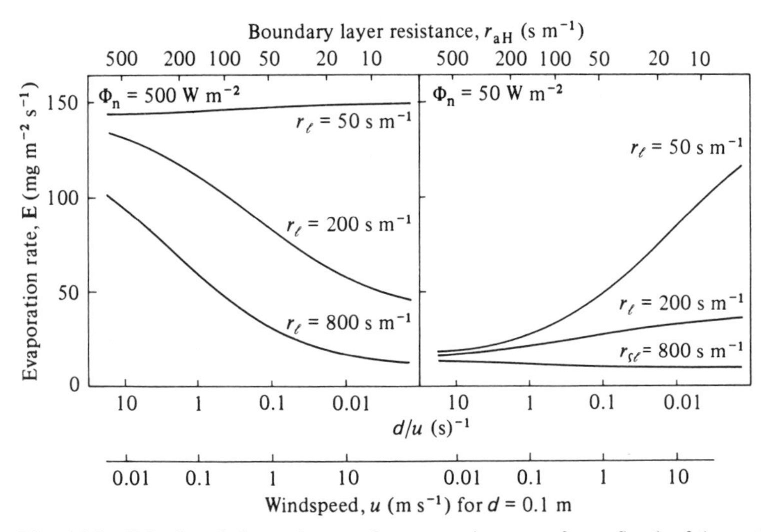

Fig. 11.2. Calculated dependence of evaporation rate for a flat leaf (equation 5.23 with $\delta e = 1$ kPa) on the boundary layer resistance (r_{aH}) or the parameter d/u (see equation 3.31) for various r_ℓ and for high or low net radiation ($\mathbf{\Phi}_n$). The equivalent windspeeds for a leaf with characteristic dimension $d = 0.1$ m are also shown.

This effect follows from the evaporation equation (equation 5.23, p. 114) where the boundary layer resistance occurs in both numerator and denominator.

It can be shown by differentiation of equation 5.23 with respect to r_a, that **E** is independent of r_a (and hence windspeed) when

$$r_\ell = \frac{\rho_a c_p \delta e[(s/\gamma)+1]}{s(\mathbf{\Phi}_n - \mathbf{G})} \tag{11.1}$$

If r_ℓ is less than the value given by this equation, **E** increases with increasing windspeed, but when r_ℓ exceeds this critical value, **E** decreases with increasing windspeed. For a free water surface, where the 'surface resistance' term is zero, **E** always increases with windspeed (u), roughly in proportion to $\sqrt{u}$. The effect of boundary layer resistance on evaporation rate in a range of conditions is illustrated in Fig. 11.2.

High windspeeds can also affect evaporation by indirect effects on either the stomatal or cuticular resistances. There are conflicting reports on the effect of wind on the stomata (see Grace 1977). In many cases it is likely that any closure in response to increased windspeed is simply a feedback or feedforward response (see Chapter 6) to increased evaporation rate. Species tolerant of more exposed habitats may have stomata that are less responsive to windspeed, though it is very difficult to generalise. For example, both *Rhododendron ferrugineum* and *Pinus cembra* occur above the treeline in the European Alps, but the former is found mainly in more sheltered microsites. Using potted plants in a wind tunnel, it was apparent that *Rhododendron* stomata were the more sensitive to raised windspeed (Caldwell 1970). However, no such differential sensitivity could be detected in the field (A. Cernusca, unpublished data). It may be that the wind tunnel results were atypical for a number of reasons, including the difficulty of growing the plants in pots, problems of natural endogenous rhythms obscuring responses, and differences in leaf size with consequent differences in r_a (Ch. Körner, personal communication). Cuticular resistances, on the other hand, may be decreased by wind. This probably results from damage to the cuticle as a result of leaves flexing in the wind and rubbing against each other.

Dwarfing, deformation and mechanical damage

Plants developing in windy environments show a range of characteristics from stunted growth and deformation to actual breakage. At sites with a predominant prevailing wind, plants often show a marked asymmetry (see e.g. Daubenmire 1974), growth being very much reduced to windward. There can be several reasons for this including 'wind-training', where the branches are oriented downwind as a result of constant wind pressure. The reduced growth on the windward side can be attributed to both physiological and mechanical effects. These include, for example, enhanced desiccation and a more extreme temperature regime for exposed tissues (particularly in alpine environments), and direct damage or damage by materials such as salt or sand transported by the wind. All of these may preferentially kill buds on the windward side. These effects also lead to a marked tendency for the whole plant stature to be smaller in more exposed situations. Ecological advantages of dwarf plants in cold environments have already been discussed in Chapter 9. For example, the reduced windspeeds in the vicinity of cushion plants enable the tissues to be much warmer than the air in high irradiance conditions.

Even quite moderate wind velocities can affect growth. Often a major factor is the impaired plant water status that results from the increased evaporation rate frequently associated with increased windspeed (see above). Where stomata are closed, cuticular damage resulting from high windspeeds can also lead to increased water loss. In many cases, however, no decline of Ψ_ℓ is observed at elevated windspeeds. In alpine environments the decreased leaf temperature, and hence decreased metabolic rate that results from the more efficient sensible heat transfer at high windspeeds, can be more important than any effects on water status. In arid environments, however, desiccation effects tend to be dominant.

As an alternative it has been suggested that the shaking effect of wind can be an important factor reducing growth. For example, Neel & Harris (1971) demonstrated that growth of *Liquidambar* could be reduced to 30% of normal when the plants were 'shaken' for only 30 s each day; while less extreme results have been reported for many other species. The detailed mechanism for this effect is not yet known, though it does not appear to be consistently related to impaired water status (as might be expected, for instance, if shaking induced cavitation in the xylem vessels). In the grass *Festuca arundinacea* shaking even increased stomatal conductance (Grace *et al.* 1982).

It is now well accepted that mild mechanical stimulation can have a major effect on plant development. This 'thigmomorphogenesis' may be an important means of adaptation to windy or other mechanically damaging environments. Mechanical stimulation tends to produce shorter and sturdier plants. A well-known example of the effects of wind and of the consequent stem bending is the production of flexure- or compression-wood in trees. Mechanical stimulation can also enhance drought tolerance, probably through the stimulation of ABA production (Biddington & Dearman 1985). At least four touch-sensitive genes have been isolated in *Arabidopsis* (Braam & Davis, 1990), whose expression is enhanced within 10 min of stimulation: the fact that many of these genes are calmodulin-related or encode calmodulin (a calcium binding protein) suggests that Ca^{++} is involved at an early stage in thigmomorphogenetic responses. The implication of Ca^{++} in mechanical stress responses is consistent with observations that mechanical injury can give rise to action potentials (Pickard 1971) and other electrical changes. There is a close relationship between the responses to touch and to ethylene: the growth responses are similar, and touch results in ethylene production (see Jaffe *et al.* 1985). The fact that ethylene stimulates only some of the touch sensitive genes suggests that this cannot be a primary sensor in the touch response.

Lodging

An important problem in high winds is that the forces exerted on the plants can lead to structural failure, either the uprooting of whole plants (particularly common with trees) or else the breaking or buckling of stems. This process of plants being laid flat by the wind is called lodging (or windthrow in trees). In addition to the obvious importance in forestry, lodging can significantly reduce yield in cereals, where the commonest form of lodging is buckling of the stem. The lodging of cereals can decrease the harvestable yield either by effects on photosynthesis (e.g. as a result of either poor light penetration in the compressed canopy or stress caused by the damaged conducting system) or else by making the grain difficult to harvest.

The occurrence of lodging depends on the forces exerted on the plant by wind, rain, etc., on the height from the ground at which they act, and on the strength of the stem (see Pinthus 1973; Grace 1977). In a cereal, for example, the force due to the wind acts primarily on the head of the plant and induces a torque (**T**) or turning moment, that increases down the stem and causes bending. The torque at a point is the product of the force × the perpendicular distance of its line of action from that point (see Fig. 11.3), so it has units of N m (or J). Thus the total torque at the base of a stem is

$$\mathbf{T} = \Sigma(\mathbf{F}_i h_i) \tag{11.2}$$

where h_i is the height of the i-th part of the plant, and $\mathbf{F}_i$ is the horizontal force due to the wind acting on that part. Once a significant bending occurs, there is an additional turning moment due to gravity $= \Sigma(m_i x_i \boldsymbol{g})$, where x_i is the displacement from the vertical of a part of mass m_i and $\boldsymbol{g}$ is the acceleration due to gravity. Therefore the total torque is

$$\mathbf{T} = \Sigma(\mathbf{F}_i h_i + m_i x_i \boldsymbol{g}) \tag{11.3}$$

The torque induces stresses (S) that cause deformation (strain), in this case stem bending, but this is resisted by the bending-resistance moment of the stem. The maximum bending-resistance moment is called the stem strength. The magnitude of the stress for a cylindrical stem is related to the torque by

$$S = \mathbf{T}d/2I \tag{11.4}$$

where d is the stem diameter, and I is the moment of inertia. I depends on the shape of the stem cross-section, being equal to $\pi d^4/64$ for a solid cylinder and $\pi(d^4 - d_i^4)/64$ for a hollow cylinder with internal diameter d_i. The maximum stress that can be withstood without irreversible deformation occurring is called the elastic limit.

The amount of bending is proportional to S (and hence **T**) and inversely proportional to the flexural rigidity or stem stiffness, given by the product $\varepsilon_Y I$, where ε_Y is a measure of linear elasticity of the stem material called Young's modulus. A large value of $\varepsilon_Y I$ represents a stiff stem. A very stiff stem, as in trees, will transfer the torque operating on it to the root system and may thus promote root failure.

The actual force exerted by the wind (**F**) results largely from form drag (cf. skin friction – see Chapter 3) and is given by

$$\mathbf{F} = c_D \tfrac{1}{2} \rho u^2 A \tag{11.5}$$

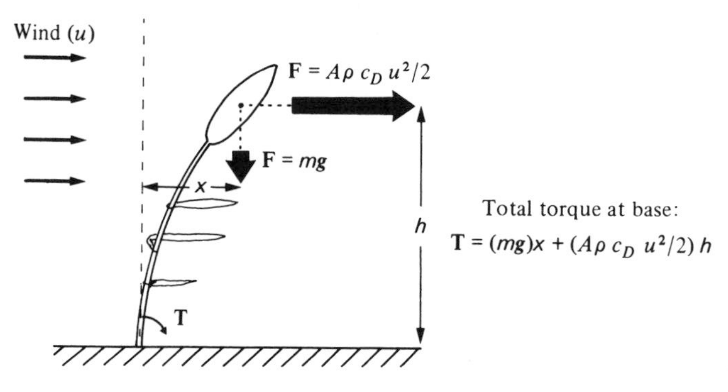

Fig. 11.3. Illustration of forces acting on a cereal ear in the wind.

where c_D is a dimensionless drag coefficient dependent on u, ρ is the air density and A is the area projected in the direction of the wind. 'Streamlining' minimises turbulence and form drag in a flowing fluid (Fig. 11.4), while a 'bluff' body such as a cube is subject to greater form drag as a result of the turbulence and lowered pressure that occurs in the object's wake. For a cylinder, form drag contributes 57% of the total at a Reynolds number (p. 59) of 10 (e.g. for $d = 0.001$ m, $u = 0.15$ m s^{-1}) increasing to 97% at a Reynolds number of 10000 (e.g. $d = 0.1$ m, $u = 1.5$ m s^{-1}). The drag on a flat leaf oriented parallel to the flow is small and largely a result of skin friction, but when normal to the flow, drag is much greater and form drag predominates. For leaves c_D is typically between 0.03 and 0.6.

For many plants increasing windspeed forces the foliage to align with the streamlines so that the value of the drag coefficient, c_D, decreases in a way that is dependent on the rigidity of the foliage. Mayhead (1973) has presented data on drag coefficients for a number of British forest trees: at windspeeds that are liable to cause windthrow (above about 30 m s^{-1}) c_D for a variety of conifers ranged between 0.14 for the very supple Western hemlock (*Tsuga heterophylla*) to 0.36 for the much more rigid Grand fir (*Abies grandis*). At a windspeed of 10 m s^{-1} the corresponding values of c_D for the two species were above 0.3 and 0.8 respectively.

If the airstream contains denser particles such as raindrops, these can increase the force exerted on the object by impacting on the plant, partly because the denser particles have greater inertia and consequently less tendency to be diverted in the streamlines round the plant (Fig. 11.4). It is possible to calculate the relative forces exerted by raindrops and wind on, for example, a cereal ear. The force due to wind is given by equation 11.5: assuming a wind velocity of 5 m s^{-1} at 2 m, the average velocity at the level of the ears might be $\simeq 1.5$ m s^{-1} (equation 3.36, p. 67) and assuming

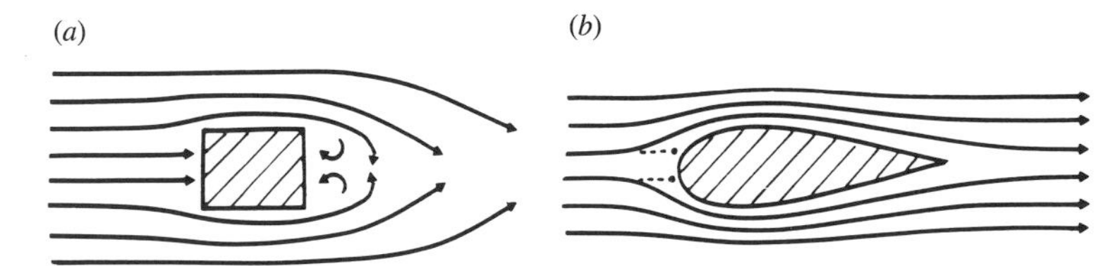

Fig. 11.4. Flow around (*a*) a bluff body (*b*) a streamlined body. The dotted lines show the trajectories of dense objects impacting on the streamlined body.

$c_D = 0.4$, this gives, for an ear of 0.01 m^2 cross-sectional area, a continuous force $\mathbf{F} = 0.4 \times 1.2 \times (1.5)^2 \times 0.01/2 = 5.4 \times 10^{-3}$ N. For a large enough object, the separate impulses of individual raindrops are equivalent to a steady force given by the mass of water impacting each second × velocity. Assuming that the mean horizontal velocity of the drops is 5 m s^{-1} this gives, for a low rate of precipitation of 1 mm h^{-1} (= 0.28 g m^{-2} s^{-1}), $\mathbf{F} = (\text{mass m}^{-2}\text{ s}^{-1}) \times u \times A = 0.28 \times 10^{-3} \times 5 \times 0.01$ N $= 1.4 \times 10^{-5}$ N. Even for a high rate of precipitation of 25 mm h^{-1}, $\mathbf{F} = 3.5 \times 10^{-4}$ N, which is more than an order of magnitude smaller than the force of the wind.

The tendency to lodge will also be enhanced if the frequency of turbulence in the wind corresponds to a natural resonant frequency of the plant, or if the stem is weakened by disease or moisture absorption. Such dynamic responses have been shown to be particularly important for storm damage in forests. Work on the dynamic behaviour of trees has been reviewed by Mayer (1987).

A reduction in the tendency to lodge has been an objective for some years in cereal breeding programmes. The turning moment may be reduced by decreasing the size of the ear, by improving its streamlining or by decreasing its height. The latter possibility has provided the main impetus for breeding the dwarf cereals which are increasingly grown throughout the world. The original problem was that the increased levels of nitrogen fertiliser that were applied to increase yield also increased stem height and therefore the tendency to lodge. Reduction of stem height conferred an improved tolerance of higher levels of nitrogen.

The alternative approach to reduction of lodging might be to increase the stem strength, for example by increasing I either by increasing the material in the stem or preferably by increasing the stem diameter (as hollow stems have a higher I for any given mass of material). Increases in the elastic limit would also be advantageous.

Shelter

Shelter belts have been used since prehistoric times to reduce wind damage to crops and livestock. It has been known empirically for many years that erection of windbreaks, whether natural belts of trees or artificial screens of slatted wood or plastic mesh, can have a very important mollifying effect in windy environments, resulting in greatly improved plant growth and yield, particularly where drought or exposure can be problems. As well as having direct effects on crop performance, shelter belts can also improve soil water distribution by collecting snow during the winter, so that the melt water is more evenly distributed than would otherwise be the case.

Micrometeorological effects of shelter. A recent review of some micrometeorological aspects of shelter may be found in McNaughton (1989). Although the detailed effects of a windbreak depend on its orientation and on the turbulence characteristics of the wind, typical airflow patterns behind a thin solid barrier such as a wall can be sketched as in Fig. 11.5*a*. In this case there is a large recirculating eddy downwind of the barrier. As the porosity of the windbreak increases, however, the recirculating eddies retreat downwind, becoming smaller and more intermittent until they finally vanish when the porosity increases above about 30%. Characteristic patterns of variation in relative wind velocity with distance behind the barrier and with porosity are shown in Fig. 11.5*b*, demonstrating that the effect can extend downwind to about 25 × the height of the barrier.

A triangular region that extends between the barrier and a line to the ground at about $8h$ behind the barrier (where h is the height of the barrier) is termed the 'quiet zone'; this remains similar in shape as barrier porosity increases. In this region, not only is mean windspeed reduced but so is the size of the turbulent eddies. Behind and above this region lies a wake zone where turbulence is enhanced. Transport processes therefore tend to be depressed in the quiet zone (where the crop becomes relatively decoupled from the overhead airstream), and enhanced in the wake.

The reduced rates of turbulent transfer immediately behind shelter tend to lead to increased air and surface temperatures during the day, as incoming radiation may be less easily dissipated than in the open field. At night, shelter may have the opposite effect on temperature, partly by preventing the break-up of inversions near the ground. It is difficult to generalise concerning the effects of shelter on air humidity and canopy evaporation because of the complexities of the energy balance (see Chapter 5), but it is usually observed that the increased canopy boundary layer resistance in the quiet zone behind the shelter reduces the rate of removal of water vapour, thus leading to a build-up of humidity and a concomitant

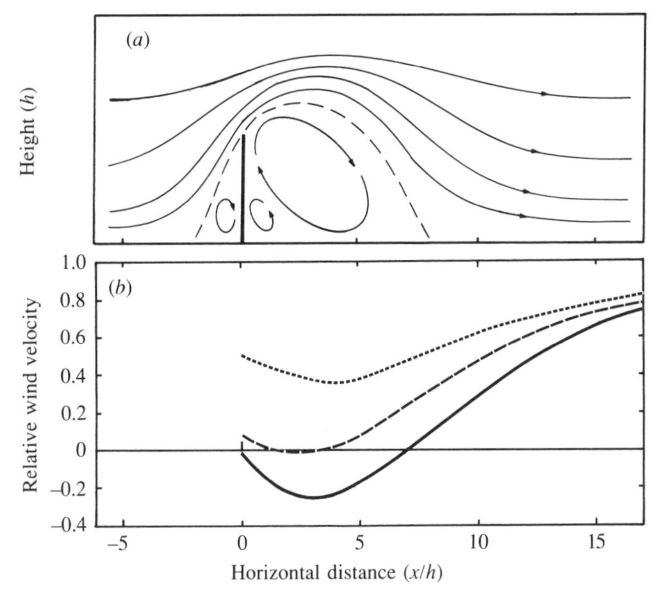

Fig. 11.5. (*a*) Typical windflow patterns caused by a windbreak of low porosity and height (h) $\simeq 100z_0$ standing normal to the windflow, showing the recirculating eddy immediately behind the barrier. The horizontal distance is represented as multiples of the barrier height and the broken line indicates the separation streamline. (*b*) Corresponding horizontal profiles of mean wind velocity at a height equal to $0.25h$ with barriers having porosities equal to zero (——), 0.3 (– – –) or 0.5 (············) (after McNaughton 1989).

decrease in crop evaporation. As a result of this reduced evaporation rate, leaf water potentials are commonly higher than in exposed sites.

Radiation is little affected by north–south oriented shelter belts, since any shading in the morning or evening is partially compensated for by increased reflection off the shelter belt. An east–west oriented shelter belt, however, tends to reduce net radiation receipt to its north (in the northern hemisphere), the distance of influence depending on height as well as on solar elevation (itself a function of latitude and time of year).

Effects of shelter on plants. In addition to shading effects, shelter belts can be harmful in that they use significant water and nutrients. For this reason, species like alders, that have nitrogen fixing capability, are often planted as shelter species. Another problem is that shelter belts may harbour both pests (including birds as well as insects) and diseases. In general, however, crop yields are usually markedly improved by shelter. Grace (1977) reviewed more than 95 experiments conducted at many sites with a wide range of species. In these experiments the average yield benefit of shelter was 23%, with fewer than a quarter of the cases giving less than 10% benefit.

The improved yields usually result from raised temperatures or from better water status that follows from decreased potential evaporation. This can lead to higher stomatal conductances and more rapid photosynthesis. Another beneficial effect of shelter is protection of soils from wind erosion. Further discussion of shelter effects may be found in texts by Grace (1977) and Rosenberg *et al.* (1983).

Dispersal of pollen, seeds and other propagules

Many species have evolved to take advantage of the wind to aid cross-fertilisation and dispersal (Daubenmire 1974). Anemophily (wind pollination) is widespread (e.g. in the Gramineae and in conifers), particularly among species of cool and cold climates. Pollination efficiencies of different types of floral structure in relation to patterns of windflow can be studied by means of scale models in a wind tunnel. In using scale models, it is necessary to adjust the windspeed to maintain the value of the Reynolds number ($= ud/\nu$ – Chapter 3) equal to that appropriate for the original floral structure in its natural environment (i.e. windflow must decrease as dimensions increase).

There are several types of wind dispersal (anemochory). These include having minute disseminules that can be easily blown long distances (e.g. the minute seeds of the Orchidaceae, as well as the spores of fungi and bryophytes). Larger seeds may have hairy 'parachutes' (as in the Compositae) or 'wings' (as in many tree species). In addition, some species (e.g. Chenopodiaceae) have papery fruits that readily roll along the ground, while others, such as tumble weed, have vegetative propagules that roll in the wind.

Altitude

Altitudinal variation of climate causes many visible changes in the composition of vegetation and in the growth habit of individual plants. Plants growing at high elevations often exhibit many characteristic morphological and physiological features including dwarf, compact habit and small, narrow or densely pubescent leaves (see e.g. Bliss 1962; Daubenmire 1974; Larcher 1983). In addition to altitude, local topography plays a major role in determining the microclimate at any site in the mountains. For example aspect and slope affect the shortwave irradiance (and hence soil temperatures – see e.g. Rorison 1981), while topography

Table 11.1 *Estimated average values of different climatic factors at* 600 *m and* 2600 *m above sea level in the Central Alps (from data compiled by Körner & Mayr* 1981). *Summer refers to June, July and August*

Height above sea level		600 m	2600 m
Atmospheric pressure	(10^4 Pa)	9.46	7.40
Average windspeed (summer)	(m s^{-1})	1	4
Average water vapour pressure (summer)	(10^2 Pa)	14.7	6.9
Maximum water vapour pressure deficit (July)	(10^2 Pa)	~ 20	~ 8
Annual number of days with fog		0–10	80
Annual sum of precipitation (±30%)	(mm)	900	1800
Average air temperature (July)	(°C)	18	5
Annual average air temperature	(°C)	8	−3
Global radiation on clear days (summer)	(%)	100	120
Global radiation on overcast days (summer)	(%)	100	260
Number of clear days (summer)		10	5
Number of sunshine hours (July)		200	160
Evaporation from low vegetation (July)[a]	(mm day^{-1})	4–6	3–4
Number of days with snow cover		80	280

[a] Ch. Körner, personal communication.

also affects the pattern of windflow and the degree of 'shelter' and, on a large scale, can even have marked effects on precipitation (high to the windward and low to the leeward of a mountain range). In this section, however, the effects of altitude will be considered only in relation to 'standard' level sites. Further information on mountain climates and microclimates may be found in texts by Barry (1981) and Geiger (1965).

Some typical climatic data for different altitudes are summarised in Table 11.1. The decreasing pressure and increasing windspeed with increasing altitude are probably the most fundamental physical effects of altitude; most other effects follow from these.

Pressure

The atmospheric pressure over the altitudinal range of botanical interest may be derived. For a thin layer of atmosphere of thickness dz, the downward pressure (force per unit area) due to this layer alone (dP) is given by the mass per unit area multiplied by the acceleration due to gravity ($\mathbf{g}$). Since mass per unit area equals the mass per unit volume (the density, ρ) multiplied by the thickness of the layer,

$$\mathrm{d}P = \rho\,\mathrm{d}z\,\mathbf{g} \qquad (11.6).$$

substituting from equation 3.6 this gives

$$dP = (PM_A/\mathscr{R}T)\,dz\,\mathbf{g} \tag{11.7}$$

Integrating this from $z = 0$ (sea level), where $P = P^o$, to any altitude z, gives the pressure at that altitude (P_z) as

$$P_z \simeq P^o \exp(-M_A \mathbf{g} z/\mathscr{R}\bar{T}) \tag{11.8}$$

where $\bar{T}$ is the mean temperature over the altitude range. The observed mean pressures at 600 and 2600 m in the Central Alps (Table 11.1) are close to the values predicted by equation 11.8.

Temperature

As a result of the reduction of atmospheric pressure with height, there is a marked tendency for temperature to decrease. The reason for this is that as a parcel of air rises, it tends to expand as the pressure decreases. This expansion requires the performance of work. If there is no heat exchange with the environment, i.e. the system is adiabatic, the energy for this expansion is extracted from the air itself, thus lowering its temperature. The rate at which air temperature changes with altitude is called the lapse rate. For dry air this theoretical temperature gradient, the dry adiabatic lapse rate, is approximately equal to 0.01 °C m^{-1}. The adiabatic lapse rate for wet air, saturated with water vapour, is much smaller, ranging from 0.003 to 0.007 °C m^{-1}, depending on air temperature. The smaller lapse rate in wet air arises because condensation liberates some heat that partially compensates for the air expansion.

In natural situations, temperatures decrease on average by about 0.006 °C m^{-1} (e.g. Table 11.1), though quite large deviations do occur. This temperature reduction with altitude means that cold tolerance becomes an increasingly important factor determining plant distribution at high elevations (see e.g. Larcher 1983). The general temperature reduction with altitude also affects the growing season; for example, in Britain the decrease can be 12–15 days per 100 m rise, or approximately 5% per 100 m rise.

Partial pressure and molecular diffusion coefficients

As the total air pressure decreases with altitude, the partial pressures of its component gases (including N_2, CO_2 and O_2) decrease in proportion. On average, water vapour partial pressure decreases rather more rapidly because of condensation as air temperature falls. These changes all affect the driving forces for diffusion to and from plant leaves.

Similarly, the changes of pressure and temperature with altitude affect diffusion coefficients according to equation 3.18 (p. 51). Although the

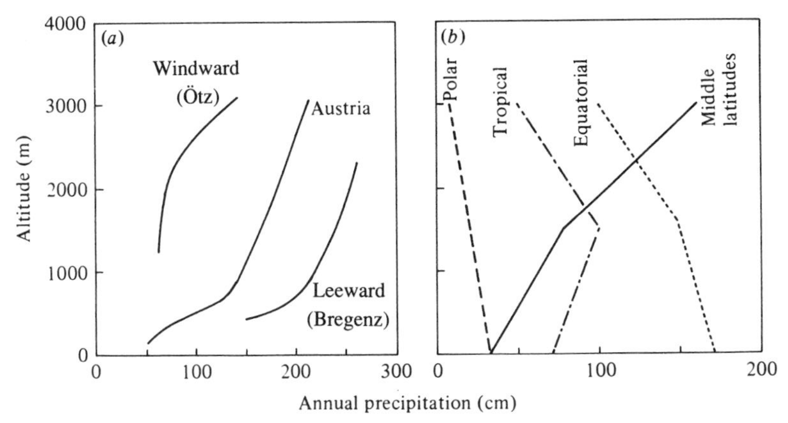

Fig. 11.6. Altitudinal profile of mean annual precipitation (*a*) in the Austrian Alps, and (*b*) in different climatic regions. (After Lauscher 1976).

effects of temperature and pressure are in opposite directions, the pressure effect is dominant so that the $\boldsymbol{D}_i$ increases with altitude.

Relative humidity and precipitation

Although water vapour pressure decreases with altitude, relative humidity tends to increase with increasing altitude as a result of adiabatic cooling. Similarly, vapour pressure deficit tends to decrease. This results in a general tendency for precipitation (and cloudiness) to increase with altitude. Detailed trends depend, however, on latitude and local topography (particularly in relation to prevailing wind direction) so there is no unique relation between altitude and precipitation (see Fig. 11.6).

Radiation

On clear days, the incoming solar radiation increases with altitude as the attenuating air mass decreases (Table 11.2; see also Chapter 2). Remembering that the air mass at any altitude is proportional to P/P°, the theoretical altitude dependence of $\mathbf{I}_{S(dir)}$ on a clear day (ignoring changes in atmospheric water vapour with altitude) can be determined using equations 11.8 and 2.10 (p. 24) (but see Gates (1980) for details).

Prediction of altitudinal variation in total (global) shortwave radiation ($\mathbf{I}_S$) is complicated by the variation in cloudiness with altitude (see entries for fog, overcast days and sunshine hours in Table 11.1). In many regions cloud cover tends to increase along with the altitudinal increases in relative humidity. The effects of this are typified by the report (Yoshino 1975) that sunshine duration in the Japanese mountains fell from around 2200 h yr^{-1}

Table 11.2. *Daily totals of diffuse and total shortwave radiation on a horizontal surface in June at three altitudes in the European Alps (data from Dirmhirn* 1964)

	Clear sky			Overcast
Altitude m	Diffuse ($MJ\ m^{-2}\ day^{-1}$)	Total	Diffuse/total (%)	Diffuse (= total) ($MJ\ m^{-2}\ day^{-1}$)
200	4.2	28.9	14.5	6.5
1500	3.3	32.6	10.0	10.3
3000	2.6	34.9	7.4	16.9

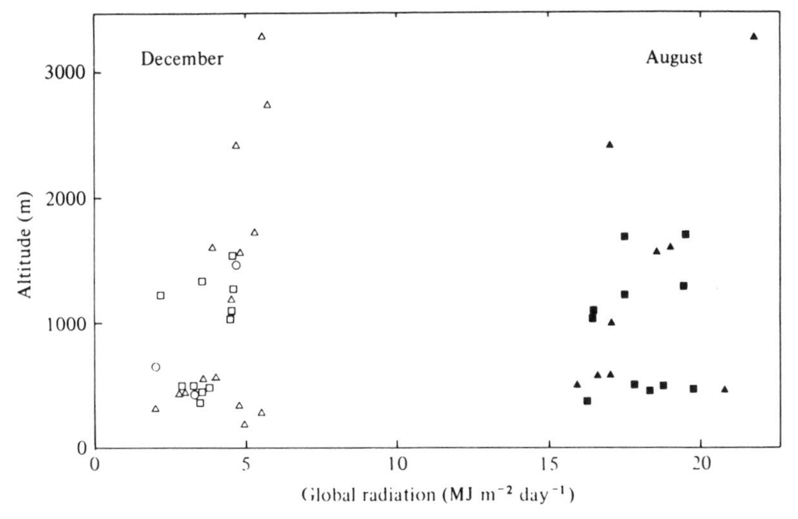

Fig. 11.7. Altitudinal variation of mean daily global radiation for August and December 1980 on a horizontal surface for different sites: ○ stations in central Switzerland; △▲ eastern Switzerland; □■ western Switzerland. (From Valko 1982.)

at heights up to 500 m to about 1300 h yr^{-1} at 1500 m. On the highest peaks, sunshine duration increased (and fog decreased) with further increases in elevation. The increased cloudiness and smaller air mass at high elevations tend to balance out in the long term so that global radiation receipts may be relatively insensitive to altitude (Fig. 11.7). With clear skies, diffuse radiation decreases with increasing altitude, but conversely in overcast conditions diffuse radiation (and global radiation) increases with altitude (Table 11.2). Becker & Boyd (1957) give further information on solar radiation variation with altitude.

Ultraviolet. Radiation increases with altitude are particularly marked in the UV (Caldwell 1981): for example Dirmhirn (1964) quotes a 4.8-fold increase of UV-B wavelengths (280–315 nm) in the solar beam on rising from 200 m to 3000 m in winter. In this context it is also worth noting that depletion of the stratospheric ozone layer as a result of atmospheric pollutants such as chlorofluorocarbons is also likely to be biologically important (UNEP 1989), so both effects will be considered together.

Even quite small changes in UV-B can have quite large effects on plant functioning (UNEP 1989; Tevini & Teramura 1989), with effects such as stomatal closure and the inhibition of growth and photosynthesis often, but not always, being observed. Assessment of the biological effects of UV-B is not simple; not only is there much variability between species and varieties in their sensitivity and their ability to acclimate (for example, by the synthesis of screening pigments such as flavonoids), but there is also great variability of natural UV-B irradiance with location and time of year, and the unnatural spectral distribution of supplementary sources often used (which results in the need to use an appropriate weighting function) causes further problems. Physiological effects of additional UV-B are often apparent for increases in UV-B irradiance of only 50–100 mW m^{-2} (equal to an additional daily dose of between about 1500 and 5000 J m^{-2} and simulating a 15–25% depletion of the ozone layer). The effects on photosynthesis have been intensively studied, with recent work having demonstrated that exposure to UV can lead to rapid (50% within 3 h) down-regulation of genes for a number of important photosynthetic proteins (Jordan *et al.* 1991).

Effects on plants

Gas exchange. The changes of photosynthesis and transpiration with altitude are difficult to predict, partly because of the complexity of the physical changes and partly because of biological responses to these changes. In this context it is necessary to distinguish physical effects of altitude (such as the effect of P on diffusion coefficients) from physiological and anatomical changes (such as alterations in stomatal dimensions and aperture).

The advantages of molar units for conductance are particularly apparent in studies of altitude effects on plant gas exchange. If one uses standard units (mm s^{-1}) for diffusive conductance (see Chapter 3) where $g_\ell = \mathbf{D}/\ell$, and substitutes for $\mathbf{D}$ from equation 3.18, it follows that g_ℓ would tend to increase with altitude (as pressure decreases) even with constant stomatal dimensions. Taking as an example the July means in Table 11.1, it is apparent that g_ℓ would increase by 18% between 600 and 2600 m (i.e. $100 \times (9.46/7.40) \times (278/291)^{1.75}$) as a result of these purely physical effects. With molar units (mol m^{-2} s^{-1}), on the other hand, where $\boldsymbol{g}_\ell = \boldsymbol{P}\boldsymbol{D}/\ell\mathcal{R}T$, the direct pressure dependence cancels the pressure dependence of $\boldsymbol{D}$, so that the net effect of moving between 600 and 2600 m is a decrease in $\boldsymbol{g}_\ell$ of only about 3% ($100 \times (278/291)^{0.75}$). The use of molar units does not, however, eliminate the need to consider the purely physical effects of altitude, because the fact that $g_a \propto \boldsymbol{D}^{\frac{2}{3}}$ means that the effect of pressure on g_a differs from that on g_ℓ. Making the appropriate substitutions it can be shown that, at constant windspeed, g_a would increase by *c.* 11% over the same altitudinal range, while the corresponding change in the molar conductance $\boldsymbol{g}_a$ would be a decrease of *c.* 9%. In addition to these effects, the altered driving forces with changed temperature and pressure and the altered g_a as a result of the windspeed gradient must also be considered.

Applying these results to evaporation, the fairly small physical effects on diffusion conductances are found to be much less important than changes in stomatal dimensions and frequency, the increase in g_a resulting from increased windspeed, and especially the tendency for vapour pressure deficit to decrease with increasing altitude. The typical summer vapour pressures in Table 11.1, for example, correspond to a decreased δe of approximately 70% from 0.59 kPa at 600 m to 0.18 kPa at 2600 m, with the corresponding water vapour mole fraction deficit ($\delta x_W = x_{Ws} - x_{Wa}$) decreasing by about 61%. Frequently this effect is dominant and tends to decrease potential evaporation with altitude (see e.g. Barry 1981). A further factor to consider is radiation, as this is also an important determinant of **E** (equation 5.23). Although $\mathbf{I}_S$ increases strongly with height in clear weather, the mean value is less sensitive to altitude. There is also evidence that, even allowing for the physical effects discussed above, stomatal conductance tends to increase with altitude (Körner & Mayr 1981). This may result partly from the stomatal humidity response, where stomata tend to open more with the smaller humidity deficits at the higher altitude, though high altitude species also tend to have a greater stomatal frequency, particularly on the upper leaf surface (Körner & Mayr 1981; Körner *et al.* 1986*a,b*). Smith & Geller (1979) have incorporated many of these factors into a model for predicting effects of altitudinal variation on **E**.

An altitude correction for potential evaporation that has been found to be appropriate for Scotland is that monthly potential evaporation decreases by approximately 21 mm $(100\ m)^{-1}$ in summer and 8 mm $(100\ m)^{-1}$ in winter (Smith 1967). The corresponding figures for England are rather less extreme at 17 and 12 mm $(100\ m)^{-1}$ respectively.

The effects of altitude on photosynthesis are just as complex as those on evaporation, with the details in any region depending on the actual lapse rates and radiation profiles so that no general predictions can be made. One difference is that the partial pressure of CO_2 consistently decreases with altitude, thus tending to decrease photosynthesis, though temperature and radiation changes are again dominant.

Plant form. The rather small effects of altitude on evaporation rate, together with evidence that the importance of leaf water potential as a rate limiting factor decreases with altitude (see Körner & Mayr 1981), suggest that the xeromorphy commonly found in plants at high altitudes is not primarily related to the aerial environment, and may even be unrelated to water stress. For this reason the term schleromorphy (from the Greek word for hard) is perhaps a better term to describe the characteristic plant type. Although the dwarfing at high altitude has an environmental component, it is primarily a genetically determined adaptation to prevailing conditions, including high windspeeds, low temperatures, etc.

Treelines. The transition between trees and dwarf shrub vegetation in mountains is characteristically sudden, with many species at such a treeline showing characteristically stunted and gnarled growth (*krummholz*). In general the level of the treeline corresponds with that where the temperature of the warmest month is less than about 10 °C. The close correlation with summer temperature suggests that an important factor may be the increased potential for raising tissue temperature above that of the air as the height of the vegetation, and hence the boundary layer conductance, decreases. The sharpness of treelines may be enhanced by frost killing of unhardened foliage or by frost drought (where trees cannot restrict transpiration losses to a level that can be replaced by water uptake from the still frozen soil, especially in late winter: Grace 1989).

Greenhouse effect

There is currently much interest in the reality and possible consequences of 'global warming' as a result of the build-up of carbon dioxide and other so-called greenhouse gases in the Earth's atmosphere. The main components of the energy balance of the Earth–atmosphere system are illustrated in Fig. 11.8, which shows the situation at equilibrium when the solar radiation absorbed equals the outgoing thermal radiation. The partial trapping of the thermal radiation by the lower 10–15 km of the atmosphere (the troposphere) is a major factor in making life on Earth possible as it increases the surface temperature by about 33 °C above what would occur

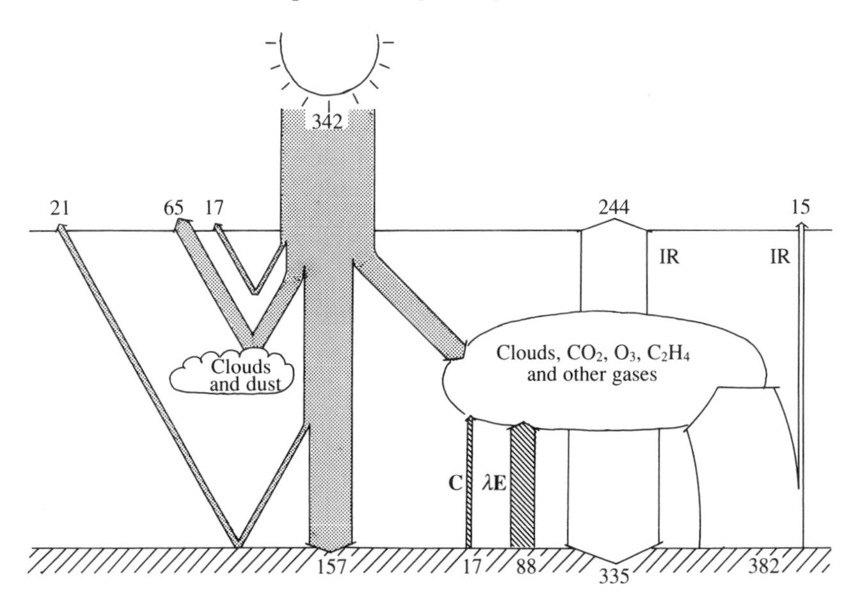

Fig. 11.8. Schematic diagram of the annual global average of different energy fluxes within the climate system. Incident shortwave radiation is illustrated by the stippled arrows on the left side of the diagram, while longwave radiative fluxes (IR) are shown as clear arrows on the right. Energy transfer from the earth to the atmosphere by sensible heat transfer (**C**) and by latent heat transfer (λ**E**) are shown diagonally hatched in the centre. The system is depicted at equilibrium, as the total absorbed solar radiation (239 W m^{-2}) equals the total thermal radiation lost from the system. All fluxes are expressed in W m^{-2}. (Data from Dickinson & Cicerone 1986.)

in the absence of an atmosphere. The tropospheric temperature increases towards the Earth's surface so that thermal radiation emission from atmospheric gases decreases with height (as a function of temperature: see equation 2.3). This results in the upward thermal radiation from the top of the atmosphere being less than that emitted from the Earth's surface. At the actual average surface temperature ($T_s = 287.5$ K) the thermal radiation emitted from the Earth would be *c.* 387 W m^{-2}, but the actual radiative flux from the top of the atmosphere is only *c.* 239 W m^{-2} (Fig. 11.8), implying a trapping of 148 W m^{-2}. (The surface temperature of the Earth that would occur in the absence of an atmosphere can be determined by substituting the value of the radiative loss (equal to the solar radiation absorbed) of 239 W m^{-2} into equation 2.3: this gives a surface temperature of only 255 K.) Even an increase in trapping of 1 W m^{-2} would significantly increase the equilibrium surface temperature.

Table 11.3. *Atmospheric concentrations in* 1985 (*mole fraction – equivalent to volume fraction*) *of various 'greenhouse gases', together with the thermal trapping* (**ΔΦ**, *defined as the change in the net radiation at the tropopause if that constituent is removed, but the atmosphere is otherwise held fixed*), *the range of possible concentration increases for each gas by* 2050 *and the consequent effects of these increases on the radiation balance at the tropopause* ($\Delta\Phi_{2050}$) (*data from Dickinson & Cicerone* 1986, *but see also Houghton* et al. 1990)

Gas	Concentration (mol mol^{-1})	Attenuation ΔΦ (W m^{-2})	Increase by 2050	Δ,Φ_{2050} (W m^{-2})
CO_2	345×10^{-6}	~ 50	16–74 %	0.9–3.2
CH_4	1.7×10^{-6}	1.7	24–135 %	0.2–0.9
O_3	$10–100 \times 10^{-9}$	0.2	15–50 %	0.2–0.6
N_2O	304×10^{-9}	1.3	15–48 %	0.1–0.3
CCl_3F	0.22×10^{-9}	0.06	200–1200 %	0.2–0.7
CCl_2F_2	0.38×10^{-9}	0.12	430–1200 %	0.6–1.4

The main natural absorbers of outgoing thermal radiation are CO_2 and especially water vapour, but they are both weak absorbers in the 8–12 μm wavelength region where there is an 'atmospheric window'. Although increasing atmospheric CO_2 is often assumed to be the major contributor to warming, the combined effect of a number of trace gases, principally methane, nitrous oxide and the chlorofluorocarbons (CFCs), though present at concentrations that are two to six orders of magnitude lower than CO_2, can rival the effect of CO_2 because, per molecule, they absorb infra-red radiation much more strongly, particularly in the critical 8–12 μm waveband. Values of the atmospheric concentrations and thermal trapping (**ΔΦ**) are tabulated for different gases in Table 11.3. Note that radiation absorption by trace gases (such as CFCs) does not necessarily follow the logarithmic relation expected from Beer's Law (Chapter 2), being approximately linearly related to concentration because of their low concentration. It is also important to note that the effects of minor gases are not necessarily directly related to their absorption coefficients because of overlap between their absorption bands and those of major absorbers such as CO_2: for example, methane and nitrous oxide lose about half of their potential trapping by such overlap.

Table 11.4 *Estimates of the magnitudes of carbon fluxes and sinks (data from Bolin* et al. 1986; *Hall* 1989)

CO_2 in atmosphere (1986)	730×10^{12} kg carbon
Carbon in aquatic biomass	3×10^{12} kg carbon
Carbon in terrestrial biomass (80 % trees)	560×10^{12} kg carbon
Carbon in soil	1515×10^{12} kg carbon
Carbon in total fossil fuel resources	6500×10^{12} kg carbon
Carbon in top 75 m of oceans	725×10^{12} kg carbon
Carbon in medium and deep oceans	38000×10^{12} kg carbon
CO_2 fossil fuel emissions	5.3×10^{12} kg carbon yr^{-1}
Net CO_2 emissions from the biosphere	$1–2.6 \times 10^{12}$ kg carbon yr^{-1}
Net primary production (terrestrial)	60×10^{12} kg carbon yr^{-1}
Net primary production (aquatic)	46×10^{12} kg carbon yr^{-1}
CO_2 uptake by oceans and freshwater	2.7×10^{12} kg carbon yr^{-1}

Changing atmospheric gas concentrations

For at least 2000 years prior to AD 1800 (and probably since the last ice age) the concentration of CO_2 remained fairly stable at around 280 ± 15 vpm, but since then it has been rising (now at a rate of about 1.5 vpm per year, though with marked seasonal cycles relating to seasonal changes in photosynthesis). Much of this rise in CO_2 concentration is attributable to the burning of fossil fuels (perhaps 5.3×10^{12} kg carbon yr^{-1}: Table 11.4), but there is also a major contribution from changes in the biosphere such as the burning of tropical rainforests and from other changes in land management and changed fertility of (and hence photosynthesis by) the oceans, with estimates of net biospheric emissions in the range of $1.0–2.6 \times 10^{12}$ kg carbon yr^{-1}. Estimates of the sizes of carbon sinks are given in Table 11.4. Although the general trends are clear, it is difficult to predict accurately future changes in atmospheric CO_2, because the rate of uptake by the oceans, for example, is not well understood.

The role of forests in the global carbon balance is somewhat controversial. Although burning and clearance of tropical forests can release large quantities of CO_2 it does not necessarily follow that planting trees will reduce CO_2 build-up in the atmosphere. In the long term the amount of carbon stored in trees and their products will tend to a steady value that depends on the species, its management and the uses of the wood products (Thompson & Matthews 1989). By making assumptions about the fate of carbon (burning, early decomposition, or preservation for centuries before decay) it is possible to calculate the total carbon accumulated (Fig.

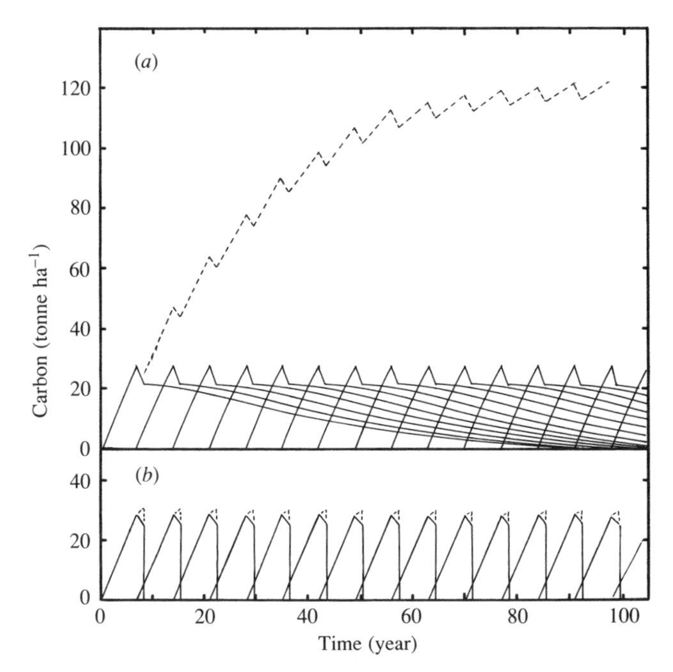

Fig. 11.9. Modelled time course of carbon accumulation in a poplar coppice system, showing the importance of final use of the wood (based on calculations of Thompson & Matthews 1989). The carbon accumulated in each rotation is given by the series of solid curves, while the total accumulated carbon at any time is shown by the dashed line. When the product is converted into a relatively long-lived product such as medium density fibreboard (*a*), the total carbon in trees and product would be expected to rise slowly to a steady value of about 120 tonne ha^{-1} after about 100 years. In contrast, where the product is rapidly used as a biomass source for energy (*b*), the maximum amount of carbon locked up in the system would be less than 30 tonne ha^{-1}.

11.9). The main point to note is that eventually a steady state will be reached, the level of which (and the time taken to reach it) will depend on the use of the timber. Where the trees are long-lived and/or the product goes into a long-life use (such as building material with a life of hundreds of years) many hundreds of tonnes ha^{-1} will accumulate (and it will take hundreds of years to attain the steady state). If, on the other hand, the product is allowed to decay rapidly, smaller amounts will accumulate (and a lower steady state will be attained rapidly). Planting more woodland, therefore, is only likely to be beneficial to the global carbon balance in the period before attainment of the steady state, or where the product is used to substitute for fossil fuel. Grassland could even be more effective at lowering net CO_2 emissions than forests (because of a potentially higher net primary productivity) if used as a biomass source to replace fossil hydrocarbons or if some of the produce is prevented from decay. Unfortunately the achievement of high productivity by grassland or any other system requires a high input of nitrogen, but denitrification of fertilisers to N_2O is one of the reasons for increasing atmospheric N_2O, which may itself be damaging to the environment. For a detailed discussion on the relative merits and economic aspects of biomass for carbon sequestration or for fossil fuel substitution as means of reducing net CO_2 emissions, see Hall *et al.* (1991).

Global concentrations of methane were also relatively stable until about 150 years ago, but are now rising rapidly at $> 1\%$ per year, and are another major contributor to potential global warming. The reasons for this rapid increase are not clear, though the contributions of rice paddies, ruminant animals, biomass burning and emission from swamps and marshes including the tundra regions have all been implicated. The effect of any global warming could even be amplified if it leads to significant releases of CH_4 from warming tundra. Atmospheric N_2O is also increasing, possibly as a result of either increased microbial nitrification and denitrification of agricultural fertilisers or combustion. Together with the CFCs (especially CCl_3F (CFC-11) and CCl_2F_2 (CFC-12)) these trace gases probably make a contribution to global warming at least equivalent to that of CO_2 (Table 11.3), particularly when one takes account of their long residence times in the atmosphere.

The other main climatic effect of CFCs on destruction of the tropospheric ozone layer, with the consequential increased transmission of damaging UV radiation to the earth's surface, has led to international agreement (the Montreal Protocol) to restrict their emission (Anon. 1987), but their long lifetimes (*c.* 120 yr for CFC-12) means that the effects of existing and known future emissions of these gases will take a long time to reverse. The candidate replacements (hydrohalocarbons) have shorter atmospheric lifetimes, so that the cumulative global warming potential from their release is much less.

Effects on climate

The simple energy balance shown in Fig. 11.8 indicates that increasing the absorption of thermal radiation in the atmosphere should lead to global warming (perhaps 1.2–1.3 K for a doubling of CO_2 concentration: Hansen *et al.* 1988). Climate, however, is the long-term result of an extremely complex and dynamic system of atmospheric motion that involves a large number of feedbacks that modify this simple prediction. Increasing temperature, for example, would be expected to increase the amount of water vapour in the atmosphere by increasing evaporation and by increasing the water holding capacity of the air; this in turn affects atmospheric absorption of thermal radiation. Another important, yet poorly understood, process is the way changing global temperatures might affect cloud cover and type; as is apparent from Fig. 11.8, many of the fluxes, especially the reflection of solar radiation from clouds, are much larger than the transmission of thermal radiation through the atmosphere, so small changes in cloud cover are critical for the overall energy balance. The polar ice caps make a major contribution to the albedo of the earth, so changes in their extent would significantly affect the amount, and distribution, of solar radiation absorbed.

Although several general circulation models (GCMs) simulating the behaviour of the atmosphere on a global scale have been developed that attempt to take account of atmospheric dynamics, there are still large uncertainties, with the temperature rises predicted for a doubling of CO_2 ranging between about 2 and 5 K. More importantly for plant production, there are likely to be large regional differences, with the largest temperature increases occurring near the poles, while effects on rainfall and the water balance are likely to be at least, if not more, important than changes in temperature. The potential complexities involved in assessing the likely effects of global warming and CO_2 changes on evaporation from vegetated areas have been outlined by Rosenberg *et al.* (1989). There may also be changes in climatic variability and the probability of extremes, though the evidence for this is unclear.

Consequences of global warming for agriculture and natural ecosystems

There are a number of effects to consider. First, there are the direct effects of increased CO_2 concentration on plant performance. The most obvious are the effects on photosynthesis, though there are many other likely effects such as increased water use efficiency. These are reasonably well understood, though most research has concentrated on the effects of short-term changes in CO_2 concentration and ignored the longer-term adaptations that can occur. Secondly, there are the effects of altered climate (temperature, humidity, rainfall, etc. and their seasonal distribution).

Effect of elevated CO_2 concentration on photosynthesis. The effects of increased CO_2 concentrations are commonly studied in controlled environment chambers, but an alternative approach is to enrich the natural environment by release of CO_2: this can either be done by the use of supply lines in the field (though this can be very expensive in CO_2 supply) or else more efficiently by the use of open-top chambers. In all cases, however, it is necessary to distinguish between the short-term effects of altered CO_2 concentration on photosynthesis and the effects of long-term exposure (for which there is much less information).

Differences in the short-term photosynthetic responses to CO_2 of plants having different photosynthetic pathways were outlined in Chapter 7. Although the net CO_2 fixation rate is often close to CO_2 saturation in C_4 plants, photosynthesis in C_3 plants tends to continue to increase with increasing CO_2 above 350 vpm, partly because CO_2 inhibits photo-respiratory loss by competing with O_2 and shifting Rubisco function towards carboxylation. This short-term response of assimilation to CO_2 occurs even though the stomata tend to close with increasing CO_2 partial pressure, and hence minimise the potential increase in p_i' (see Chapter 6). With longer-term exposure to elevated CO_2, however, it is frequently observed that a decreased photosynthetic capacity at light saturation may offset any gain from decreased photorespiration (Sage *et al.* 1989). The acclimation may involve morphological (including effects on dry matter partitioning, leaf cell packing and stomatal distribution), physiological (including changes in stomatal aperture) and biochemical changes.

An example of the way in which the photosynthetic capacity of individual leaves can acclimate to atmospheric CO_2 concentration during ontogeny is shown in Fig. 11.10*a*. For the very young leaves, the light-saturated assimilation rate was found to be similar for leaves grown in low (300 vpm) or high (1000 vpm) concentrations of CO_2, whatever CO_2 concentration was used for measurement. As the leaves matured, the light-saturated assimilation rate for leaves grown in 300 vpm CO_2 became increasingly sensitive to the CO_2 concentration during the measurement, with $\mathbf{P}_n$ being approximately double when measured at 1000 vpm compared with 300 vpm. The picture was very different, however, if the plants were grown continuously in high CO_2: in this case $\mathbf{P}_n$ was similar for fully expanded leaves grown and measured at 300 vpm and for those grown and measured at 1000 vpm. The acclimation involved a particularly rapid decline of the activity (and also amount) of Rubisco in the leaves of those plants grown at elevated CO_2 (Fig. 11.10*b*).

Increasing the partial pressure of CO_2 for growth above the present ambient does not have consistent effects on stomatal frequency but tends to increase leaf thickness (at least in C_3 plants). Decreasing the CO_2 partial pressure below ambient, on the other hand, tends to increase stomatal frequency (Thomas & Harvey 1983; Woodward & Bazzaz 1988). It seems reasonable to suggest that the variability of the morphological responses to

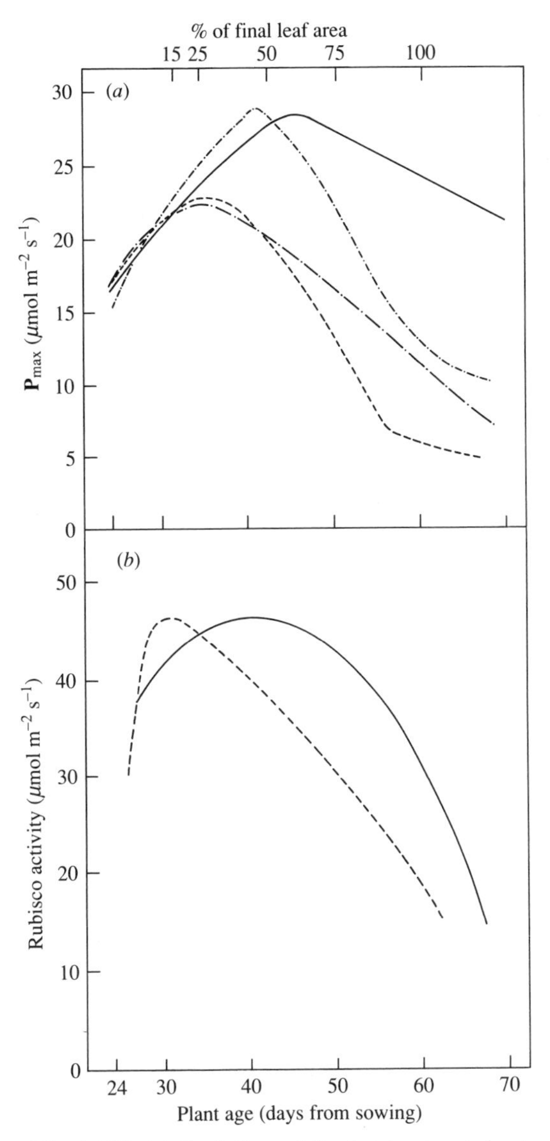

Fig. 11.10. (*a*) Variation of the light-saturated rate of photosynthesis of tomato leaves during ontogeny, either when grown in 340 vpm CO_2 and measured at either 300 vpm (—·—) or 1000 vpm (———), or when grown in 1000 vpm CO_2 and measured at either 300 vpm (–·–·–·–) or 1000 vpm (–––––).
(*b*) Corresponding variation in RuBP carboxylase activity for the leaves of plants grown in either 340 vpm CO_2 (———) or 1000 vpm CO_2 (–––––). (Curves from Besford *et al.* 1990.)

CO_2 arises because they are largely indirect responses determined by differences in plant water relations.

In contrast to the acclimation commonly observed in the light-saturated situation, there is less evidence for acclimation of photosynthesis at low irradiance. For example, *Scirpus olneyi* when grown in open-top chambers in the field at different atmospheric CO_2 concentrations showed acclimation of neither quantum yield nor the light compensation point to growth CO_2 concentration (Long & Drake 1990), even though doubling the CO_2 concentration for measurement from 350 to 700 vpm increased the quantum yield from 0.078 to 0.095, and decreased the light compensation point from 48 to 30 μmol m^{-2} s^{-1} (Fig. 11.11).

The benefits of CO_2 enrichment for crop growth are well established for a wide range of greenhouse crops (Mortensen 1987), though, because CO_2 is often only supplemented in daylight hours, full acclimation may not occur and the results are not necessarily applicable to a future higher CO_2 environment outdoors. There are large differences in CO_2-responsiveness of growth: in a wide range of C_3 species (Hunt *et al.* 1991), the response to a doubling of CO_2 ranged from a mean of only 5% in four stress-tolerant or ruderal species to 43% for eleven species of near-competitive strategy. As expected, maize (C_4) was unresponsive. These results imply that changing atmospheric CO_2 concentrations are likely to have large effects on the relative competitive ability of species and hence on plant distribution.

Effects on water use efficiency. An important effect of rising CO_2 concentration is expected to be an increase in WUE, because raising CO_2 tends to close stomata without decreasing **P**. Morphological and anatomical changes (e.g. stomatal density) as a consequence of altered CO_2 would also influence WUE. Changes in WUE are likely to be particularly important in the main agricultural regions of the world where alterations in water availability are probably more crucial for plant production than are changes in mean temperatures in any CO_2-induced climate change.

Indirect effects through altered climate. The general effects of temperature, water availability and other aspects such as climate variability have been discussed in other chapters, and the basic information is available to predict the effects of a given climate change on natural vegetation and on crop production. Their use for predicting the consequences of climate change is limited by the quality of the climatic predictions, which for any particular area cannot provide reliable predictions of the exact combinations of weather conditions that are likely to occur. Although simple models of plant growth can be used where there is a clear single most limiting environmental constraint, there is still a long way to go in developing

Table 11.5. *Estimated total global emissions of some important atmospheric pollutants from anthropogenic and natural sources and typical background concentrations (data from Freedman* 1989; *Cicerone & Oremland* 1988; *Rasmussen & Khalil* 1988)

	Anthropogenic (10^6 tonne ann^{-1})	Natural	Background (mol $mol^{-1} \times 10^9$)
SO_2	146–187	5	0.2
H_2S	3	100	0.2
NO		430	0.2–2
NO_2	53	650	0.5–4
N_2O		590	250
NH_3	4	1160	6–20
O_3			25
Isoprene[a]		400–470	
CH_4		400–640	1700

[a] 2-methyl-1,3-butadiene.

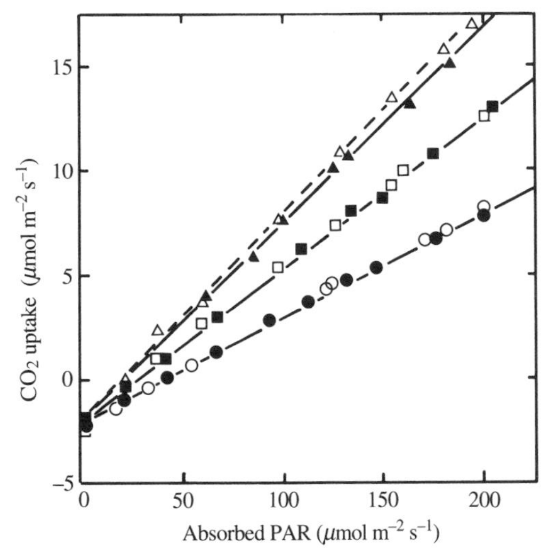

Fig. 11.11. The initial slope of the response of CO_2 uptake to absorbed photon flux density for a shoot of *Scirpus olneyi* grown in 350 vpm CO_2 (open symbols) and for one grown in 700 vpm CO_2 (closed symbols). For each shoot photosynthesis was measured in normal air with either 350 vpm CO_2 (circles), or 700 vpm CO_2 (squares) or else with 350 vpm CO_2 and 1 % O_2 (triangles). (Data from Long & Drake 1990.)

models that fully take into account the seasonal variation in weather and its variability. This is essential if real progress is to be made, as environmental extremes such as frosts and droughts are probably the main constraints on crop and vegetation distribution patterns.

When considering the implications of climate change it is necessary to remember that climate will also affect in a complex manner many other factors and their interactions, such as the population and virulence of pests and pathogens.

Modelling of agricultural consequences. There have been a number of attempts to estimate the effects of climate change on agricultural production, but they inevitably involve a lot of uncertainty. Adams *et al.* (1990), for example, have attempted to predict the consequences of a doubling of atmospheric CO_2 (with the associated changes in temperature and rainfall), not only on the production of major crops such as maize, wheat and soybeans throughout the USA, but also on the requirement for irrigation, and, by incorporation of economic models, on the US economy. Although these authors used the best available crop simulation models, critical uncertainties remain, principally in relation to the direct plant response to CO_2, and even more importantly related to uncertainties in the climate predictions by the GCMs used.

Atmospheric pollutants

In the past two decades, there has been an increasing recognition of the importance of atmospheric pollutants for plant growth and distribution. Acidic pollutants are well known to lead to soil and freshwater acidification, while the widespread occurrence of forest decline in Europe (especially silver fir, Norway spruce and beech) and in Eastern North America (red spruce) has led to a major effort to improve our understanding of the mechanisms involved. These studies have emphasised the complexity of the plant responses involved and the interactions of pollutants with other environmental factors such as temperature and drought.

An enormous number of atmospheric constituents can have detrimental effects on plants. These include the various nitrogen oxides (NO_x) sulphur dioxide (SO_2) and other acidic gases predominantly arising from combustion of fossil fuels, photochemical oxidants such as ozone (O_3) and peroxyacetyl-nitrate (PAN), important constituents of photochemical smog, and, particularly in confined environments such as glasshouses, a range of highly toxic compounds such as di-*n*-butyl phthalate (used as a plasticiser).

Plants themselves can be important sources of compounds that are either toxic themselves or are involved in the generation of atmospheric pollutants. For example the volatile hydrocarbon isoprene (2-methyl-1,3-butadiene) can be emitted at high rates from many plants, while various monoterpenes are commonly emitted by conifers. Rates of between 0.2 and 1 μg C m^{-2} s^{-1} have been reported for isoprene emission from leaves of the grass *Arundo donax* (Hewitt *et al.* 1990) with global emissions having been estimated at up to 400×10^{12} g C yr^{-1} (Rasmussen & Khalil 1988), a rate that is of the same order as global methane emissions. Through its role in tropospheric photochemistry this compound contributes significantly to the generation of ozone in rural and some urban areas, and to the formation of carbon monoxide, oxygenated hydrocarbons and organic acids. Although isoprene emissions have been linked to photorespiration, the strongest evidence links them to photosynthesis.

Sources. Many important pollutants have both anthropogenic and natural sources. Estimated global emissions and background atmospheric concentrations of a number of pollutants are summarised in Table 11.5.

Uptake and deposition processes

Uptake of pollutants occurs via a number of routes: dry deposition (which refers to the absorption of gases and capture of particles by plants), wet deposition in precipitation (rain or snow) and 'occult' deposition in intercepted cloud, fog or mist.

Gases. Fluxes of important gaseous pollutants such as SO_2, HNO_3, HCl, O_3 and NH_3 can be determined using standard micrometeorological techniques (Chapter 3) or by the exposure of plants or natural communities to pollutants in enclosures where gas fluxes can be measured. The usual mass transfer theory is applicable, and it is possible to identify transfer resistances corresponding to transfer in the boundary layer and a surface component that depends on the parallel paths of uptake through stomata and direct to leaf and other surfaces (such as the soil). With the multiple sinks for a pollutant in a canopy it is not possible completely to separate the atmospheric and surface processes into a single canopy boundary layer resistance (r_A) and a single physiological resistance (r_C) (see Fig. 5.4), but as the turbulent transfer resistances within the canopy are usually much less than the surface resistances they are usually ignored.

If the concentration at the sink can be assumed to be zero, the total resistance to uptake of a pollutant gas X is given (see Table 3.1) by

$$\Sigma r_X = r_{AX} + r_{CX} = c_X / \mathbf{J}_X \quad (11.9)$$

where c_X is the concentration of X in the air. In pollutant studies the reciprocal of the uptake resistance (the conductance) is often termed the deposition velocity; this can also be regarded as the flux normalised for atmospheric concentration.

The surface resistance for uptake of reactive gases such as HNO_3 and HCl is normally considered to be negligible (i.e. these pollutants deposit on leaf and soil surfaces at rates determined by r_{AX} which itself depends on atmospheric turbulence), while that for NH_3 is neglible for wet surfaces but can become appreciable when the vegetation is dry. Deposition of these gases is therefore primarily dependent on their atmospheric concentrations and on atmospheric transfer processes.

For gases such as SO_2 and O_3, uptake depends on dissolution or reaction within the substomatal cavity, and deposition is largely determined by the canopy stomatal resistance, which is often > 90% of the total resistance to uptake. Deposition of these gases, therefore, is greatest either during daylight when the stomata are open, or when the surface is wet. In addition, because the rate of deposition of these gases is not very sensitive to r_{AX}, deposition of these gases is normally less dependent on canopy type (i.e. forest versus short crop) than is the case for HNO_3 and HCl.

Both NO and NO_2 can be absorbed or emitted by vegetation, though fluxes of NO tend to be predominantly away from the surface as a result of denitrification in anaerobic soil.

Particles, cloud and rain. The deposition of pollutants as particles and in droplets is a major mechanism for pollutant deposition. The mechanisms of deposition, which are equally applicable to transport of pollutants and of dusts, spores and bacteria, depend on particle size, changing from predominantly Brownian diffusion for particles less than about 1 μm, to impaction and increasingly sedimentation, with increasing size (see Chamberlain & Little 1981). Much of the particulate sulphate and nitrogen in the atmosphere, as well as lead from car exhausts, occurs as aerosols of between 0.1 and 1 μm diameter. Mist and fog droplets, on the other hand, average about 20 μm, and dust particles up to 100 μm.

Sedimentation is the downward movement of particles under the influence of gravity. The rate of this settling, the sedimentation velocity (v_s) increases with the square of the particle diameter and is given by *Stokes Law* for particles with diameters in the range $0.1 < d_p < 50$ μm and ρ about 1000 kg m^{-3} as

$$v_s = (d_p{}^2 \boldsymbol{g} \rho_p)/(18 \rho_a \mathbf{v}) \qquad (11.10)$$

where ρ_p and ρ_a are the densities of the particle and of air, $\boldsymbol{g}$ is the acceleration due to gravity and $\mathbf{v}$ is the kinematic viscosity of air. Even for particles as large as 30 μm and with a density of 1000 kg m^{-3}, equation

11.10 predicts sedimentation velocities of only about 27 mm s^{-1} ($=((30 \times 10^{-6})^2 \times 9.8067 \times 1000)/(18 \times 1.2 \times 15.1 \times 10^{-6})$). This is much less than typical velocities of turbulent eddies in the atmosphere, except within very dense canopies, so it is clear that sedimentation is only important for large dust particles and water droplets.

When air flows round obstacles, entrained particles tend to continue in nearly straight lines because of their inertia leading to the potential for impaction. The impaction efficiency on an object in an airstream can be defined as the number of particles striking the obstacle divided by the number that would have passed through the space occupied by it if the object had not been there. The impaction efficiency is related to the **Stokes number** which is equal to the stopping distance of the particle divided by the effective diameter of the obstacle (where the stopping distance is the horizontal distance travelled by a particle in still air when given an initial velocity u and is approximated by $v_s u/g$). Impaction efficiency therefore increases with windspeed, with particle size and with decreasing size of obstacle.

The particulate sulphate and nitrate in the atmosphere is only very inefficiently intercepted by vegetation because of the small size of the particles; transfer of these particles is therefore predominantly by Brownian motion, with some impaction and interception by micro-roughness elements such as leaf hairs. This size of particle tends to stick firmly once attached. Wind-tunnel studies suggest deposition velocities of less than 1 mm s^{-1} to forest with wind speeds as high as 5 m s^{-1} for 0.5 μm particles. This is much less than the corresponding rates of uptake of gaseous sources of S and N such as SO_2, NH_3 and HNO_3 (Fowler *et al.* 1989).

Sulphate or nitrate aerosols may act as nucleation centres for the formation of fog or cloud droplets, especially during the formation of orographic cloud when air is cooled as it rises over mountain ranges. Cloud droplets, therefore, tend to contain quite high concentrations of sulphate and nitrate. Because typical droplet sizes in orographic cloud are of the order of 20 μm in diameter (Fowler *et al.* 1989) they are collected relatively efficiently by impaction and sedimentation processes on vegetation, with deposition velocities of perhaps 20–50 mm s^{-1} for moorland and up to 200 mm s^{-1} for forest in the atmospheric conditions that apply in upland Britain. These deposition rates are one to two orders of magnitude greater than for the original aerosols.

This occult deposition is a particularly important means of pollutant deposition at higher elevations where vegetation may be bathed in cloud for long periods. Because the concentrations of major ions in cloud water often exceed those in rainwater by two to three-fold (up to six-fold), deposition of pollutant chemicals in cloud can in certain circumstances be more important than deposition in rainfall. Cloud deposition is greater onto aerodynamically rough surfaces such as forest, than onto crops or moorland, and is much greater than the deposition of small aerosol particles. Cloud deposition also has hydrological significance as an important source of water for plants in montane environments.

The effects of factors influencing the deposition of different pollutants are illustrated in Table 11.6. The data for the Kielder Forest area (a large area of recent afforestation in northern England at about 300 m above sea level) show particularly well the differential effect of forest (15 m trees) as compared with moorland on deposition. The presence of trees increases sulphur input by about 30% (as a result of increased cloud-water interception), but increases nitrogen input by about 90% (as a result of both the increased cloud-water interception and the increased dry deposition rate that occurs for gases such as HNO_3 and NH_3).

Effects of pollutants on vegetation

The effects of pollutants depend not only on the sensitivity of the plants, but on the concentration and on the duration of exposure. Short episodes at high concentration may be more damaging than continuous exposure to lower concentrations that give rise to the same long-term time-averaged concentration. Lichens and bryophytes are often particularly sensitive to atmospheric pollution and have been used to provide a sensitive bioassay of pollution in many situations (e.g. Henderson 1987), though it is often difficult to relate these results to the specific chemical problem that exists in any situation.

The effects of pollutants on plants can be direct as a result of uptake by leaf tissues, but in many cases the influences may be through processes such as soil acidification or even more subtle effects on pathogens or competitors. Research has emphasised effects on carbon assimilation and partitioning, water relations and nutrition. Detailed discussion and examples may be found in Koziol & Whatley (1984), Roberts (1984), Schulte-Hostede *et al.* (1988), Heck *et al.* (1988) and Mathy (1988). These studies show that there are large differences between species, and even cultivars, in their sensitivities to different atmospheric pollutants, with typical thresholds for damage by SO_2, for example, being of the order of > 50 vppb (volume parts per 10^9) for the appearance of chronic effects (e.g. chlorosis and premature senescence) with long-term exposure, > 180 vppb for episodic exposure and > 500 vppb for the appearance of acute damage such as leaf necrosis. Ozone, on the other hand, can cause yield losses at 50–100 vppb, concentrations that are much closer to the normal background (Table 11.5).

Table 11.6. *Representative rates of sulphur and nitrogen deposition (kg ha^{-1} y^{-1}) as different forms in different environments (data from Irving,* 1988; *Fowler* et al. 1989)

	Dry		Wet	Occult	Total
	Gas	Particle			
Sulphur					
Kielder (UK)					
– coniferous forest	3.1		13	6.5	22.6
– cotton grass	3.1		13	1.3	17.4
Oak Ridge (USA)					
– deciduous forest	7.8 ± 1.3	2.1 ± 0.3	8.5 ± 2.0	n.a.[a]	> 18.4
Nitrogen (NO_3^- and NH_4^+)					
Kielder					
– forest	13.5		8	1.9	23.4
– cotton grass	4.0		8	0.4	12.4

[a] n.a. = data not available.

Table 11.7 *Effects of fumigation on mean relative growth rate (RGR) of* Phleum pratense *when maintained well-watered or unwatered after a prior 40-d exposure to SO_2 together with NO_2 (data from Wright* et al. 1986)

	Relative growth rate (RGR, g g^{-1} d^{-1})	
Pretreatment	Watered daily	Unwatered
Control	0.024	0.013
90 vppb $SO_2 + NO_2$	0.027	0.005[a]

[a] Significant difference from control at $P < 0.05$.

Methods of study. Pollution effects on vegetation are most commonly studied in fumigation chambers with more or less sophisticated environmental control and which may be either of the closed or of the 'open-top' type. In some cases open-field fumigation systems, which have the advantage of maintaining natural microclimatic conditions, are used (McLeod *et al.* 1985).

Interactions. Pollutants interact both with each other and with biotic and abiotic stresses in their effects on plants in a complex manner. As an example of the type of effect that is commonly observed, Ting & Dugger (1968) reported that exposure of bean leaves to a combination of 0.3 vpm O_3 and 0.4 vpm SO_2 for four hours and no apparent detrimental effect, while exposure to the O_3 alone caused heavy damage. This has been explained in terms of stomatal closure in response to the SO_2 preventing entry of the O_3, though, in other experiments, low levels of SO_2 have been reported to cause stomatal opening (see Unsworth *et al.* 1972), and hence potentially enhanced damage.

The effects of ozone on plant cells provide a good example of the complexity of the reactions involved in response to pollutants. Ozone can damage cell membranes by reacting directly either with proteins (oxidising cysteine and methionine residues) or with the unsaturated fatty acids. Ozone is also quite soluble in water and may decompose in aqueous cell compartments to produce highly reactive and damaging radicals such as hydroxyl ($OH^{\bullet}$), peroxy ($HO_2^{\bullet}$) and the superoxide ion ($O_2^{-\bullet}$). In addition to these direct effects, however, the impact may be enhanced by the formation of highly reactive hydroperoxides on reaction with gaseous alkenes produced within the leaf (Hewitt *et al.* 1990*a*). These biogenic alkenes include ethylene (induced in response to a number of stresses) as well as isoprene or monoterpenes in appropriate species. The effect may even be amplified by positive feedback as tissue damage releases even more alkene.

An example of another important type of interaction occurs where exposure to pollutants enhances sensitivity to other stresses. Typical of this effect is an experiment where growth of *Phleum pratense* seedlings *after* exposure to a mixture of SO_2 and NO_2 at concentrations up to 90 vppb for 40 days was compared under well-watered and drought conditions (Wright *et al.* 1986). Fumigation had little effect on the subsequent relative growth rate over a 23 day period when the plants were maintained well-watered (Table 11.7). When, however, water was withheld over the same period, the relative growth rate of the non-fumigated control plants decreased by about 50%, but the inhibition was much greater for the previously fumigated plants. This enhanced sensitivity to water deficit was related to a greater water use by the fumigated plants.

Pollutants such as O_3 and acidic mists can also markedly enhance sensitivity to frost (e.g. Brown *et al.* 1987; Wolfenden & Mansfield 1990), and this appears to be a factor in some types of forest decline. There is, for example, strong evidence that the enhanced sensitivity of spruce to frost damage that is caused by pollutants results from a delay in the onset of hardening in autumn, rather than from an effect on the final degree of hardiness. Exposure to pollutants during the dormant period can also enhance winter damage, possibly by lowering the efficiency of control of water loss by the stomata or the cuticle.

Another important type of interaction that should not be forgotten, but is relatively poorly understood, is that caused by the effects of pollutants on pests and disease organisms.

12

Physiology and yield improvement

This chapter introduces some of the ways in which information of the type discussed in earlier chapters can be applied to the improvement of crop yields. Crop yields have been improving slowly over hundreds of years, though increases in yield have been particularly rapid only in the last 50 years or so (Fig. 12.1). These yield increases have resulted both from the introduction of new varieties and from advances in crop management (agronomy), including the widespread use of chemical fertilisers, precision drilling, combine harvesters, the introduction of herbicides, pesticides and fungicides and (particularly in horticultural crops) the use of chemical plant growth regulators. In addition to their increased yield potential, the new varieties that have been developed by plant breeders often incorporate improved pest or disease resistance and the ability to benefit from increased levels of fertilisation.

An indication of the relative contribution made by new varieties and improved management to yield increases can be obtained from direct comparisons of yields of old and new varieties in one trial. Any increase in yield not attributable to varieties can be ascribed to agronomy. A study based on the national average wheat yields in the UK (Fig. 12.2) has indicated that agronomy and breeding have made approximately similar contributions to the doubling of national cereal yields over the 30 years to 1978, though almost all the yield advance since 1967 is apparently due to variety. It is not possible to partition precisely these contributions to yield improvement, partly because of what are called genotype × environment interactions, where different varieties respond differently (e.g. the newer semi-dwarf cereal varieties can respond better to high nitrogen input than can tall varieties that become liable to lodging).

In some crops, breeding has had a much smaller impact than in cereals. For example, the major varieties of both dessert and culinary apple currently grown in the UK were discovered in the nineteenth century. Another consideration is that the long-term trend of yield improvements for any one crop can vary greatly between different parts of the world. Maize yields, for example, have more than doubled in the USA over the

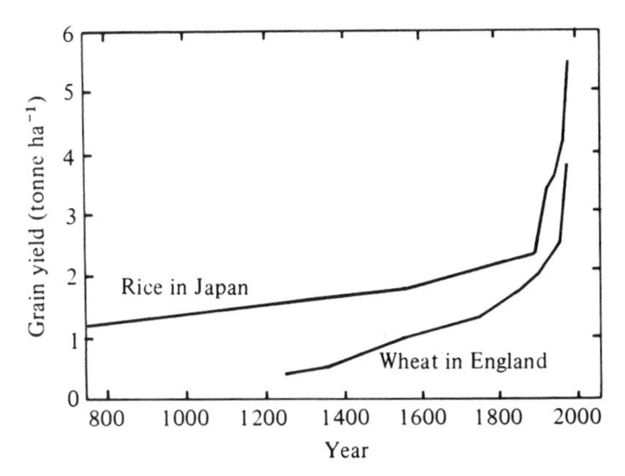

Fig. 12.1. Historical trends in grain yields of rice in Japan and of wheat in England. (From data collected by Evans 1975.)

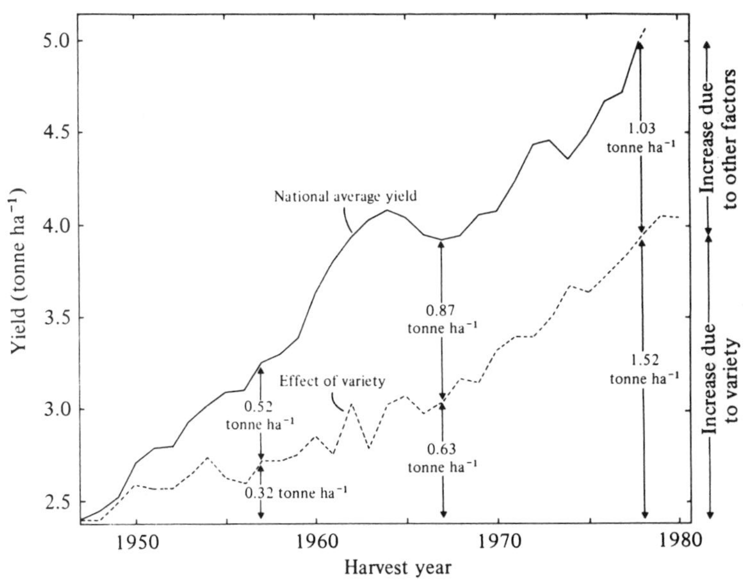

Fig. 12.2. Variation of the five-year moving average of national yields of wheat in England and Wales 1947–78. The dotted line represents the contribution of the variety grown. (After Silvey 1981.)

past 30 years, increasing from about 2.3 to 6.0 tonne ha^{-1} (largely as a result of the introduction of F1 hybrid varieties), but yields have remained nearly constant at about 0.7 tonne ha^{-1} in Africa (Leng 1982).

An important question in relation to attempts both to identify the environmental factors such as radiation, temperature or water, that limit the yield of any crop and to overcome these constraints by breeding or management, is how closely the current yields approach the theoretical maximum. There is no evidence from the UK wheat yield data shown in Fig. 12.2 that yields have been levelling off in recent years. On the other hand, some other crops such as sorghum in the USA do show indications of a plateau being approached. The fact that there is still ample scope for improvement, at least on a farm scale, is implicit from the observation that average yields are often only about half the best yields (5 tonne ha^{-1} compared with approximately 12 tonne ha^{-1} for wheat in the UK). The latter value is in fact close to the potential yield calculated on the basis of the observed behaviour of a wheat crop in terms of leaf area development and the assumption that virtually all the dry matter produced during the 40-day grain filling period goes into the grain (Austin 1978). An alternative estimate of the potential yield can be obtained using the type of efficiency calculation described in Chapter 7, where it was shown that 3% efficiency of conversion of incident solar radiation to dry matter should be attainable. If one assumes an input of approximately 3 GJ of solar radiation during the life of a winter wheat crop in the UK, this corresponds to a production of about 5100 g dry matter m^{-2}, or about 23 tonne ha^{-1} of grain (assuming a harvest index of 45%). Even if one only considers the 2 GJ available during the summer, 15 tonne ha^{-1} should be possible. It is likely, however, that yields of this order could only be achieved by quite radical alterations in the phenology of the wheat crop.

Discussion of yield processes in crops and the role of physiology in increasing crop yields may be found in texts by Evans (1975), Milthorpe & Moorby (1979) and Johnson (1981). There are many areas where physiology and an understanding of plant interactions with the aerial environment have made important contributions to crop production. For example, the understanding of photoperiodic control of flowering has enabled the development of widely adapted photoperiod-insensitive varieties (e.g. in cereals) that can be grown successfully over a wide geographic range, and has allowed the artificial control of flowering in horticulture. Similarly, knowledge of the natural control of plant growth and development has led to the successful use of plant growth regulators in fruit production (see Luckwill 1981) and in propagation *in vitro*. Some other potential applications of physiology to yield improvement are discussed in this chapter.

Variety improvement

The traditional approach of plant breeders to improving crop performance is to choose, as parents, plants known to have the required characters such as high yield or disease resistance, to cross them and then select those progeny that have the required combination of characters. There are, however, many practical problems involved, a full discussion of which may be found in texts such as those by Allard (1960) and Mayo (1980).

One problem is caused by the vast number of ways in which even a relatively small number of genes can recombine. For example, if the parents are heterozygous (that is they carry different alleles) at n loci, there are 2^n possible combinations. That is, over a thousand possibilities for a mere ten genes. When one considers that a large number of plant characters are multigenic, being determined by many genes, the number of possible combinations is clearly astronomical. This means that breeders must be able to evaluate or screen very large numbers of progeny. Furthermore, except where clonal propagation is possible, selection has to be repeated for several generations in order to ensure that the plants are homozygous and breed true to type.

Other problems with particular crops include the fact that there may be a long period between making a cross and being able to test the progeny. In fruit trees, for example, it may be several years before any fruit is produced that can be tested. Even in annual crops, the number of plants tested would be greatest if all characters could be tested using seedlings.

Because of the large numbers of plants required in traditional breeding, breeders have in the past had to rely on visual selection methods, even though it is notoriously difficult visually to rank plants for yield. Many new tools that will help to speed the process are now becoming available to breeders. Foremost among these are the new technologies opened up by recent rapid advances in plant molecular biology and genetic manipulation (see e.g. Grierson & Covey 1988). For example it is now possible to isolate individual genes from one organism and to transfer them to unrelated species using techniques that include direct DNA injection, virus and *Agrobacterium tumefaciens*-mediated transformation. Even without direct gene transfer, a number of new approaches are already contributing to plant improvement: for example restriction fragment length polymorphism (RFLP) markers closely linked to useful genes can be used to select useful recombinants, advances in tissue culture have enabled the development of embryo rescue to permit wide crosses, the generation of somaclonal variants, somatic hybridisation, and the use of protoplasts together with high-intensity cell selection techniques (e.g. the selection of adapted cell lines: see Rains 1989). Other methods for short-circuiting the breeding process include the use of mutagenesis by X-rays or chemical mutagens to increase genetic variability, and the use of selection of haploid plants (derived, for example, from antherculture) with subsequent chromosome doubling.

Genetic engineering approaches have already led to the successful insertion of a number of genes into crop plants including those conferring herbicide resistance and insect resistance. In a few cases single genes may confer stress tolerance (e.g. the salinity resistance in wheat that has been associated with chromosome 4 of the D genome and which has been related to efficient exclusion of Na^+ may be a single gene character: Gorham *et al.* 1987) and there are hopes that other single genes could be isolated and incorporated into new cultivars to improve crop yield or environmental tolerance, but in most cases improvement of crop adaptation is likely to require the identification and insertion of complexes of genes (see e.g. Hughes *et al.* 1989). The complexity of adaptive and compensatory mechanisms in whole plants means that any individual modification will almost certainly require additional changes for its optimal expression, while the insertion of new biosynthetic pathways may involve the insertion of many genes.

In the next few years it is likely that the major contribution of molecular biology to the improvement of environmental adaptation will be to the isolation of appropriate genes and regulatory sequences. These can then be used to study the detailed mechanisms of adaptation, and to help in defining the particular characters that are likely to be of most use to plant breeders. Unfortunately screening by molecular or traditional techniques for genes related to physiological characters (as opposed to herbicide tolerance, for example) is limited by the availability of rapid screening tests or of techniques for applying high selection pressure.

In principle, a detailed knowledge of the mechanisms of yield production should enable us to short-circuit the traditional empirical breeding approach of simply selecting those plants with the highest yield. In particular one could pin-point those processes that are most limiting yield so that effort can be concentrated on them and so perhaps even indicate ways of overcoming the limitations. A next stage would be to help devise rapid screening techniques that could be used by the breeders. The objectives of physiology in a breeding programme can be summarised: (1) to define the ideal plant 'ideotype' for a particular situation and (2) to devise rapid screening procedures.

Definition of the ideotype

The determination of the final yield of any crop is a complex process that depends on the cumulative effects of environment over the whole life interacting with genotypically determined developmental sequences (Fig. 12.3). For grain crops the life cycle can be divided into the vegetative and reproductive phases, where the amount of growth during the vegetative phase sets a limit to the yield achievable.

The early attempts to gain an insight into the way that environment and

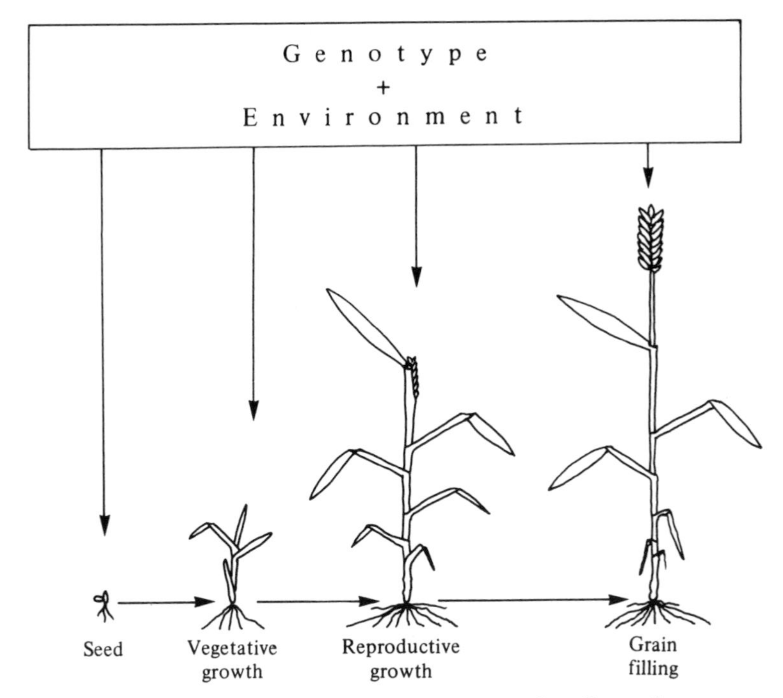

Fig. 12.3. Final grain yield for a cereal integrates the effects of genotype and environment over the whole growth cycle.

plant genotype acted to determine cereal yields were based on yield component analysis (Engledow & Wadham 1923). Engledow & Wadham suggested that 'Theoretically the procedure would be to find out the plant characters which control yield per acre and by a synthetic series of hybridization to accumulate into one plant form the optimum combination of characters'. In this approach the total yield per unit ground area can be treated as the product of several individual components; i.e. for a cereal

$$\text{Yield} = \text{plants m}^{-2} \times \text{ears plant}^{-1} \times \text{grains ear}^{-1} \times \text{weight grain}^{-1}$$

The idea was that effects on these different components could then be studied separately.

Later, the techniques of growth analysis were adopted (e.g. Watson 1958), with a change of emphasis from a leaf area basis (as in the net assimilation rate: Chapter 7), to a unit ground area basis (giving a crop growth rate). Using growth analysis, the dynamics of partitioning of carbohydrate could be analysed in more detail. Realisation of the importance of photosynthesis in productivity led to emphasis on leaf area index and its time integral, the leaf area duration. Additionally it became clear that the economic yield depended not only on the total dry matter production, but also on the harvest index, that is, the proportion of total dry matter in the harvestable product.

The importance of the concept of harvest index was recognised as long as 70 years ago by the barley breeder, E. S. Beaven (see Donald & Hamblin 1976), though he called it a 'migration coefficient'. The importance of harvest index in varietal improvements is well illustrated by a comparison of the wheat varieties Little Joss and Holdfast, that were widely grown in the UK in the late 1940s, with varieties such as Maris Huntsman, Maris Kinsman and Hobbit, that were available in the 1970s (see Table 12.1). A large proportion of the increased yield in the modern semi-dwarf varieties is attributable to the increased harvest index, which itself is related to the smaller requirement for dry matter in the shorter stems. Similarly, yield responses to environmental factors such as water supply or nitrogen fertilisation partly depend on altered harvest index, as illustrated in Fig. 12.4.

Perhaps the greatest difficulty with attempts to define an ideal plant is that most plants have a great capacity for yield compensation. For example, it might be thought from study of the yield components that simply increasing ear number per plant would be a good way of increasing yield. Unfortunately such a simple approach does not necessarily work because the size of each ear is likely to decrease to compensate. In fact quite divergent strategies can result in similar yields. For example, two-row barley varieties tend to have many relatively small ears but they produce yields similar to those of the six-row varieties, that have relatively few large ears. Compensation can occur at all stages of the life cycle. For example, if some seeds fail to germinate, neighbouring plants may compensate by producing more shoots and a greater leaf area.

Another problem is that the relative performance of different cultivars depends on the environment in which they are grown. A typical example of such genotype × environment interaction is shown in Fig. 12.5. Usually cultivars with higher potential yield tend to be most affected by stress (in this case drought). Analysis of genotype × environment interactions has been discussed by Yates & Cochran (1938), Finlay & Wilkinson (1963) and Eberhardt & Russell (1966).

These concepts have provided the basis of what is often called crop physiology. More recently, new biochemical and physical techniques including those described earlier in this book have become available for investigating yield mechanisms. Combined with the development of mathematical models, these now allow us to attempt to determine the optimal plant responses at any stage and to define the crop ideotype.

Table 12.1. *Yield characteristics of some winter wheat varieties* (*data from Austin* 1978)

Variety	Date released	Height to base of ear (cm)	Relative yield	Harvest index (%)
Little Joss	1908	130	100	30
Holdfast	1935	112	94	31
Maris Huntsman	1970	95	148	40
Maris Kinsman	1975	82	145	38
Hobbit	1975	67	166	45

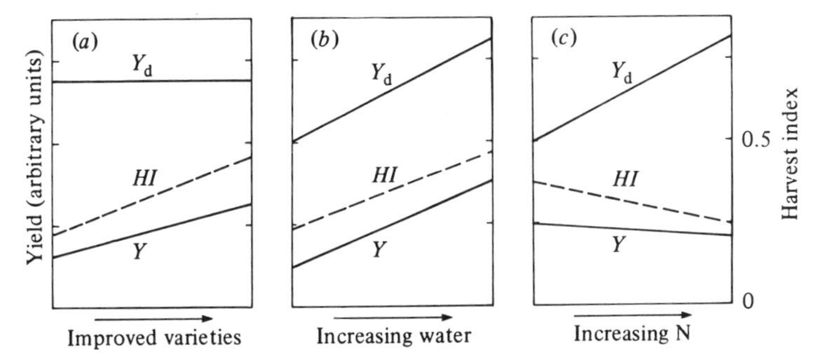

Fig. 12.4. Contribution of changes in total dry matter production (Y_d) and harvest index (*HI*) to economic yield (*Y*): (*a*) with improved varieties (largely increased *HI*); (*b*) increasing water (both Y_d and *HI* increase); (*c*) increasing nitrogen fertilisation (Y_d increases but *HI* decreases). (After Donald & Hamblin 1976.)

Modelling and determination of crop ideotype

Several examples where modelling techniques have been used to investigate the consequences of changes in plant morphology or of physiological response in different environments have already been discussed: the models for water use efficiency that were described in Chapter 10 can be used to determine optimum combinations of leaf size and stomatal behaviour for particular environments, while the possibility of an optimum leaf angle for maximal radiation interception was pointed out in Chapter 2. Such models can be used in attempts to explain plant distribution; or else the same approach may be adopted to determine crop ideotypes for maximum yield in certain environments.

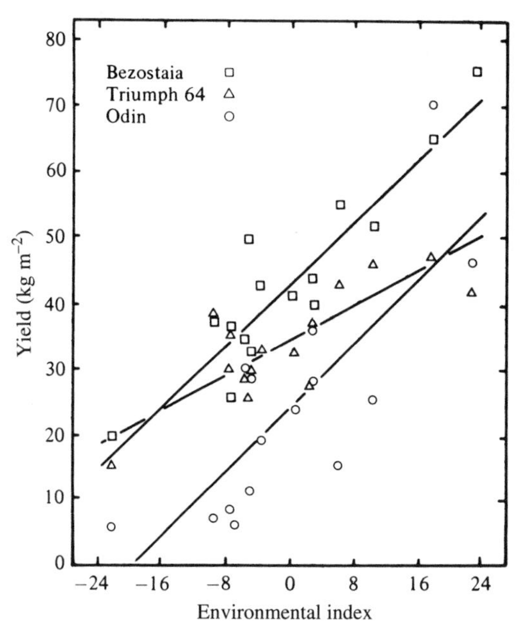

Fig. 12.5. Genotype × environment interactions in wheat yields in a multi-site international yield trial. Yield is plotted against an environmental index, the site mean yield determined over all varieties. (After Stroike & Johnson 1972.)

In this section there is a very simple example of optimisation to illustrate the principles of how modelling can be used in the determination of a crop ideotype. In general this approach involves the use of some type of optimisation procedure.

Timing of switch from vegetative to reproductive development

In an interesting example of the use of optimisation theory, Cohen (1971) showed that for any given environment, the maximum grain yield will be achieved by a plant that switches suddenly from a vegetative phase, where all available photosynthate is used for vegetative growth, to a reproductive phase where all resources are used for grain growth (see also Paltridge & Denholm 1974). Many plants, for example cereals such as wheat, in fact approximate this type of response, where flowering on all shoots is approximately synchronous, occurring after vegetative growth ceases. On the other hand, there are many species, for example most legumes, that have an 'indeterminate' flowering pattern. That is, they continue to grow and produce flowers while the earliest formed seeds are ripening. This indeterminate behaviour is probably an adaptation to a relatively unpredictable environment with some seed being produced even if the growing season is very short, without sacrificing the potential for further seed production if water supply is maintained.

A question which might face a plant breeder breeding new plants for a particular environment, or an agronomist choosing a suitable genotype, is 'when is the optimum time for this vegetative to floral switch?'. To answer this one has first to set up an appropriate model for the crop, and then find the optimum by an appropriate method, which may be graphical, analytical or by an iterative process of trial and error, usually using a computer. Methods are outlined by Thornley & Johnson (1990) and Gold (1977).

To start with a very simple model, one can assume that any photosynthate can go either into grain growth or into expansion of leaf area and that there is a fixed growing season of T days available (for ease of calculation assume $T = 10$ 'days', though a more realistic number can easily be substituted). If it is further assumed that the mass of photosynthate (m, in CO_2 equivalents) that gives one unit of leaf area (A) equals b, it follows that in the vegetative phase when all photosynthate is diverted into leaf area

$$\mathrm{d}A = \mathrm{d}m/b \tag{12.1}$$

The value of b depends on leaf thickness and the amount of photosynthate that is required for supporting structures such as stems and roots. Similarly one might assume that the rate of production of photosynthate ($\mathbf{P} = \mathrm{d}m/\mathrm{d}t$) is proportional to the leaf area (though see below), i.e.

$$\mathbf{P} = aA \tag{12.2}$$

where a is a constant. It is also necessary to assume an initial leaf area (A_0). On this basis various models can be derived:

(*a*) *Discontinuous model.* Perhaps the simplest approach is to assume that increases in leaf area only occur at night, so that the growth in leaf area over the season is given by the discontinuous curve in Figure 12.6.

If the vegetative to floral switch occurs early in the season there will be a long period available for filling grain, but the rate of grain filling will be slow because of the small leaf area available (dotted line in Fig. 12.6*a*). If, however, the switch occurs later, the area (and consequently photosynthetic rate) is larger but the time available shorter.

The total photosynthate available for grain filling is given by the product of the photosynthetic rate for the appropriate area × time left. If the area after t days is denoted by A_t, the potential yield of grain dry matter (Y_d, again in CO_2 equivalents) is given by

$$Y_d = aA_t(T - t) \tag{12.3}$$

The optimum time for the switch is that which gives a maximum Y_d, and may be determined by solving equation 12.3 for all t between 0 and T, as

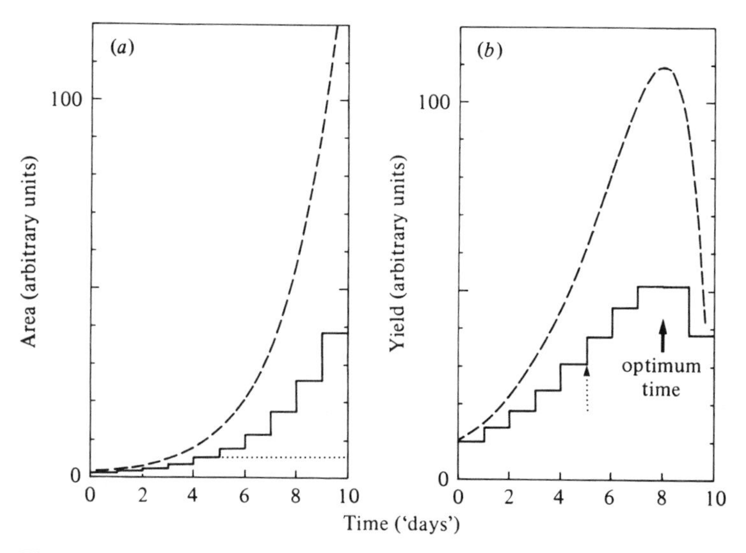

Fig. 12.6. Yield models for a cereal (see text). (*a*) Increase in leaf area with time for discontinuous model (solid line) or continuous model (dashed line). The dotted line illustrates the time course for leaf area for the case where the switch to flowering occurs after 5 'days'. (*b*) Dependence of final yield for the two models on the time of the switch from vegetative to floral development. The dotted arrow indicates the yield corresponding to a flowering switch after 5 days.

done in Fig. 12.6*b*. It is clear from this that for the particular values chosen ($A_0 = 1$, $a = 1$, $b = 2$), the optimum switch occurs after 8 days.

(*b*) *Continuous model.* If photosynthate can be used immediately to generate new leaf area according to equation 12.1, the continuous exponential growth curve shown dashed in Fig. 12.6*a* is obtained. In this case, the rate of change of area can be written

$$dA/\mathrm{d}t = (1/b)\,\mathrm{d}m/\mathrm{d}t = \mathbf{P}/b \qquad (12.4)$$

Substituting from equation 12.2

$$\mathrm{d}A/\mathrm{d}t - (a/b)\,A = 0 \qquad (12.5)$$

This is a first-order differential equation of the form already seen above (e.g. equation 4.30, p. 100), the solution being

$$A = c\mathrm{e}^{(a/b)t} \qquad (12.6)$$

where c is an arbitrary constant. The value of c may be obtained by substituting particular known values for the variables (= boundary conditions), i.e. at $t = 0$, $A = A_o$ which gives

$$c = A_o \qquad (12.7)$$

Substituting back into equation 12.6, a function for the time dependence of A is given by

$$A_t = A_o e^{(a/b)t} \qquad (12.8)$$

As before, one can solve to find the optimum value for t by calculating Y_d at various values of t and obtaining the optimum by graphical interpolation (Fig. 12.6*b*).

A more elegant solution is to make use of calculus. Equation 12.8 can be substituted into equation 12.3 to give

$$Y_d = aA_o e^{(a/b)t}(T-t) \qquad (12.9)$$

It can be seen from Fig. 12.6*b* that at the maximum Y_d, the slope of the curve relating Y_d and t is horizontal (i.e. $dY_d/dt = 0$). Therefore the optimum value for t can be determined by differentiating

$$dY_d/dt = aA_o e^{(a/b)t}[(a/b)T-(a/b)t-1] \qquad (12.10)$$

then setting (dY_d/dt) equal to zero, and solving for t. This gives $t = T-(b/a)$, which can easily be shown to be an optimum rather than a minimum. As b increases, the optimum time gets earlier.

(*c*) *More complex models.* Although the above models are useful for illustrative purposes, they include many major simplifications, such as that involved in equation 12.2. In practice, a more appropriate equation takes account of the diminishing returns as leaf area increases beyond a leaf area index of about 3, so that equation 12.2 can be replaced by, for example, a rectangular hyperbola (see Chapter 7):

$$\mathbf{P} = \mathbf{P}_{max} A(k+A) \qquad (12.11)$$

Unfortunately, the differential equations resulting from the use of this or other more realistic assumptions rapidly become too complicated to be solved analytically, so that solution of more realistic models generally requires the use of numerical procedures on a computer using packages such as ACSL (Advanced Computer Simulation Language). Other refinements could include treatment of seasonal and daily changes in environmental conditions, adequate modelling of stomatal behaviour and so on.

In fact many large-scale dynamic simulation models have been developed for particular crops such as sorghum (Arkin *et al.* 1976), sugar beet (Fick *et al.* 1975) and apple (Landsberg 1981). These models usually incorporate a collection of subsystems that describe individual crop processes such as light interception, photosynthesis, respiration, water use and dry matter partitioning. It is then possible to investigate the consequences of changing any component in isolation (e.g. time of flowering), to determine the optimum value. Sensitivity analysis of these large-scale simulation models can be used to pin-point deficiencies in our understanding of processes controlling yield and to determine those yield-determining processes that are most amenable to alteration by breeding or management. Large-scale ecosystem models can also be used in a similar fashion to investigate the consequences of changes in biotic or physical environment.

In spite of the greater realism in many of these large simulation models, there is a lot to be said for the approach with which this illustration began, that is, of using the simplest possible model for the purpose, since this can frequently give a better and more comprehensible picture of the system behaviour than the more complex models with all the interactions that occur.

The simple 'trial-and-error' optimisation approach described above can be usefully applied to a wide range of problems. I have found that the use of exercises of this type based on extremely simple assumptions can provide instructive insights into optimisation and how different factors interact in determining plant yield. Suitable examples for class projects include investigation of optimal sowing density (balancing the cost of extra seed against the earlier canopy closure at high seed rates), determining optimal stomatal behaviour (see Chapter 10) or determining the optimal partitioning of photosynthate between roots and shoots when water is limited.

Validation of the ideotype

An essential stage in development of an ideotype for breeders is its validation. In its simplest form this involves a positive correlation within a set of genotypes between expression of the character and yield under appropriate conditions. A better approach is to use isogenic lines which differ only in the character under investigation. Unfortunately preparation of isogenic lines usually requires a complete breeding programme itself as it involves a long series of back-crosses, unless the character appears as a mutant within a commercial variety. For this reason very few characters have been tested using isogenic lines. Examples include stomatal frequency (see below) and the uniculm (single stem) mutant in barley.

Development of screening tests

Rapid screening tests have been successfully used particularly in breeding for disease and pest resistance. In addition, seedling survival tests are used for screening large numbers of plants for tolerance of cold, heat, drought or salt stress. Physiological knowledge has also proved valuable in selecting plants with appropriate daylength or vernalisation requirements for particular environments.

Many other physiological tests have been proposed as a means of screening for particular characters though they have rarely been successful in yield improvement, often because of compensation effects. A typical example of a physiological mass screening test is the compensation point test for C_4 plants described in Chapter 7.

There has been much interest in selecting cultured protoplasts for tolerance of stresses such as osmotica or salinity, because of the opportunities that this system provides for applying high selection pressures. The aim is that selected cell lines can be used directly for the clonal regeneration of improved genotypes, but these studies have not yet lived up to expectation, partly because of the very different tolerance mechanisms commonly found in intact plants as compared with isolated cells (Rains 1989). Once particular gene sequences or gene products have been identified that are either involved in determining particular required characters, or else are closely linked to them, molecular and immunological techniques will then allow rapid screening of recombinants at an early stage.

To be of value, a useful screening test must satisfy a number of criteria:

1. The character must be easier to assess than yield itself.
2. There must be correlation (preferably causal) between the character and yield in the field.
3. The test should be simple, rapid, cheap and preferably capable of being used on seedlings at any time of year.
4. There must be heritable variation in the character.
5. A test involving a single measurement is likely to be better than the more complex tests usually required for estimates of responses.

Attributes for which screening can be carried out fall into four classes (Austin & Jones 1975):

1. *Morphological and anatomical*, for example plant height, leaf size or stomatal frequency. In practice, these are the easiest and most widely used by breeders.
2. *Compositional.* Screening for grain composition, such as protein or lysine content, is widespread. Also included in this class is screening for hormone content, such as concentration of abscisic acid as a test for drought tolerance (see below).
3. *Process rates*, for example photosynthesis, respiration or vernalisation.
4. *Process control*, for example enzyme activity or stomatal aperture and its behaviour.

For many characters it is necessary to test the performance of the selected genotype on a crop scale, because as we have seen earlier (e.g. Chapter 5 for sensitivity of evaporation and water use efficiency to stomatal behaviour), the performance of the genotype depends on its interaction with neighbours and with the environment itself. Models can, however, provide a useful basis for predicting behaviour and limiting wasted effort on the development of inappropriate ideotypes.

Examples of applications

In the past, many successful applications of environmental physiology have been largely 'explanatory', though there have been important applications in the area of crop management including the use of evaporation models in irrigation scheduling and light penetration models in the evaluation of pruning systems for fruit tree management. In this section some examples of physiological methods that have potential for inclusion in plant breeding programmes are examined.

Breeding for drought tolerance

Drought is a major factor limiting yields of crops in many areas, but irrigation is often either not possible or uneconomic, so that much effort has been (and is being) devoted to breeding drought-tolerant cultivars. There are many possible characters for drought tolerance (Chapter 10) but, as was seen there, the best combination depends on the crop, the climate and even on the farming system. A list of characters that have been suggested for inclusion in a particular drought-tolerance breeding programme for sorghum (a crop predominantly grown in semi-arid areas) is presented in Table 12.2. I have indicated with + + + those characters that with current breeding technology are most likely to provide significant improvements in sorghum drought tolerance. A number of other screening tests, such as the use of chlorophyll *a* fluorescence (Chapter 7; Havaux & Lannoye 1985) or carbon isotope discrimination (Chapter 10), have been proposed and could perhaps be added to this list. Although there is evidence that there may be significant genetic variation in many metabolic characters (e.g. respiration) or responsive characters (e.g. developmental plasticity or stomatal closure), their use is at present speculative and really needs improved screening techniques. The use of heat and desiccation tolerance tests applicable to large numbers of seedlings holds rather more promise, but general validation in terms of correlation with yield under drought conditions is still awaited.

Use of stomatal characters

Because of the central role that stomata have in the control of water loss, much effort has concentrated on the use of stomatal characters in breeding for drought tolerance (see e.g. Table 12.2, numbers 10 and 14). Physiological

Table 12.2. *Characters for drought tolerance of possible use in a sorghum improvement programme (modified from Seetharama* et al. 1982)

	Techniques	Genetic variability	Prospects for breeding
Morphological/phenological			
1. Maturity	Visual	+++	+++
2. Developmental plasticity	Visual	+?	+?
3. Glossy leaf	Visual	+++	+++
4. Leaf number, size, shape	Visual	++	++
Physiological – constitutive			
5. Desiccation tolerance	Survival tests	+++	+++
6. Heat tolerance	Survival tests, ion leakage	+++	+++
7. High growth rates	Visual, growth analysis	++?	++
8. Low respiration	Gas exchange	++?	+?
9. Recovery after stress	Visual	+++	+++
10. Anatomical (e.g. stomatal density)	Microscope	++?	?
11. Root/shoot ratio	Growth analysis	++	+
12. Liquid phase resistance	Pressure chamber, anatomy	?	?
13. Deep roots	Root boxes	++	+?
Physiological – facultative			
14. Stomatal closure	Leaf temperature, porometer	++	?
15. Leaf rolling	Visual	+++	++
16. Epidermal wax production	Visual, chemical analysis	++	++?
17. Leaf area increase	Visual	++	+?
18. Leaf senescence	Visual	+++	+++
19. Remobilisation of stem reserves	^{14}C, growth analysis	++?	+++?
20. Osmotic adjustment	Psychrometry, chemical analysis	++?	+?
21. Relative increase in root growth	Root box, growth analysis	++	++

The number of +'s indicates the variability that exists for any character and the possibilities for breeding.

thinking, along the lines described in Chapter 10, has led to the widely held view that drought tolerance would be favoured by reduced rates of water use, at least in some situations. This should be attainable by reducing stomatal conductance. Several approaches have therefore been tried by physiologists and breeders to select for stomatal characters (see Jones 1979*b* for a review of possible methods and of the practical problems).

(i) *Stomatal frequency.* Because conductance might be expected to decrease as stomatal frequency decreases (Chapter 6), there have been many attempts to select for reduced numbers of stomata per unit area of leaf. There is a fair degree of genetic variation for stomatal frequency and some of these studies have been successful in their initial objective. Although there are reports that the reduced frequency can reduce water use or increase water use efficiency, it can also be disadvantageous. In an attempt to validate the idea that a reduced stomatal frequency per unit area should improve drought tolerance, isogenic lines for stomatal frequency were developed in barley. Unfortunately it was discovered that, contrary to expectations, the low frequency lines transpired water up to 6% faster than the high frequency lines (Table 12.3). Further analysis showed that although stomatal frequency had been successfully reduced by selection, this had been offset by increases in pore size (so that leaf conductance was unchanged) and in total leaf area. In fact the increase in leaf area had dominated in its effect on transpiration (Table 12.3). Future attempts to breed for low stomatal frequency must therefore take account of the negative correlation between leaf (and cell) size and stomatal frequency.

(ii) *Stomatal conductance.* Direct measurements of stomatal conductance should, in principle, be better than indirect characters such as stomatal frequency. Unfortunately, though such measurements can be rapid with modern porometers, biological and environmental variability (particularly in the field) means that large numbers of measurements are required to distinguish different genotypes (see Jones 1979*b*).

(iii) *Stomatal response.* A potential disadvantage of selection for low stomatal frequency or low conductance is that it might limit potential photosynthesis where water is not limiting. Therefore it should, in theory, be better to select a plant that is effective at closing its stomata in *response* to drought (Table 12.2). Unfortunately it is more difficult to measure a response than a steady state because one needs to make at least twice as many measurements, while it may also be necessary to measure Ψ_ℓ. A further complication is that both mean conductance and mean response tend to change during ontogeny (see Jones 1979*b*).

(iv) *Abscisic acid production.* Because of the difficulty of selecting directly for stomatal response and because it is known that the plant growth regulator abscisic acid (ABA) is both produced in stress and closes stomata,

Table 12.3. *Effect of selection for stomatal frequency on water use by barley (data from Jones 1977b). Transpiration rate quoted is for 45 days after sowing*

	Whole plant		Flag leaf			
	Transpiration ($g\ day^{-1}$)	Leaf area (cm^2)	Stomatal frequency (mm^{-2})	Pore length (μm)	Leaf area (cm^2)	Leaf conductance ($mm\ s^{-1}$)
High frequency lines:						
Minn. 92–43	73	329	83.4	16.7	24.9	4.6
CI 5064	75	412	97.6	17.7	18.3	4.6
Low frequency lines:						
Minn. 161–16	110	513	67.8	19.1	29.0	5.1
CI 4176	128	522	65.8	20.4	21.7	4.7

Table 12.4. *Breeding for ABA accumulation. Interpretation of differences in ABA accumulation in response to stress in different species*

High ABA accumulation	
(*a*) leads to good *control* of stress	– Tolerant varieties of maize, sorghum
or	
(*b*) is a *measure* of stress	– Susceptible varieties of wheat
Low ABA accumulation	
(*a*) because good avoider of stress	– Tolerant varieties of wheat
(*b*) because poor ability to synthesise ABA	– Susceptible varieties of maize, sorghum

it has been suggested that selection for a high production of ABA in response to drought would be a good short-cut to drought-tolerant varieties. Unfortunately such an approach is clearly too naïve, as illustrated in Table 12.4, because high levels of ABA may either indicate good stomatal control and therefore avoidance of stress (though with a corresponding potential reduction in photosynthesis – e.g. in maize: Larqué-Saavedra & Wain 1974), or they may occur *because* the plant is particularly stressed (as may be the case in wheat: Quarrie & Jones 1979). Breeding experiments for a number of species have now confirmed the heritability of ABA production; in at least one case (Innes *et al.* 1984), the high ABA accumulators have tended to have the highest yields (see Quarrie 1991 for a detailed discussion of breeding for different ABA accumulation capacity).

Photosynthesis and crop yield

Photosynthesis is clearly essential for crop yield and there is wide variation between plants in leaf photosynthetic rates. It might be expected, therefore, that increases in photosynthesis should be one way of increasing yield. Although there has been controversy as to whether yields are limited by 'source' processes (i.e. photosynthesis) or by 'sink processes' (e.g. capacity for grain growth), the general consensus is that both are important and co-limiting.

With this background, many studies have demonstrated varietal or species differences in photosynthesis at various levels including activity of Rubisco, ^{14}C fixation by protoplasts, leaf slices or discs, or CO_2 fixation at the leaf level. There has, however, rarely been an indication from any of these studies of a clear positive association between leaf photosynthetic rate and yield. In fact the reverse tends to be true. For example, Dunstone *et al.* (1973) in a good early study, the results of which have been confirmed several times since, showed that in wheat the highest photosynthetic rates tend to occur in small-leaved primitive diploid species, and the lowest rates in the high-yielding modern hexaploid varieties. It seems that total leaf area, and leaf area display, are much more crucial determinants of yield. This has been confirmed by many studies showing a close relationship between dry matter productivity and leaf area or intercepted radiation (see Chapter 7).

Notwithstanding the failure so far to relate crop yield to leaf photosynthesis, many workers are still optimistic, and it is probably true that such a relationship would hold if *other things could be kept equal.* This is the crux of the problem of attempting to incorporate any single character into a new variety and remains a major challenge for the future.

An ideotype for wheat in Britain

A final example of the range of physiological attributes that have been proposed for one crop is shown in Table 12.5. In this Table, attention has been concentrated on those attributes for which it should be possible to undertake some screening with present technology. Selection is already being practised for many of these characters. Although proposed some years ago, the main suggestions are still generally applicable as there has been rather little progress in developing biochemical screens, other than for characters such as protein quality of the grain. The Table emphasises the fact that a delicate balance must be maintained, as many characters proposed have potential disadvantages as well as advantages under certain conditions. Some characters (such as high harvest index) will be of value in any environment, while others (such as low root hydraulic conductivity) might be expected to be of value only in specific situations, such as the driest parts of the country. There is often a conflict between the requirements for wide adaptability and those for specific environments. Furthermore, many desirable characters are mutually exclusive, or at least partly so (e.g. large ears and high tiller number) and the extent to which these are inherent in the physiology or are a consequence of genetic linkage is often not known.

Table 12.5. *Physiological attributes of the ideal model wheat plant for British conditions* (*modified from Austin & Jones* 1975)

Attribute	Possible benefits	Possible disadvantages
Seed		
Large seed	Rapid seedling emergence. Reduced sensitivity to sowing depth	Unknown
High concentration of seed protein	High grain quality and seedling vigour	May be low yield of seed per hectare
Root growth		
Adequate investment in root growth	Plants less susceptible to drought	Rapid depletion of soil water. Root growth may occur at expense of shoot growth
Low root hydraulic conductivity	Water conservation in drought	Adverse effects in conditions of ample water supply
Continued root growth during the critical phase of ear development and during grain growth	Supply of water and nitrogen assured. May delay demand for nitrogenous compounds from upper leaves, and so prolong their photosynthetic life	May be adverse competitive effects on ear development and grain growth
Nitrogen economy		
Ability to take up and reduce nitrate rapidly; ability to store nitrate and reduced nitrogen compounds	Minimises loss of soil nitrogen by leaching and denitrification	Possible toxic accumulation of nitrogen compounds at high levels of nitrogen fertilisation
Rapid export of nitrogen from leaves when senescing	Particularly important if soil nitrogen or water is limiting	Unknown

Tillering and development		
Early and near synchronous formation of tillers; absence of late-tiller production	No competition from late-formed tillers. Avoidance of 'wasteful' use of water, minerals and assimilates	Reduced ability to compensate for death of main shoot and tillers
Homeostasis of tiller number per unit area of ground (perhaps 800 m^{-2} in winter wheat)	Compensation for variation in seeding rates and seedling survival	Homeostasis mechanism may be dependent on tiller death
In spring wheats, photoperiod and temperature responses giving constant ear size but smaller tiller number with later sowings	Yield per hectare can be maintained in later sowing by increased seeding rate	Need for information on how seeding rates should be adjusted to allow for variation in sowing date
Response to photoperiod and temperature to give slow growth rates during cold spells and in midwinter	Favours development of winter-hardiness	Small plants, possibly with smaller ear primordia and fewer, or later developing, tillers
Morphology		
Canopy structure (e.g. erect upper leaves and lax lower leaves) to give maximum interception of light	Maximises canopy photosynthesis rates	Erect leaves give poor interception if leaf area index is less than 3
Dwarf habit	Confers resistance to lodging. High harvest index	May adversely affect crop microclimate and have agronomic disadvantages
High stem density (dry weight per unit stem length) and resistance to flexing	Confers resistance to lodging. Stem reserves can provide some 'insurance' against shortfall in assimilate during late grain filling (caused by drought or disease)	May be adverse competitive effects on ear development. Too extensive mobilisation of stem reserves could lead to lodging and brackling

Table 12.5 (*cont.*)

Attribute	Possible benefits	Possible disadvantages
Highest possible dry weight per unit ground area at anthesis, concentrated in the ear-bearing tillers	Favours maximum uptake of nitrogen and, in turn, grain protein yield per hectare	May be adverse competitive effects and reduced efficiency of water use. May be associated with late anthesis
Photosynthesis and gas exchange		
High rates of photosynthesis per unit leaf area	Increases potential for dry matter production	May be compensating changes in leaf size and thickness which offset any benefits
May be achieved by		
(*a*) low mesophyll resistance to CO_2 uptake	Increases efficiency of water use, especially important in drought	Unknown
(*b*) low photorespiration rate	Particularly advantageous at high temperatures and in drought	Unknown
(*c*) low stomatal resistance. High stomatal frequency	Useful only in conditions of ample water supply	In dry situations, there may be an unacceptably high rate of water loss
Stomata, sensitive to water stress, but reopening quickly after stress is relieved	Prevents irreversible damage in periods of drought	Undesirable with short periods of water stress
High cuticular resistance to water loss	Reduces transpiration rate. Particularly advantageous in drought	Unknown

APPENDIX 1

Units and conversion factors

The International System of Units (SI) is followed. The following table defines the derived units in terms of SI base units and gives some useful conversions to other units in common use.

Quantity	SI base units	SI derived units	Equivalent forms of SI units	Equivalents in other units
Mass	1 kg			= 2.2046 pounds
Length	1 m			= 3.2808 feet
Time	1 s	(or min, h, etc.)		
Temperature	1 K			= 1 °C
Electric current	1 A			
Amount of substance	1 mol			
Energy	1 kg m^2 s^{-2}	joule (J)	N m	= 10^7 erg = 0.2388 calorie
Force	1 kg m s^{-2}	newton (N)	J m^{-1}	= 10^5 dyne
Pressure	1 kg m^{-1} s^{-2}	pascal (Pa)	N m^{-2}; J m^{-3}	= 10 dyne cm^{-2} = 10^{-5} bar = 0.9869×10^{-5} atmosphere = 7.5×10^{-3} mm Hg
Power	1 kg m^2 s^{-3}	watt (W)	J s^{-1}	= 10^7 erg s^{-1}
Electric charge	1 s A	coulomb (C)		
Electric potential difference	1 m^2 kg s^{-3} A^{-1}	volt (V)	J C^{-1}	
Kinematic viscosity	1 m^2 s^{-1}			= 10^4 stokes
Dynamic viscosity	1 Pa s			= 10 poise

APPENDIX 2

Mutual diffusion coefficients of binary mixtures containing air or water at 20 °C

Values of $\boldsymbol{D}$ in air may be corrected for temperature (giving values with less than 1 % error over the range 0–45 °C) by multiplying by $(T/293.2)^{1.75}$ (data for solutes from Weast 1969, others mostly from Monteith & Unsworth 1990).

		In air ($mm^2\ s^{-1}$)	In water ($mm^2\ s^{-1}$)
Water	– $\boldsymbol{D}_W$	24.2	0.0024[a]
Carbon dioxide	– $\boldsymbol{D}_C$	14.7	0.0018
Oxygen	– $\boldsymbol{D}_O$	20.2	0.0020
Heat	– $\boldsymbol{D}_H$ (= thermal diffusivity)	21.5	0.144
Momentum	– $\boldsymbol{D}_M$ (= kinematic viscosity, $\boldsymbol{\nu}$)	15.1	1.01
Sucrose (0.38 % solution)		—	0.52×10^{-3}
Glycine (dilute)		—	1.06×10^{-3}
$CaCl_2$ ($10\ mol\ m^{-3}$)		—	1.12×10^{-3}
NaCl ($10\ mol\ m^{-3}$)		—	1.55×10^{-3}
KCl ($10\ mol\ m^{-3}$)		—	1.92×10^{-3}

[a] Coefficient of self-diffusion.

APPENDIX 3

Some temperature-dependent properties of air and water

Density of dry air (ρ_a), density of air saturated with water vapour (ρ_{as}), psychrometer constant ($\gamma = Pc_p/0.622\,\lambda$), latent heat of vapourisation of water (λ), radiative resistance ($r_R = \rho c_p/4\varepsilon\sigma T^3$), the factor converting conductance in units of mm s^{-1} to mmol m^{-2} s^{-1} ($\mathcal{g}/g = g^m/g = P/\mathcal{R}T$) and kinematic viscosity of water (ν). (At 100 kPa where appropriate.)

T (°C)	ρ_a (kg m^{-3})	ρ_{as} (kg m^{-3})	γ (Pa K^{-1})	λ (MJ kg^{-1})	r_R (s m^{-1})	$\mathcal{g}/g$ $\left(\frac{\text{mmol m}^{-2}\text{ s}^{-1}}{\text{mm s}^{-1}}\right)$	ν (mm^2 s^{-1})
−5	1.316	1.314	64.6	2.513	304	44.8	—
0	1.292	1.289	64.9	2.501	282	44.0	1.79
5	1.269	1.265	65.2	2.489	263	43.2	—
10	1.246	1.240	65.6	2.477	244	42.5	1.31
15	1.225	1.217	65.9	2.465	228	41.7	—
20	1.204	1.194	66.1	2.454	213	41.0	1.01
25	1.183	1.169	66.5	2.442	199	40.3	—
30	1.164	1.145	66.8	2.430	186	39.7	0.80
35	1.146	1.121	67.2	2.418	174	39.0	—
40	1.128	1.096	67.5	2.406	164	38.4	0.66
45	1.110	1.068	67.8	2.394	154	37.8	—

APPENDIX 4

Temperature dependence of air humidity and associated quantities

Saturation water vapour pressure (e_s), saturation water vapour concentration (c_{sW}), slope of saturation vapour pressure curve (s) and the ratio of the increase of latent heat content to increase of sensible heat content of saturated air ($\varepsilon = s/\gamma$).

T (°C)	e_s (Pa)	c_{sW} (g m^{-3})	s (Pa °C^{-1})	ε	T (°C)	e_s (Pa)	c_{sW} (g m^{-3})	s (Pa °C^{-1})	ε
−5	421 (402)[a]	3.41	32	0.50	20	2337	17.30	145	2.20
−4	455 (437)[a]	3.66	34	0.53	21	2486	18.34	153	2.31
−3	490 (476)[a]	3.93	37	0.57	22	2643	19.43	162	2.44
−2	528 (517)[a]	4.22	39	0.60	23	2809	20.58	170	2.56
−1	568 (562)[a]	4.52	42	0.65	24	2983	21.78	179	2.69
0	611	4.85	45	0.69	25	3167	23.05	189	2.84
1	657	5.19	48	0.74	26	3361	24.38	199	2.99
2	705	5.56	51	0.78	27	3565	25.78	210	3.15
3	758	5.95	54	0.83	28	3780	27.24	221	3.31
4	813	6.36	57	0.88	29	4005	28.78	232	3.48
5	872	6.79	61	0.94	30	4243	30.38	244	3.66
6	935	7.26	65	1.00	31	4493	32.07	257	3.84
7	1002	7.75	69	1.06	32	4755	33.83	269	4.02
8	1072	8.27	73	1.12	33	5031	35.68	283	4.22
9	1147	8.82	78	1.19	34	5320	37.61	297	4.43
10	1227	9.40	83	1.26	35	5624	39.63	312	4.65
11	1312	10.01	88	1.34	36	5942	41.75	327	4.86
12	1402	10.66	93	1.42	37	6276	43.96	343	5.09
13	1497	11.35	98	1.49	38	6626	46.26	357	5.33
14	1598	12.07	104	1.58	39	6993	48.67	376	5.58
15	1704	12.83	110	1.67	40	7378	51.19	394	5.84
16	1817	13.63	117	1.77	41	7780	53.82	413	6.11
17	1937	14.48	123	1.86	42	8202	56.56	432	6.39
18	2063	15.37	130	1.97	43	8642	59.41	452	6.68
19	2196	16.31	137	2.07	44	9103	62.39	473	6.98

[a] Saturation vapour pressure over ice.

The value of $e_{s(T)}$ (the saturation vapour pressure of moist air over water) is approximated over the normal range of environmental temperatures by the following empirical relationship (Buck 1981):

$$e_{s(T)} = f\left(\mathrm{a}\exp\left\{\frac{\mathrm{b}T}{\mathrm{c}+T}\right\}\right)$$

where the values of the various coefficients are

$\mathrm{a} = 611.21$
$\mathrm{b} = 17.502$
$\mathrm{c} = 240.97$
$f = 1.0007 + 3.46 \times 10^{-8} P$

and T is in °C, and $e_{s(T)}$ and P are in Pa. The second term in this equation gives the saturation pressure of pure water vapour over water, while the first term (f) represents an 'enhancement factor' to convert the saturation pressure of pure water vapour to the saturation partial pressure of water vapour in moist air. Because the enhancement factor represents a correction of only about 0.4% at normal atmospheric pressure it is usual to incorporate it into a as a constant (equation 5.12).

APPENDIX 5

Thermal properties and densities of various materials and tissues (at 20 °C)

Selected principally from Herrington (1969), Weast (1969), Leyton (1975) and Edwards *et al.* (1979).

	Specific heat capacity, c_p (J kg^{-1} K^{-1})	Thermal conductivity, k (W m^{-1} K^{-1})	Density, ρ (kg m^{-3})
Air	1010	0.0257	1.204
Aluminium	896	237.0	2710
Cellulose	2500	—	1270–1610
Glucose	1260	—	1560
Plant leaves	3500–4000	0.24–0.57	530–910
Seasoned oak	2400	0.21–0.35	820
Fresh red pine	1960–3130	0.15–0.38	360–490
Polyethylene (high density)	2090	0.33	960
Polyvinyl chloride	1050	0.092	1714
Clay soil: dry	890	0.25	1600
Clay soil: wet (40% water)	1550	1.58	2000
Peat: dry	300	0.06	1920
Peat: wet (40% water)	1100	0.50	3650
Water	4182	0.59	998.2[a]

[a] Rising to a maximum of 1000 kg m^{-3} at 4 °C.

APPENDIX 6

Physical constants and other quantities

Acceleration due to gravity ($\boldsymbol{g}$):	9.8067 m s^{-2} (at sea level, latitude 45°)
Avogadro's number:	6.022×10^{23} particles mol^{-1}
Gas constant ($\mathscr{R}$):	8.3144 J K^{-1} mol^{-1}
Planck's constant (h):	6.6262×10^{-34} J s
Solar constant ($\boldsymbol{\Phi}_{pA}$):	1370 W m^{-2}
Speed of light ($\boldsymbol{c}$):	2.998×10^{8} m s^{-1}
Stefan–Boltzmann constant (σ)	5.6703×10^{-8} W m^{-2} K^{-4}
Molar volume of ideal gas at 0 °C:	2.27106×10^{-2} m^3 mol^{-1} (at 100 kPa)
	2.241×10^{-2} m^3 mol^{-1} (at 101.3 kPa)
Molecular weight of air (M_A):	28.964×10^{-3} kg mol^{-1}
Specific heat of air (c_p):	1012 J kg^{-1} K^{-1}
Water – dielectric constant ($\mathscr{D}$):	80.2 (at 20 °C)
Water – dynamic viscosity ($\eta = \rho \boldsymbol{v}$):	1.008×10^{-3} N s m^{-2} (at 20 °C) (= Pa s)
Water – latent heat of fusion:	334 kJ kg^{-1} or 6.01 kJ mol^{-1}
Water – partial molal volume ($\bar{V}_W$):	18.05×10^{-6} m^3 mol^{-1} (at 20 °C)
Water – surface tension against air ($\boldsymbol{\sigma}$):	74.2×10^{-3} N m^{-1} (at 10 °C)
	72.8×10^{-3} N m^{-1} (at 20 °C)
	71.2×10^{-3} N m^{-1} (at 30 °C)

APPENDIX 7

Solar geometry and irradiance

Useful relationships for calculating irradiance for modelling purposes may be derived from spherical geometry. These include the following (expressing all angles in degrees):

1. *Solar elevation.* The solar elevation at any site is given by:

$$\sin\beta = \cos\theta = \sin\lambda\sin\delta + \cos\lambda\cos\delta\cos h \tag{A7.1}$$

where β is the solar elevation above the horizontal, θ is the zenith angle of the sun (the complement of β), λ is the latitude of the observer, δ is the angle between the sun's rays and the equatorial plane of the earth (solar declination) and is a function only of the time of year (see Table A7.1), h is the hour angle of the sun (the angular distance from the meridian of the observer) and is given by 15 $(t-t_0)$ where t is the time in hours and t_0 is the time at solar noon. Unfortunately the time of solar noon varies during the year by an amount that is given by 'the equation of time' (Table A7.1). In the Western hemisphere, the standard time at local apparent noon = 12.00 – (equation of time) – 4 × (longitude in degree). As an example of the method of calculation, standard time at local apparent noon at New York (74° W) on 1 February would be 07 h 12.3 min (i.e. 12 h + [17.3 – 296] min), which is equal to a local time (Eastern Standard Time) of 12 h 12.3 min because Eastern Standard Time is 5 h before the Greenwich standard. Further details of the calculation of solar noon on any date for any given longitude and latitude are described e.g. by Šesták *et al.* (1971).

2. *Daylength.* The daylength (N), that is the number of hours that the sun is above the horizon, may be obtained by solving equation A7.1 for $\beta = 0$.

This gives the hour angle of the sun, h, at sunrise or sunset as

$$\cos h = -\tan\lambda\tan\delta \tag{A7.2}$$

so that the daylength in hours equals $2\,h/15$.

Table A7.1. *Solar declination (δ, degree) and the equation of time (e, min) on the first day of each month*

For simulation purposes δ may be obtained from

$\delta = -23.4\cos[360(t_d + 10)/365]$

where t_d is the number of the day in the year.

Month	δ	e	Month	δ	e
January	−23.1	−3	July	+23.2	−4
February	−17.3	−14	August	+18.3	−6
March	−8.0	−13	September	+8.6	0
April	+4.1	−4	October	−2.8	+10
May	+14.8	+3	November	−14.1	+16
June	+21.9	+2	December	−21.6	+11

3. *Angle between any surface and the sun.* This is given by:

$$\cos\xi = [(\sin\lambda\cos h)(-\cos\alpha\sin\chi) - \sin h(\sin\alpha\sin\chi) + (\cos\lambda\cos h)\cos\chi]\cos\delta + [\cos\lambda(\cos\alpha\sin\chi) + \sin\lambda\cos\chi]\sin\delta \quad \text{(A7.3)}$$

where ξ is the angle between the sun's rays and the normal to the surface, χ is the zenith angle (= slope) of the surface, α is the azimuth or aspect of the surface (measured east from north). This equation can be used to calculate irradiance at sloping sites or on leaves of any orientation (see e.g. Fig. 2.9).

An example of application: Estimate the direct irradiance on a horizontal surface at solar noon on 1 April at a site at sea level and 45° N latitude: Substituting in equation A7.1 gives

$$\sin\beta = \sin 45 \sin 4.1 + \cos 45 \cos 4.1 \cos 0 = 0.756$$

Using equation 2.10 (p. 24) with $m = 1/\sin\beta$ (equation 2.9), and assuming an atmospheric transmittance of 0.7, gives

$$\mathbf{I}_{S(dir)} = 1370 \times 0.7^{1.32} \times 0.756 = 646 \text{ W m}^{-2}.$$

4. *Distance to the sun.* The value of the irradiance incident at the top of the atmosphere varies by up to about 3% as a result of seasonal variation in the distance between the Earth and the sun.

APPENDIX 8

Measurement of leaf boundary layer conductance

As an alternative to the use of equations such as 3.31–3.33 (p. 63) for estimating g_a, it is often better to measure leaf conductance directly. This is particularly true for very irregular-shaped leaves or for leaves in gas exchange chambers. Two main approaches are available, based on measurements of either the evaporation rate or the energy balance of model leaves.

1. *From evaporation rate.* The boundary layer conductance to water loss, g_{aH}, may be determined directly from the rate of evaporation (**E**) from a 'wet' model (having no surface resistance analogous to cuticular or stomatal components) of the same dimensions and surface characteristics and exposed in the same situation, by using equation 5.17 (p. 112):

$$g_{aW} = \mathbf{E}/[(0.622\rho_a/P)(e_{s(T_s)} - e_a)] \qquad \text{(A8.1)}$$

where $e_{s(T_s)}$ is the saturation vapour pressure at 'leaf' temperature and e_a is the water vapour pressure in the bulk air.

Adequate models can be made from wet blotting paper, though the exact surface characteristics of real leaves may be difficult to mimic. **E** is usually estimated gravimetrically, but can also be estimated by gas-exchange techniques in a leaf chamber (see Chapter 6) if the purpose is to determine g_a inside a particular gas-exchange cuvette.

2. *From heat transfer properties.* An alternative approach is to measure the heat transfer properties of model leaves in a radiation environment where net radiation is zero and where there is no evaporative cooling (e.g. aluminium model leaves in the dark). Real leaves can also be used if evaporation is prevented by covering the surface with a material such as petroleum jelly to prevent transpiration. In either case the only significant mode of energy exchange with the environment is by 'sensible' heat transfer (that is convection and conduction), and the rate of heat loss is proportional to the leaf–air temperature difference (ΔT) as predicted by equation 3.29 (p. 60). This is an example of Newton's Law of Cooling and provides a convenient method for estimation of g_{aH}.

The technique is to follow the time course of the change of 'leaf' temperature (T_ℓ) after the model has been heated above air temperature. Leaf temperature approaches air temperature asymptotically as shown in Fig. A8.1. The instantaneous rate of heat loss per unit area by the 'leaf' is given by the rate of change of T_ℓ multiplied by its thermal capacity per unit area, i.e.

$$\mathbf{C} = -\rho^* c_p^* \ell^* \frac{\mathrm{d}T_\ell}{\mathrm{d}t} = -\rho^* c_p^* \ell^* \frac{\mathrm{d}(\Delta T)}{\mathrm{d}t} \tag{A8.2}$$

where ρ^*, c_p^* and ℓ^* are, respectively, the density, specific heat and thickness of the 'leaf', and ΔT is the air–leaf temperature difference. Where there is no other significant form of energy exchange (i.e. net radiation is zero), this can be equated to the rate of sensible heat loss given by equation 3.29 to give the differential equation

$$(\rho^* c_p^* \ell^*) \frac{\mathrm{d}(\Delta T)}{\mathrm{d}t} + g_{\mathrm{aH}} \rho_{\mathrm{a}} c_p (\Delta T) = 0 \tag{A8.3}$$

where ρ_{a} and c_p are the density and specific heat of air. Solution of equation A8.3 and rearrangement gives

$$t_2 - t_1 = (\rho^* c_p^* \ell^* / g_{\mathrm{aH}} \rho_{\mathrm{a}} c_p) \ln(\Delta T_1 / \Delta T_2) \tag{A8.4}$$

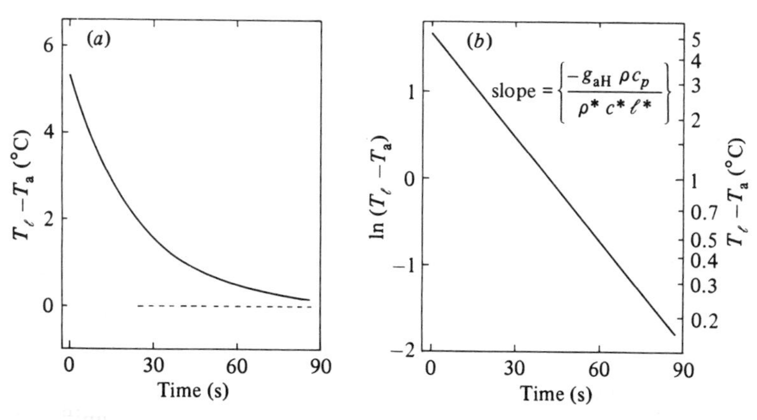

Fig. A8.1. Estimation of g_{aH} for an aluminium leaf model ($\ell^* = 0.001$ m, $c_p^* = 899$ J kg^{-1}, $\rho^* = 2.702 \times 10^3$ kg m^{-3}) from a cooling curve. The time course of 'leaf'–air temperature difference ($T_\ell - T_{\mathrm{a}}$) in (*a*) is transformed by plotting $\ln(T_\ell - T_{\mathrm{a}})$ against time in (*b*). The slope in this example is -0.04 s^{-1} $= -g_{\mathrm{aH}} \rho_{\mathrm{a}} c_p / \rho^* c_p^* \ell^*$. Substituting values for ρ^*, c_p^*, ℓ^*, ρ_{a} and c_p gives $g_{\mathrm{aH}} = 0.04 \times 0.001 \times 899 \times 2702/1010 \times 1.204 = 80$ mm s^{-1}.

where t_1 and t_2 are two times when the temperature differences are given by ΔT_1 and ΔT_2. The value of g_{aH} can then be determined from the slope of a plot of t against $\ln(\Delta T)$ as shown in Fig. A8.1. Further examples of the use of leaf energy balance are presented in Chapters 5 and 9.

APPENDIX 9

Derivation of equation 9.9

When the leaf temperature is at the equilibrium temperature for a particular environment ($T_\ell = T_e$),

$$\mathbf{S} = 0 = \mathbf{\Phi}_n - \mathbf{C} - \lambda\mathbf{E} \tag{A9.1}$$

From equation 5.8 (p. 108) it is known that

$$\mathbf{\Phi}_n = \mathbf{\Phi}_{ni} - \frac{\rho_a c_p}{r_R}(T_e - T_a) \tag{A9.2}$$

and from 9.3 (p. 232) that

$$\mathbf{C} = \frac{\rho_a c_p}{r_{aH}}(T_e - T_a) \tag{A9.3}$$

and from 9.4 that

$$\lambda\mathbf{E} = \frac{\lambda \rho_a 0.622}{P(r_{aW} + r_{\ell W})}(e_{s(T_e)} - e_a) \tag{A9.4}$$

this can be expanded, making use of equation 5.18 (p. 112) to

$$\begin{aligned}\lambda\mathbf{E} &= \frac{\lambda \rho_a 0.622}{P(r_{aW} + r_{\ell W})}(\delta e + s(T_e - T_a)) \\ &= \frac{\rho_a c_p}{\gamma(r_{aW} + r_{\ell W})}(\delta e + s(T_e - T_a))\end{aligned} \tag{A9.5}$$

Substitution of A9.2, A9.3 and A9.5 into A9.1 gives

$$\begin{aligned}0 &= \mathbf{\Phi}_{ni} - \rho_a c_p (T_e - T_a)\{(1/r_{aH}) + (1/r_R) + [s/\gamma(r_{aW} + r_{\ell W})]\} \\ &\quad - \rho_a c_p \delta e/\gamma(r_{aW} + r_{\ell W}) \\ &= \mathbf{\Phi}_{ni} - \rho_a c_p (T_e - T_a)\{(1/r_{HR}) + [s/\gamma(r_{aW} + r_{\ell W})]\} \\ &\quad - \rho_a c_p \delta e/\gamma(r_{aW} + r_{\ell W})\end{aligned} \tag{A9.6}$$

If, however, $T_\ell \neq T_e$,

$$\mathbf{S} = \mathbf{\Phi}_{\text{ni}} - \rho_{\text{a}} c_p (T_\ell - T_{\text{a}}) \{(1/r_{\text{HR}}) + [s/\gamma (r_{\text{aw}} + r_{\ell\text{W}})]\} \\ - \rho_{\text{a}} c_p \, \delta e / \gamma (r_{\text{aw}} + r_{\ell\text{W}}) \tag{A9.7}$$

Subtracting A9.6 from A9.7 gives

$$\mathbf{S} = \rho_{\text{a}} c_p (T_{\text{e}} - T_\ell) \{(1/r_{\text{HR}}) + [s/\gamma (r_{\text{aw}} + r_{\ell\text{W}})]\} \tag{A9.8}$$

Substituting this into equation 9.8 gives 9.9, i.e.

$$\frac{\text{d}T_\ell}{\text{d}t} = \frac{\rho_{\text{a}} c_p}{\rho^* c_p^* \ell^*} (T_{\text{e}} - T_\ell) \{(1/r_{\text{HR}}) + [s/\gamma (r_{\text{aw}} + r_{\ell\text{W}})]\} \tag{A9.9}$$

APPENDIX 10

Answers to selected problems

2.1 (i) (a) The total shortwave radiation absorbed is the sum over all wavebands 0.3–3.0 μm of $\alpha\mathbf{I}$, i.e. $(0.85\times450)+(0.20\times380)+(0.65\times70)=504$ W m^{-2}.
(b) The shortwave absorption coefficient is the ratio of the total shortwave radiation absorbed (504 W m^{-2}) to the total incident shortwave ($450+380+70=900$ W m^{-2}) $=0.56$.
(c) As there is no sensible or latent heat exchange, the energy absorbed ($\mathbf{I}_{S(absorbed)}+\mathbf{I}_{L(absorbed)}$) = the thermal radiation given off (which from equation 2.3 $=\varepsilon\sigma T^4$). $\mathbf{I}_{L(absorbed)}=\varepsilon\sigma T_{environent}{}^{4}$. Assuming $\varepsilon=1$, substituting 293 K for $T_{environment}$, and rearranging gives
$\mathbf{I}_{S(absorbed)}=\sigma(T^4-293^4)$
therefore, with further rearrangement, $T^4=\mathbf{I}_{S(absorbed)}/(\sigma+293^4)=$ 504 (W m^{-2})/5.6703 × 10^{-8} (W m^{-2} K^{-4})+293^4 (K^4), so that taking the fourth root gives leaf temperature as 357 K or 84 °C.
(ii) Sensible heat exchange has not been allowed for.

2.2 (i) The net radiation absorbed is the difference between that absorbed ($\mathbf{I}_{Sd}+\mathbf{I}_{Su}+\mathbf{I}_{Ld}+\mathbf{I}_{Lu}$) and the thermal radiation emitted from the two sides of the leaf, i.e. $(0.5\times500)+(0.3\times0.5\times500)+(\sigma\ 268^4)+(\sigma\ 297^4)-(2\ \sigma\ 293^4)=223$ W m^{-2}.
(ii) We have had to assume that emissivities equal 1.

2.3 (i) From equation 2.1, $\mathbf{E}=hc\lambda=(6.6262\times10^{-34}\text{ J s})\times(2.998\times10^{8}\text{ m s}^{-1})/(500\times10^{-9}\text{ m})=3.97\times10^{-19}$ J photon^{-1} for green light.
A similar calculation for infra-red gives 0.993×10^{-19} J photon^{-1}.
(ii) The wavenumber $=\lambda^{-1}$ (in cm^{-1}), so for green light the wavenumber is $1/(500\times10^{-7}\text{ cm})=20000$ cm^{-1}, and for infra-red light it is given by $1/(2000\times10^{-7}\text{ cm})=5000$ cm^{-1}.
(iii) Since energy per photon is proportional to $1/\lambda$, the number of photons per unit energy is proportional to λ, so that there are 4 × as many photons at 2000 nm as for equal energy at 500 nm.

2.4 (i) The fraction of ground area sunlit (which for opaque leaves is $\mathbf{I}/\mathbf{I}_o$) equals e^{-L} with a horizontal leaved canopy (equation 2.15) so that (*a*) for $L = 1$, this equals $e^{-1} = 0.368$, and (*b*) for $L = 5$, this equals $e^{-5} = 0.0067$.
(ii) For a horizontal leaved canopy, the sunlit leaf area index (L_{sunlit}) $= 1 - e^{-L}$ (equation 2.16), so the corresponding values for L_{sunlit} are 0.632 and 0.993.
(iii) For randomly oriented leaves $L_{sunlit} = (1 - e^{-kL})/k$, where $k = 0.5 \operatorname{cosec} \beta = 0.5/\sin\beta$. For $\beta = 40°$, $k = 0.5/0.6428 = 0.7779$. Therefore for $L = 1$, $L_{sunlit} = (1 - \exp(-1 \times 0.7779))/0.7779 = 0.695$. Similarly for $L = 5$, $L_{sunlit} = 1.259$.

2.5 $\mathbf{I}/\mathbf{I}_o = 0.25$, which, from equation 2.17 $= e^{-kL}$, so $L = -\{\ln(0.25)\}/k$, and using Table 2.5 to calculate k: (i) for $k = 1/(2\sin(60))$, $L = 2.401$; (ii) for $k = 1$, $L = 1.386$.

3.1 (i) From equation 3.20, $\mathbf{J}_W = (24.2 \times 10^{-6}\ \mathrm{mm^2\ s^{-1}}) \times ((17.3 - 11)\ \mathrm{g\ m^{-3}})/(0.1\ \mathrm{m}) = 1.525 \times 10^{-3}\ \mathrm{g\ m^{-2}\ s^{-1}} = 1.525\ \mathrm{mg\ m^{-2}\ s^{-1}}$.
(ii) From equation 3.21, $g_W = \boldsymbol{D}_W/\ell\ (24.2\ \mathrm{mm^2\ s^{-1}})/(100\ \mathrm{mm}) = 0.242\ \mathrm{mm\ s^{-1}}$.
(iii) $\mathbf{J}_W^m = \mathbf{J}_W/M_W = (1.525/18) = 0.085\ \mathrm{mmol\ m^{-2}\ s^{-1}}$.
(iv) From Appendix 3, $g_W = 41 \times g_W = 9.92\ \mathrm{mmol\ m^{-2}\ s^{-1}}$.

3.2 (i) (*a*) If one assumes that the characteristic dimension is $0.9 \times$ diameter and that for a leaf g_{aH} is $1.5 \times$ the value given by equation 3.31, $g_{aH} = 1.5 \times 6.62\,((1\ \mathrm{m\ s^{-1}})/(0.9 \times 0.02\ \mathrm{m}))^{0.5} = 74\ \mathrm{mm\ s^{-1}}$.
(*b*) From Table 3.2, $g_{aW} = 1.08 \times 74 = 79.9\ \mathrm{mm\ s^{-1}}$.
(*c*) $g_{aM} = 0.8 \times 74 = 59.2\ \mathrm{mm\ s^{-1}}$.
(*d*) δ for momentum is given by $2 \times \boldsymbol{D}_M/g_M = (2 \times 15.1\ \mathrm{mm^2\ s^{-1}})/(59.2\ \mathrm{mm\ s^{-1}}) = 0.51\ \mathrm{mm}$.
(ii) For heat the surface is now the top of the tomentum, so g_{aH} is still $74\ \mathrm{mm\ s^{-1}}$; for water vapour, on the other hand, there is an extra resistance approximately equal to $\ell/\boldsymbol{D}_W = (1\ \mathrm{mm})/(24.2\ \mathrm{mm^2\ s^{-1}})$, so by the rules for two resistances in series $g_{aW} = 1/\{(1/24.2) + (1/79.9)\} = 18.6\ \mathrm{mm\ s^{-1}}$; for momentum, the conductance is still unchanged at $59.2\ \mathrm{mm\ s^{-1}}$.
(iii) The Reynolds number $(ud/v) = (1 \times 10^3\ \mathrm{mm\ s^{-1}}) \times (0.9 \times 20\ \mathrm{mm})/(15.1\ \mathrm{mm^2\ s^{-1}}) = 1192$, which is in the range where a laminar boundary layer breaks down.

3.3 (i) If one assumes that $d = 0.64\,h = 0.512$ m, and $z_o = 0.13\,h = 0.104$ m, and substitutes into equation 3.36, $u_* = (4 \times 0.41)/\{\ln[2 - 0.512)/0.104]\} = 0.616\ \mathrm{m\ s^{-1}}$.

(ii) Also substituting into equation 3.36, $u_{0.8} = (0.616/0.41) \ln\{(0.8-0.512)/0.104\} = 1.53$ m s^{-1}.
(iii) Substituting in equation 3.40, $\tau = (1.204$ kg m$^{-3}) \times (0.616$ m s$^{-1})^2 = 0.457$ kg m^{-1} s^{-2}.
(iv) From equation 3.41, $g_{AM} = (0.616$ m s$^{-1})^2/(4$ m s$^{-1}) = 0.095$ m s^{-1}.

4.1 (i) (*a*) From equation 4.1 (assuming that $\alpha = 0$, $T = 20$ °C), $h = \{2 \times (7.28 \times 10^{-2}$ N m$^{-1}) \times \cos 0°\}/\{0.5 \times 10^{-3}$ (m) $\times 998.2$ (kg m$^{-3}) \times 9.8$ (m s^{-2})$] = 0.0298$ m $= 2.98$ cm.
(*b*) 2.98 cm (it still rises to the same height, but further along the capillary).
(*c*) $2.98 \cos\alpha$ (cm) $= 1.92$ cm.
(*d*) As for (*a*), but substituting 0.5 μm for 0.5 mm, gives 29.8 m.
(ii) Pressure required to prevent capillary rise $= 2\sigma\cos\alpha/r = 0.291$ MPa.

4.2 (i) $\Psi_p = \Psi - \Psi_\pi = -1 + 1.5$ MPA $= 0.5$ MPa.
(ii) The solute concentration decreases from c_s to $c_s/1.25$, therefore since Ψ_π is proportional to $-c_s$ (equation 4.8), the new value of $\Psi_\pi = \Psi_{\pi o}/1.25 = -1.5/1.25$ MPA $= -1.2$ MPa.
(iii) The new $\Psi_p = -0.5 + 1.2$ MPa $= 0.7$ MPa.
(iv) If Ψ is linearly related to volume, when Ψ reaches 0, the volume will be $1.5 \times V_o$. The initial relative water content of the cell will equal initial volume/turgid volume $= 1/1.5 = 0.666$ (assuming the whole volume is water).
(v) The easiest way to determine ε_B is graphically giving the value at full turgor as 2 MPa.

4.3 (i) From equation 4.24, $\mathbf{J}_v = \{(0.1 \times 10^{-3}$ (m))$^2/\{8 \times (1.008 \times 10^{-3}$ (N s m$^{-2})) \times 1$ (m)$\}\} \times (5 \times 10^3$ Pa) $= 6.2004 \times 10^{-3}$ m s^{-1}. The volume flow per pipe therefore $= \mathbf{J}_v \times \pi r^2 = 1.95 \times 10^{-10}$ m^3 s^{-1}.
(ii) From equation 4.22, $L = 6.2004 \times 10^{-3}$ (m s$^{-1}) \times 1$ (m)/ $(5 \times 10^3$ (Pa)) $= 1.24 \times 10^{-6}$ m^2 s^{-1} Pa^{-1}; from equation 4.23, $L_p = 1.24 \times 10^{-6}$ m s^{-1} Pa^{-1}; $R = 1/L_p = 8.06 \times 10^{-5}$ Pa s m^{-1}.
(iii) Rearranging equation 4.24 and substituting appropriate values gives $r^2 = (6.2004 \times 10^{-3}$ (m s$^{-1})) \times (8 \times 1.008 \times 10^{-3}$ (N s m$^{-2}))/(1 \times 10^3$ (Pa)), $= 5.00 \times 10^{-8}$ m^2, therefore $r = 2.24 \times 10^{-4}$ m, and $d = 0.448$ mm.

4.4 (i) From equation 4.25, (*a*) $= (1.1$ (MPa))$/(0.1 \times 10^{-6}$ (m^3 m^{-2} s^{-1})) $= 1.1 \times 10^7$ MPa s m^{-1}.
(*b*) $= (1.1$ (MPa))$/\{(0.1/10) \times 10^{-6}$ (m^3 plant^{-1} s^{-1})$\} = 1.1 \times 10^8$ MPa s m^{-3}.
(*c*) $= (1.1$ (MPa))$/\{30 \times (0.1/10) \times 10^{-6}$ (m^3 m^{-2} s^{-1})$\} = 3.67 \times 10^6$ MPa s m^{-1}.

(ii) Assuming that all resistances stay the same, and that Ψ_{soil} stays at -0.1 MPa, the potential drop across the plant will double (as flow per plant has doubled) to give $\Psi_\ell = -2.3$ MPa.
(iii) If half the shoots are removed, the shoot resistance doubles so that the total plant resistance $= R+\frac{1}{2}R = 1\frac{1}{2}R$, therefore the potential drop across the plant $= 1.5\times1.1 = 1.65$ MPa, giving $\Psi_\ell = -1.75$ MPa.

5.1 (i) $e_s = 4243$ Pa (Appendix 4).
(ii) $e = 0.4\times4243$ Pa $= 1697$ Pa.
(iii) $c_W = 0.4\times30.38 = 12.15$ g m^{-3} (using Appendix 4).
(iv) $\delta e = 4243 - 1697$ Pa $= 2546$ Pa.
(v) From Fig. 5.2 (see equation 5.15), $T_w = 20$ °C.
(vi) $T_d = 15$ °C (i.e. the air temperature from Appendix 4 at which e_s equals 1697 Pa).
(vii) From equation 3.7, $m_W = 12.15$ (g m^{-3})/(1.164$-${(1.164–1.145)$\times$ 0.4} kg m^{-3}) $= 1.05\times10^{-2}$.
(viii) For $P = 101.3$ kPa, $x_W = 1697/(101.3\times10^3) = 1.68\times10^{-2}$.
(ix) From equation 5.11, $\Psi = \{8.3144\,(\text{J K}^{-1}\,\text{mol}^{-1})\times303\,(\text{K})/(18.05\times10^{-6}\,(\text{m}^3\,\text{mol}^{-1}))\}\ln(0.4) = -127.9$ MPa.

5.2 (i) Assuming that $\varepsilon = 1$ and substituting in equation 5.4, $\mathbf{\Phi}_{\text{ni}} = 430+4\times(5.6703\times10^{-8})\times(295^4-292^4)$ W m^{-2} $= 447$ W m^{-2}.
(ii) From equation 5.7, $g_R = (4\times5.6703\times10^{-8}\times292^3)/(1012\times1.204) = 4.64\times10^{-3}$ m s^{-1} $= 4.64$ mm s^{-1}.

5.3 (i) Substituting into equation 5.25: (*a*) When the surface is wet $g_A = g_W$, so that $\mathbf{E}$ (for forest) $= \{145$ Pa K$^{-1}\times400$ W m$^{-2}+1.204$ kg m$^{-3}\times$ 1010 J kg^{-1} K$^{-1}\times0.2$ m s$^{-1}\times1000$ Pa$\}/\{2.454\times10^6$ J kg$^{-1}\times(145$ Pa K^{-1} $+(66.1$ Pa K$^{-1}\times0.2/0.2)\} = 0.581$ g m^{-2} s^{-1}.
Similarly for short grass, but substituting $g_A = 0.01$ m s^{-1} gives, $\mathbf{E} = 0.135$ g m^{-2} s^{-1}.
(*b*) $g_W = ((g_A)^{-1}+(g_L)^{-1})^{-1}$ so that for forest ($g_A = 0.2$ m s^{-1}) and where $g_L = 0.03$ m s^{-1}, $g_W = ((0.2)^{-1}+(0.03)^{-1})^{-1} = 0.02609$ m s^{-1}. A similar calculation for grassland gives $g_W = 0.0075$ m s^{-1}. Substitution of these values into equation 5.23 gives $\mathbf{E}$ of 0.188 g m^{-2} s^{-1} and 0.123 g m^{-2} s^{-1}, respectively for forest and grass.
(ii) The Bowen ratio, $\boldsymbol{\beta} = \mathbf{C}/\lambda\mathbf{E}$. Making use of the energy balance (equation 5.1) and remembering that $\mathbf{M}$ and $\mathbf{S}$ are zero at the steady state, enables one to write $\boldsymbol{\beta} = ((\mathbf{\Phi}_n-\mathbf{G})-\lambda\mathbf{E})/\lambda\mathbf{E}$. Using the answers from (i) above enables one to write for wet forest, for example, $\boldsymbol{\beta} = (400-2454\times0.581)/(2454\times0.581) = -0.72$. Corresponding answers for the other cases are 0.2, -0.133 and 0.33.

5.4 (i) From equation 5.28, $\Omega = (2.20+1)/(2.20+1+15/5) = 0.516$.

(ii) From equation 5.29 the relative reduction in transpiration $(d\mathbf{E}/\mathbf{E}) = (1-\Omega)\, dg_\ell/g_\ell = 0.484 \times 0.5 = 0.242$.
(ii) Equation 5.29 is only strictly true for *small* changes in g_ℓ, so expressing the changes with respect to the initial **E** and g_ℓ is only an approximation.

6.1 (i) $r_{aW} = (c_{W(leaf)} - c_{W(air)})/\mathbf{E}_{(blotting paper)} = (17.30\ \text{g m}^{-3} - 0.2 \times 17.30\ \text{g m}^{-3})/0.230\ \text{g m}^{-2}\ \text{s}^{-1} = 60.2\ \text{s m}^{-1}$.
(ii) The cuticular resistance, r_c, is given by the difference between the total resistance to water loss with the stomata closed and the boundary layer resistance (i.e. assuming that the final rate is achieved when the stomata close completely) $= \{(17.30 \times 0.8\ \text{g m}^{-3})/(0.002\ \text{g m}^{-2}\ \text{s}^{-1})\} - 60.2\ \text{s m}^{-1}) = 6860\ \text{s m}^{-1}$.
(iii) For the initial rate of water loss, $r_{\ell W} = \{(17.30 \times 0.8\ \text{g m}^{-3})/(0.08\ \text{g m}^{-2}\ \text{s}^{-1})\} - 60.2\ \text{s m}^{-1} = 112.8\ \text{s m}^{-1}$. Since the stomatal and cuticular resistances are in parallel, the stomatal resistance is given by $((r_{\ell W})^{-1} - (r_c)^{-1})^{-1} = 114.7\ \text{s m}^{-1}$.

6.2 (i) From equation 6.3, r_{sW} (for one surface) $= [10 \times 10^{-6}\ \text{m} + (\pi \times 2.5 \times 10^{-6}\ \text{m}/4)]/(200 \times 10^{6}\ \text{m}^{-2} \times \pi \times (2.5 \times 10^{-6}\ \text{m})^2 \times 24.2 \times 10^{-6}\ \text{m}^2\ \text{s}^{-1}) = 125.9\ \text{s m}^{-1}$; therefore for both surfaces, $r_{sW} = 125.9/2 = 62.9\ \text{s m}^{-1}$.
(ii) $g_{sW} = 1/(62.9\ \text{s m}^{-1}) = 0.0159\ \text{m s}^{-1} = 15.9\ \text{mm s}^{-1}$.

6.3 (i) As $T_\ell = T_a$ one can use equation 6.10. Assuming that $T_a = 25\ °\text{C}$, the molar volume of air $= 0.02241\ \text{m}^3\ \text{mol}^{-1} \times 298\ \text{K}/273\ \text{K} = 0.024462\ \text{m}^3\ \text{mol}^{-1}$, so that $u_e = (2 \times 10^{-6}\ \text{m}^3\ \text{s}^{-1})/(0.024462\ \text{m}^3\ \text{mol}^{-1}) = 8.176 \times 10^{-5}\ \text{mol s}^{-1}$, so that $g_W^{-1} = \{[(1/0.35) - 1] \times 1.5 \times 10^{-4}\ \text{m}^2/(8.176 \times 10^{-5}\ \text{mol s}^{-1})\}$, therefore $g_W = 0.293\ \text{mol m}^{-2}\ \text{s}^{-1}$.
(ii) When the system is non-isothermal one needs to use equation 6.9. $x_{Ws} = e_{s(T\ell)}/P = (3167\ \text{Pa})/(1.013 \times 10^5\ \text{Pa}) = 0.03126$, $x_{Wo} = (0.35 \times 3565\ \text{Pa})/(1.013 \times 10^5) = 0.01232$, and $x_{We} = 0$, so that $g_W = (8.176 \times 10^{-5}\ \text{mol s}^{-1} \times 0.01232)/(1.5 \times 10^{-4}\ \text{m}^2 \times (0.03126 - 0.01232) \times (1 - 0.01232)) = 0.359\ \text{mol m}^{-2}\ \text{s}^{-1}$.

6.4 (i) The stomatal response to δe can be written as $g_\ell = 10(1 - (1/3)\delta e)$ (e in kPa), so for $\delta e = 1$ kPa, $g_\ell = 6.6666\ \text{mm s}^{-1}$. In mass units, $\mathbf{E} = (c_{Ws} - c_{Wa}) \times g_W$, so, using equation 5.9 to convert δe to the corresponding water vapour concentration difference, one obtains (assuming a T of 293 K), $\mathbf{E} = (2.17/293)\ \text{g m}^{-3}\ \text{Pa}^{-1} \times 1000\ \text{Pa} \times 0.006666\ \text{m s}^{-1} = 0.0494\ \text{g m}^{-2}\ \text{s}^{-1}$.
(ii) Assuming a linear response of g_ℓ to Ψ, g_ℓ can be written as $g_{\ell o} \times (1 + 0.5\,\Psi)$, but as $\Psi = -10\,\mathbf{E}$, combining gives $g_{\ell o}\ (1 - 5\,\mathbf{E})$, which together with the humidity response gives g_ℓ (mm s^{-1}) $= 10(1 - (1/3))$

$\times(1-5\,\mathbf{E})$, therefore $\mathbf{E} = (2.17/293)$ g m^{-2} Pa$^{-1} \times 1000$ (Pa) $\times 0.006666$ (m s^{-1}) $\times (1-5\,\mathbf{E})$. Rearranging and solving for $\mathbf{E}$ gives 0.040 g m^{-2} s^{-1}.

7.1 Using the formulae in Table 7.2:
(i) $F_v/F_m = 2.7/3.7 = 0.730$.
(ii) $q_o = (1-0.9)/1 = 0.1$.
(iii) $q_P = (3.2-1.2)/(3.2-0.9) = 0.870$.
(iv) $q_N = (3.7-3.2/0.9)/(3.7-1) = 0.053$.
(iv) The quantum yield of electron transport through PS II is given by equation 7.17 as $(3.2-1.2)/3.2 = 0.625$.

7.2 (i) $u_e = (5\times10^{-6}$ m^3 s$^{-1})/(0.0227107$ m^3 mol$^{-1} \times 296$ K$/273$ K$) = 2.031 \times 10^{-4}$ mol s^{-1}.
(ii) From equation 6.6*b*, $u_o = 2.031\times10^{-4}$ mol s$^{-1} \times (1-0.5/100)(1-1.5/100) = 2.052\times10^{-4}$ mol s^{-1}.
(iii) The CO_2 concentration is 0.6 g m^{-3}, dividing by M_C (= 44), gives 0.01364 mol m^{-3}. The number of moles of gas per m^3 = $1/(0.0227107$ m^3 mol$^{-1} \times 296/273) = 40.611$ mol m^{-3}, therefore $x'_e = 0.01364/40.611 = 335.8\times10^{-6}$ mol mol^{-1}, or 335.8 ppm. (Alternatively one could use equations 3.5 and 3.6.)
(iv) From equation 7.9, $\mathbf{P}^m = \{(2.031\times10^{-4}$ mol s$^{-1} \times 335.8\times10^{-6}) - (2.052\times10^{-4}$ mol s$^{-1} \times 335.8\times10^{-6} \times 450/600)\}(10\times10^{-4}$ m$^2) = 16.52$ μmol m^{-2} s^{-1}.
(v) From equation 6.9, $r_W = (10^{-3}$ m$^2)\{(2.809/100)-(1.5/100)\}\times\{1-(1.5/100)\}/\{(2.031\times10^{-4}$ mol s$^{-1})\times(1.5/100-0.5/100)\} = 6.348$ m^2 s mol^{-1}, therefore $g_W = 0.158$ mol m^{-2} s^{-1}, therefore from Table 3.2 assuming largely still air $g' = (0.68/1.12)\times0.158$ mol m^{-2} s^{-1} = 0.096 mol m^{-2} s^{-1}.

7.3 Estimating slopes from Fig. 7.16, gives ℓ'_g for the different methods:
(i) 7.5 m^2 s μmol^{-1}/(7.5 m^2 s μmol^{-1} + 7.0 m^2 s μmol^{-1}) = 0.52.
(ii) (21.5 μmol m^{-2} s^{-1} − 18.9 μmol m^{-2} s^{-1})/(21.5 μmol m^{-2} s^{-1}) = 0.12.
(iii) 7.5 m^2 s μmol^{-1}/(7.5 m^2 s μmol^{-1} + 22.1 m^2 s μmol^{-1}) = 0.25.

7.4 Since the relative masses of CO_2 and sucrose per mol C are 44/30, and assuming an energy equivalent of 16 kJ g^{-1} in sucrose, this is equivalent to $16\times30/44$ kJ (g CO_2)$^{-1}$ = 10.9 kJ g^{-1}. Therefore
(i) For May the efficiency of gross photosynthesis is ((288/7) × 10.9 kJ (g CO_2)$^{-1}$ m^{-2} day^{-1})/(14.5 × 10^3 kJ m^{-2} day^{-1}) = 3.1%. The corresponding figure for July = ((304/7) × 10.9/(17 × 10^3)) = 2.8%. Assuming that the energy in PAR is 50% of that in global radiation the corresponding efficiencies in terms of incident PAR are 6.2% and 5.6%, respectively.

(ii) For net productivity one needs to use the average energy per g in dry matter (17.5 kJ g^{-1}), so that the efficiency in May = $((167/7) \times 17.5 \times (30/44))/14.5 \times 10^3 = 2.0\%$, and in July $((101/7) \times 17.5 \times (30/44))/17 \times 10^3 = 1.0\%$, with the corresponding values in terms of PAR being 3.9% and 2%.

8.1 (i) From equation 8.3, (*a*) $\zeta = 0.904 \times 1.1 = 0.994$, (*b*) $\zeta = 0.904 \times 1.1 \times 0.08/0.35 = 0.227$.
(ii) Estimating from Fig. 8.2 gives ϕ as 0.52 and 0.28.

9.1 (i) Substituting into equation 9.5:
$T_\ell = 25\ °C + \{(40 \times (40/1.08 + 200)\ s\ m^{-1} \times 66.5\ Pa\ K^{-1} \times 400\ W\ m^{-2})/(1010\ J\ kg^{-1}\ K^{-1} \times 1.204\ kg\ m^{-3} \times [66.5\ Pa\ K^{-1} \times (237.03\ s\ m^{-1}) + (189\ Pa\ K^{-1}) \times 40\ s\ m^{-1})]\} - \{(40\ s\ m^{-1} \times 0.6 \times 3167\ Pa)/[2.332 \times 10^4\ Pa\ s\ m^{-1}\ K^{-1}] = 30.6\ °C$.
(ii) From equation 9.11, $\tau = (2.7 \times 10^6\ J\ m^{-3} \times 1 \times 10^{-3}\ m)/\{[1010\ J\ kg^{-1}\ K^{-1} \times 1.204\ kg\ m^{-3}] \times \{((1/40) + (4 \times 5.6703 \times 10^{-8} \times 298^3/1010 \times 1.204)\ s\ m^{-1}) + [189\ Pa\ K^{-1}/(66.5\ Pa\ K^{-1} \times 237.03\ s\ m^{-1})]\}\} = 53\ s$.
(iii) Similarly, when $r_{\ell W} = \infty$, $\tau = 74$ s.
(iv) Substituting $r_{\ell W} = 0$ into equation 9.5 gives $T = 20.7\ °C$.
(v) 20.8 s.

9.2 (i) From equation 9.16, $Q_{10} \simeq (0.19/0.1)\exp(10/6) = 2.91$.
(ii) Rearranging equation 9.15: $E_a = \mathscr{R}T(T+10) \times \ln(Q_{10}) = 75.4$ kJ mol^{-1}.

9.3 25 °C.

REFERENCES

Acevedo, E., Hsiao, T. C. & Henderson, D. W. (1971). Immediate and subsequent growth responses of maize leaves to changes in water status. *Plant Physiology*, **48,** 631–6.

Acock, B., Charles-Edwards, D. A., Fitter, D. J., Hand, D. W., Ludwig, J., Warren Wilson, J. & Withers, A. C. (1978). The contribution of leaves from different levels within a tomato crop to canopy net photosynthesis: an experimental examination of two canopy models. *Journal of Experimental Botany*, **29,** 815–27.

Acock, B. & Grange, R. I. (1981). Equilibrium models of leaf water relations. In Rose & Charles-Edwards. (1981), pp. 29–47.

Adams, R. M., Rosenzweig, C., Peart, R. M., Ritchie, J. T., McCarl, B. A., Glyer, J. D., Curry, R. B., Jones, J. W., Boote, K. J. & Allen, L. H. Jr. (1990). Global climate change and US agriculture. *Nature*, **325,** 219–24.

Addicott, F. T. (ed.) (1983). *Abscisic acid.* New York: Praeger.

Allard, R. W. (1960). *Principles of plant breeding*. New York: Wiley.

Allaway, W. G. (1973). Accumulation of malate in guard cells of *Vicia faba* during stomatal opening. *Planta*, **110,** 63–70.

Allaway, W. G., Austin, B. & Slatyer, R. O. (1974). Carbon dioxide and water vapour exchange parameters of photosynthesis in a Crassulacean plant. *Kalanchoe daigremontiana. Australian Journal of Plant Physiology*, **1,** 397–405.

Allaway, W. G. & Hsiao, T. C. (1973). Preparation of rolled epidermis of *Vicia faba* L. so that stomata are the only viable cells: analysis of guard cell potassium by flame photometry. *Australian Journal of Biological Science*, **26,** 309–18.

Allen, L. H., Stewart, D. W. & Lemon, E. R. (1974). Photosynthesis in plant canopies: effect of light response curves and radiation source geometry. *Photosynthetica*, **8,** 184–207.

Alscher, R. G. & Cumming, J. R. (1990). *Stress responses in plants: adaptation and acclimation mechanisms*. New York. Wiley–Liss.

Amthor, J. S. (1989). *Respiration and crop productivity*. New York: Springer.

Angus, J. F. & Moncur, M. W. (1977). Water stress and phenology in wheat. *Australian Journal of Agricultural Research*, **28,** 177–81.

Anon. (1964). *Mean daily solar radiation, monthly and annual.* Washington: US Department of Commerce.

Anon. (1980). *Solar radiation data for the United Kingdom 1951–1975.* Met. O. 919. Bracknell: Meteorological Office.

Anon. (1987). *Montreal protocol on substances that deplete the ozone layer.* Nairobi: United Nations Environment Programme.

Ansari, A. Q. & Loomis, W. E. (1959). Leaf temperatures. *American Journal of Botany*, **46,** 713–17.

Arkin, G. F., Vanderlip, R. L. & Ritchie, J. T. (1976). A dynamic grain sorghum growth model. *Transactions of the American Society of Agricultural Engineers*, **19,** 622–30.

Armond, P. A. & Mooney, H. A. (1978). Correlation of photosynthetic unit size and density with photosynthetic capacity. *Carnegie Institution Year Book*, **77,** 234–7.

Atkins, P. W. (1990). *Physical chemistry*, 4th edn. Oxford: Oxford University Press.

Austin, R. B. (1978). Actual and potential yields of wheat and barley in the United Kingdom. *Agricultural Development and Advisory Service Quarterly Review*, **29,** 76–87.

Austin, R. B. & Jones, H. G. (1975). The physiology of wheat. *Cambridge Plant Breeding Institute Annual Report* 1975, pp. 20–73.

Bainbridge, R., Evans, G. C. & Rackham, O. (eds) (1966). *Light as an ecological factor.* Oxford: Blackwell.

Baker, E. A. (1974). The influence of environment on leaf wax development in *Brassica oleracea* var. *gemmifera. New Phytologist*, **73,** 955–66.

Baldocchi, D. D., Verma, S. B. & Rosenberg, N. J. (1985). Water use efficiency in a soybean field: influence of plant water stress. *Agricultural and Forest Meteorology*, **34,** 53–65.

Ball, J. T. (1987). Calculations related to gas exchange in *Stomatal Function*, ed. E. Zeiger, G. D. Farquhar & I. R. Cowan, pp. 445–76.

Ball, J. T., Woodrow, I. E. & Berry, J. A. (1986). A Model predicting stomatal conductance and its contribution to the control of photosynthesis under different environmental conditions. *Proceedings of VII International Photosynthesis Congress*, ed. J. Biggins. Dordrecht: Martinus Nijhoff.

Balling, A. & Zimmermann, U. (1990). Comparative measurements of the xylem pressure of *Nicotiana* plants by means of the pressure bomb and pressure probe. *Planta*, **182,** 325–38.

Bangerth, F. (1979). Calcium-related physiological disorders of plants. *Annual Review of Phytopathology*, **17,** 97–122.

Barrs, H. D. (1968). Determination of water deficits in plant tissues. In Kozlowski (1968), pp. 235–368.

Barry, B. G. (1981). *Mountain weather and climate.* London: Methuen.

Bartels, D., Schneider, K., Terstappen, G., Piatkowski, D. & Salamini, F. (1990). Molecular cloning of abscisic acid-modulated genes which are induced during desiccation of the resurrection plant *Ceratostigma plantagineum*. *Planta*, **181,** 27–34.

Bates, L. M. & Hall, A. E. (1981). Stomatal closure with soil moisture depletion not associated with changes in bulk water status. *Oecologia*, **50,** 62–5.

Beakbane, A. B. & Mujamder, P. K. (1975). A relationship between stomatal density and growth potential in apple rootstocks. *Journal of Horticultural Science*, **50,** 285–9.

Beardsell, M. F., Jarvis, P. G. & Davidson, B. (1972). A null-balance diffusion porometer suitable for use with leaves of many shapes. *Journal of Applied Ecology*, **9,** 677–90.

Beck, E., Senser, M., Scheibe, R., Steiger, H.-M. & Pongratz, P. (1982). Frost avoidance and freezing tolerance in Afroalpine 'giant rosette' plants. *Plant, Cell and Environment*, **5,** 212–22.

Becker, C. F. & Boyd, J. S. (1957). Solar radiation availability on surfaces in the United States as affected by season, orientation, latitude, altitude and cloudiness. *Solar Energy*, **1,** 13–21.

Bell, C. J. & Rose, D. A. (1981). Light measurement and the terminology of flow. *Plant, Cell and Environment*, **4,** 89–96.

Berlyn, M. B., Zelitch, I. & Beaudette, P. D. (1978). Photosynthetic characteristics of photoautotrophically grown tobacco callus cells. *Plant Physiology*, **61,** 606–10.

Besford, R. T., Ludwig, L. J. & Withers, A. C. (1990). The greenhouse effect: Acclimation of tomato plants growing in high CO_2, photosynthesis and ribulose-1,5-*bis*phosphate carboxylase protein. *Journal of Experimental Botany*, **41,** 925–31.

Biddington, N. L. & Dearman, J. A. (1985). The effects of mechanically-induced stress on water loss and drought resistance in lettuce, cauliflower and celery seedlings. *Annals of Botany*, **56,** 795–802.

Bierhuizen, J. F. & Slatyer, R. O. (1965). Effect of atmospheric concentration of water vapour and CO_2 in determining transpiration-photosynthesis relationships of cotton leaves. *Agricultural Meteorology*, **2,** 259–70.

Bilger, W., Schreiber, U. & Lange, O. (1984). Determination of leaf heat resistance: comparative investigation of chlorophyll fluorescence changes and tissue necrosis methods. *Oecologia*, **63,** 156–62.

Bilger, W. & Schreiber, U. (1986). Energy-dependent quenching of dark-level chlorophyll fluorescence in intact leaves. *Photosynthesis Research*, **10,** 303–8.

Biscoe, P. V., Gallagher, J. N., Littleton, E. J., Monteith, J. L. & Scott, R. K. (1975*a*). Barley and its environment. IV. Sources of assimilate for the grain. *Journal of Applied Ecology*, **12,** 295–318.

Biscoe, P. V., Scott, R. K. & Monteith, J. L. (1975*b*). Barley and its environment. III. Carbon budget of the stand. *Journal of Applied Ecology* **12,** 269–93.

Björkman, O., Badger, M. R. & Armond, P. A. (1980). Response and adaptations of photosynthesis to high temperatures. In Turner & Kramer (1980), pp. 233–49.

Björkman, O., Boardman, N. K., Anderson, J. M., Thorne, S. W., Goodchild, D. J. & Pyliotis, N. A. (1972*a*). Effect of light intensity during growth of *Atriplex patula* on the capacity of photosynthetic reactions, chloroplast components and structure. *Carnegie Institution Year Book*, **71,** 115–35.
Björkman, O. & Demmig, B. (1987). Photon yield of O_2 evolution and chlorophyll fluorescence characteristics at 77K among vascular plants of diverse origins. *Planta*, **170,** 489–504.
Björkman, O., Gauhl, E. & Nobs, M. A. (1970). Comparative studies of *Atriplex* species with and without β-carboxylation photosynthesis and their first-generation hybrid. *Carnegie Institution Year Book*, **68,** 620–33.
Björkman, O., Ludlow, M. M. & Morrow, P. A. (1972*b*). Photosynthetic performance of two rainforest species in their native habitat and analysis of their gas exchange. *Carnegie Institution Year Book*, **71,** 94–102.
Björkman, O., Mooney, H. A. & Ehleringer, J. (1975). Comparison of photosynthetic characteristics of intact plants. *Carnegie Institution Year Book*, **74,** 743–8.
Björkman, O., Nobs, M. A., Berry, J. A., Mooney, H. A., Nicholson, F. & Catanzaro, B. (1973). Physiological adaptation to diverse environments: approaches and facilities to study plant responses to contrasting thermal and water regimes. *Carnegie Institution Year Book*, **72,** 393–403.
Blackman, F. F. (1905). Optima and limiting factors. *Annals of Botany*, **19,** 281–95.
Blackman, P. G. & Davies, W. J. (1985). Root to shoot communication in maize plants of the effects of soil drying. *Journal of Experimental Botany*, **36,** 39–48.
Bliss, D. & Smith, H. (1985). Penetration of light into soil and its role in the control of seed germination. *Plant, Cell and Environment*, **8,** 475–83.
Bliss, L. C. (1962). Adaptation of arctic and alpine plants to environmental conditions. *Arctic*, **15,** 117–44.
Blum, A. (1988). *Plant breeding for stress environments*. Boca Raton, Florida: CRC Press.
Boardman, N. K. (1977). Comparative photosynthesis of sun and shade plants. *Annual Review of Plant Physiology*, **28,** 355–77.
Boardman, N. K., Björkman, O., Anderson, J. M., Goodchild, D. J. & Thorne, S. W. (1975). Photosynthetic adaptation of higher plants to light intensity: relationship between chloroplast structure, composition of the photosystems and photosynthetic rates. In *Proceedings of the third international congress on photosynthesis*, ed. M. Avron, pp. 1809–27. Amsterdam; Elsevier.
Bolin, B., Döös, B. R., Jäger, J. & Warrick, R. A. (eds) (1986). *The greenhouse effect, climatic change and ecosystems. A synthesis of present knowledge. SCOPE 29*. Chichester: Wiley.
Box, G. E. P., Hunter, W. G. & Hunter, J. S. (1978). *Statistics for experimenters. An introduction to design, data analysis and model building*. New York: Wiley.
Boyer, J. S. (1976). Photosynthesis at low water potentials. *Philosophical Transactions of the Royal Society, London B*, **273,** 501–12.

Boyer, J. S. (1985). Water transport. *Annual Review of Plant Physiology*, **36,** 473–516.

Braam, J. & Davis, R. W. (1990). Rain-, wind-, and touch-induced expression of calmodulin and calmodulin-related genes in *Arabidopsis*. *Cell*, **60,** 357–64.

Brandrup, J. & Immergut, E. H. (eds) (1975). *Polymer handbook*, 2nd edn. New York: Wiley.

Briggs, W. R. (ed.) (1989). *Photosynthesis*. New York: Alan R. Liss.

Brown, K. W., Jordan, W. R. & Thomas, J. C. (1976). Water stress induced alterations of the stomatal response to decreases in leaf water potential. *Physiologia Plantarum*, **37,** 1–5.

Brown, K. A., Roberts, T. M. & Blank, L. W. (1987). Interaction between ozone and cold sensitivity in Norway spruce: a factor contributing to the forest decline in Central Europe? *New Phytologist*, **105,** 149–55.

Brown, K. W. & Rosenberg, N. J. (1970). Influence of leaf age, illumination, and upper and lower surface differences on stomatal resistance of sugar beet (*Beta vulgaris*) leaves. *Agronomy Journal*, **62,** 20–4.

Brown, R. W. & van Haveren, B. P. (1972). *Psychrometry in water relations research*. Logan, Utah: Utah Agricultural Experimental Station.

Brun, L. J., Kanemasu, E. T. & Powers, W. L. (1972). Evapotranspiration from soybean and sorghum fields. *Agronomy Journal*, **64,** 145–8.

Bryson, R. A. (1974). A perspective on climatic change. *Science*, **184,** 753–60.

Buck, A. L. (1981). New equations for computing vapour pressure and enhancement factor. *Journal of Applied Meteorology*, **20,** 1527–32.

Bunce, J. A. & Miller, L. N. (1976). Differential effects of water stress on respiration in the light in woody plants from wet and dry habitats. *Canadian Journal of Botany*, **54,** 2457–64.

Burke, M. J. & Stushnoff, C. (1979). Frost hardiness: a discussion of possible molecular causes of injury with particular reference to deep supercooling of water. In Mussell & Staples (1979), pp. 198–225.

Burrage, S. W. (1972). Dew on wheat. *Agricultural Meteorology*, **10,** 3–12.

Businger, J. A. (1975). Aerodynamics of vegetated surfaces. In *Heat and mass transfer in the biosphere. I. Transfer processes in the plant environment*, eds D. A. de Vries & N. H. Afgan, pp. 139–65. Washington: Scripta.

Bykov, O. D., Koshkin, V. A. & Čatský, J. (1981). Carbon dioxide compensation of C_3 and C_4 plants: dependence on temperature. *Photosynthetica*, **15,** 114–21.

Calder, I. R. (1976). The measurement of water losses from a forested area using a 'natural' lysimeter. *Journal of Hydrology*, **30,** 311–25.

Caldwell, M. M. (1970). Plant gas exchange at high wind speeds. *Plant Physiology*, **46,** 535–7.

Caldwell, M. M. (1982). Plant response to ultraviolet radiation. In Lange *et al.* (1982), vol. I, pp. 169–97.

Camacho-B, S. E., Hall, A. E. & Kaufmann, M. R. (1974). Efficiency and regulation of water transport in some woody and herbaceous species. *Plant Physiology*, **54,** 169–72.

Campbell, G. S. (1977). *An introduction to environmental biophysics*. New York: Springer-Verlag.

Campbell, G. S. (1986). Extinction coefficients for radiation in plant canopies calculated using an ellipsoidal inclination angle distribution. *Agricultural and Forest Meteorology*, **36,** 317–21.

Campbell, G. S. & Norman, J. M. (1989). Plant canopy structure. In *Plant canopies: their growth, form and function*, eds G. Russell, B. Marshall & P. G. Jarvis, pp. 1–19. Cambridge: Cambridge University Press.

Canny, M. (1990). What becomes of the transpiration stream? *New Phytologist*, **114,** 341–68.

Čermák, J. & Kucera, J. (1981). The compensation of natural temperature gradients at the measuring point during the sap flow rate determination in trees. *Biologia Plantarum*, **23,** 469–71.

Chamberlain, A. C. & Little, P. (1981). Transport and capture of particles by vegetation. In Grace *et al.* (1981), pp. 147–73.

Cheung, Y. N. S., Tyree, M. T. & Dainty, J. (1975). Water relations parameters on single leaves obtained in a pressure bomb, and some ecological interpretations. *Canadian Journal of Botany*, **53,** 1342–6.

Cicerone, R. J. & Oremland, R. S. (1988). Biogeochemical aspects of atmospheric methane. *Global Biogeochemical Cycles*, **2,** 299–327.

Ciha, A. J. & Brun, W. A. (1975). Stomatal size and frequency in soybean. *Crop Science*, **15,** 309–13.

Clayton, R. K. (1981). *Photosynthesis*. Cambridge: Cambridge University Press.

Cohen, A. & Bray, E. A. (1990). Characterization of three mRNAs that accumulate in wilted tomato leaves in response to elevated levels of endogenous abscisic acid. *Planta*, **182,** 27–33.

Cohen, D. (1971). Maximising final yield when growth is limited by time or by limiting resources. *Journal of Theoretical Biology*, **33,** 299–307.

Cohen, S. & Fuchs, M. (1987). The distribution of leaf area, radiation, photosynthesis and transpiration in a Shamouti orange hedgerow orchard. I. Leaf area and radiation. *Agricultural and Forest Meteorology*, **40,** 123–44.

Collatz, J., Ferrar, P. J. & Slatyer, R. O. (1976). Effects of water stress and differential hardening treatments on photosynthetic characteristics of a xeromorphic shrub. *Eucalyptus socialis* F. Muell. *Oecologia*, **23,** 95–105.

Cooper, J. P. (1970). Potential production and energy conversion in temperate and tropical grasses. *Herbage Abstracts*, **40,** 1–15.

Cooper, J. P. (ed.) (1975). *Photosynthesis and productivity in different environments*. Cambridge: Cambridge University Press.

Cosgrove, D. J. (1986). Biophysical control of plant cell growth. *Annual Review of Plant Physiology*, **37,** 377–405.

Coulson, K. L. (1975). *Solar and terrestrial radiation*. New York: Academic Press.

Covell, S., Ellis, R. H., Roberts, E. H. & Summerfield, R. J. (1986). The influence of temperature on seed germination rate in grain legumes. I. A

comparison of chickpea, lentil, soyabean and cowpea at constant temperatures. *Journal of Experimental Botany*, **37,** 705–15.
Cowan, I. R. (1977). Stomatal behaviour and environment. *Advances in Botanical Research*, **4,** 117–228.
Cowan, I. R. (1982). Water use and optimisation of carbon assimilation. In Lange *et al.* (1982), vol. II, pp. 581–613.
Cowan, I. R. (1986). Economics of carbon fixation in higher plants. In *On the economy of plant form and function*, ed. T. J. Givnish, pp. 133–70. Cambridge: Cambridge University Press.
Cowan, I. R. & Farquhar, G. D. (1977). Stomatal function in relation to leaf metabolism and environment. *Symposium of the Society for Experimental Biology*, **31,** 471–505.
Craig, E. A. (1986). The heat shock responses. *CRC Review of Biochemistry*, **18,** 239–80.
Crank, J. (1975). *The mathematics of diffusion.* Oxford: Clarendon Press.
Crombie, D. S., Milburn, J. A. & Hipkins, M. F. (1985). Maximum sustainable xylem sap tensions in *Rhododendron and* other species. *Planta*, **163,** 27–33.
Curran, P. J. (1985). *Principles of remote sensing*. London: Longman.
Cutler, J. M., Rains, D. W. & Loomis, R. S. (1977). The importance of cell size in the water relations of plants. *Physiologia Plantarum*, **40,** 255–60.
Daniel, C. & Wood, F. S. (1980). *Fitting equations to data*, 2nd edn. New York: Wiley.
Darwin, C. (1980). *The power of movement in plants.* London: John Murray.
Darwin, F. & Pertz, D. F. M. (1911). On a new method of estimating the aperture of stomata. *Proceedings of the Royal Society*, *London B*, **84,** 136–54.
Daubenmire, R. (1974). *Plants and environment: a textbook of plant autecology*, 3rd edn. New York: Wiley.
Davies, D. D. (ed.) (1980). *The biochemistry of plants*, vol. II: *Metabolism and respiration.* New York: Academic Press.
Davies, W. J. & Jeffcoat, B. (1990). *Importance of root to shoot communication in the response to environmental stress.* BSPGR monograph 21. Long Ashton Research Station, Bristol: British Society for Plant Growth Regulation.
Davies, W. J. & Jones, H. G. (eds) (1991). *Physiology and biochemistry of abscisic acid.* Oxford: Bios.
Day, W. (1977). A direct reading continuous flow porometer. *Agricultural Meteorology*, **18,** 81–9.
Day, W., Legg, B. J., French, B. K., Johnston, A. E., Lawlor, D. W. & Jeffers, W. de C. (1978). A drought experiment using mobile shelters: the effect of drought on barley yield, water use and nutrient uptake. *Journal of Agricultural Science, Cambridge*, **91,** 599–623.
de Jong, D. W. (1973). Effect of temperature and daylength on peroxidase and malate (NAD) dehydrogenase isozyme composition in tobacco leaf extracts. *American Journal of Botany*, **60,** 846–52.
de Silva, D. L. R., Cox, R. C., Hetherington, A. M. & Mansfield, T. A. (1985).

Suggested involvement of calcium and calmodulin in the responses of stomata to abscisic acid. *New Phytologist*, **101,** 555–63.
de Wit, C. T. (1958). Transpiration and crop yields. *Verslagen van Landbouwkundige Onderzoekingen*, **64,** 1–88.
de Wit, C. T. (1965). *Photosynthesis of leaf canopies*. Agricultural Research Report no. 663. Wageningen: PUDOC.
de Wit, C. T. & Goudriaan, J. (1978). *Simulation of ecological processes*. Wageningen: PUDOC.
Delieu, T. J. & Walker, D. A. (1983). Simultaneous measurement of oxygen evolution and chlorophyll fluorescence from leaf pieces. *Plant Physiology*, **73,** 534–41.
Denmead, O. T. (1969). Comparative micrometeorology of a wheat field and a forest of *Pinus radiata*. *Agricultural Meteorology*, **6,** 357–71.
Denmead, O. T. & Bradley, E. F. (1987). On scalar transport in plant canopies. *Irrigation Science*, **8,** 131–49.
Denmead, O. T. & McIlroy, I. C. (1970). Measurements of non-potential evaporation from wheat. *Agricultural Meteorology* **7,** 285–302.
Denmead, O. T. & Millar, B. D. (1976). Water transport in wheat plants in the field. *Agronomy Journal*, **68,** 297–303.
Denmead, O. T. & Shaw, R. H. (1962). Availability of soil water to plants as affected by soil moisture content and meteorological conditions. *Agronomy Journal*, **45,** 385–90.
Dickinson, R. E. & Cicerone, R. J. (1986). Future global warming from atmospheric trace gases. *Nature*, **319,** 109–14.
Dirmhirn, I. (1964). *Das Strahlungsfeld im Lebensraum*. Frankfurt am Main: Akademische Verlagsgesellschaft.
Donald, C. M. & Hamblin, J. (1976). The biological yield and harvest index of cereals as agronomic and plant breeding criteria. *Advances in Agronomy*, **28,** 361–405.
Doorenbos, J. & Pruitt, W. O. (1984). *Guidelines for predicting crop water requirements*. FAO Irrigation and Drainage Paper 24. Rome: Food and Agriculture Organisation of the United Nations.
Downs, R. J. & Hellmers, H. (1975). *Environment and the experimental control of plant growth*. London: Academic Press.
Downton, W. J. S., Björkman, P. & Pike, C. (1980). Consequences of increased atmospheric concentrations of carbon dioxide for growth and photosynthesis of higher plants. In Pearman (1980), pp. 143–52.
Duncan, W. G., Loomis, R. S., Williams, W. A. & Hanau, R. (1967). A model for simulating photosynthesis in plant communities. *Hilgardia*, **38,** 181–205.
Dunstone, R. L., Gifford, R. M. & Evans, L. T. (1973). Photosynthetic characteristics of modern and primitive wheat species in relation to ontogeny and adaptation to light. *Australian Journal of Biological Sciences*, **26,** 295–307.
Dyer, T. G. J. & Gillooly, J. F. (1977). On a technique to describe crop and weather relationships. *Agricultural Meteorology*, **18,** 197–202.

Eberhardt, S. A. & Russell, W. A. (1966). Stability parameters for comparing varieties. *Crop Science*, **6,** 36–40.

Edwards, D. K., Denny, V. E. & Mills, A. F. (1979). *Transfer processes*, 2nd edn. New York: McGraw-Hill.

Edwards, G. E. & Ku, M. S. B. (1987). Biochemistry of C_3–C_4 intermediates. In *The biochemistry of plants. A comprehensive treatise*, eds M. D. Hatch & N. K. Boardman, pp. 275–325. San Diego: Academic Press.

Ehleringer, J. R. (1978). Implications of quantum yield differences on the distributions of C_3 and C_4 grasses. *Oecologia*, **31,** 255–67.

Ehleringer, J. R. (1980). Leaf morphology and reflectance in relation to water and temperature stress. In Turner & Kramer (1980), pp. 295–308.

Ehleringer, J. & Björkman, O. (1977). Quantum yields for CO_2 uptake in C_3 and C_4 plants. *Plant Physiology*, **59,** 86–90.

Ehleringer, J. & Forseth, I. (1980). Solar tracking by plants. *Science*, **210,** 1094–8.

Ehleringer, J. R. & Osmond, C. B. (1989). Stable isotopes. In Pearcy *et al.* (1989), pp. 281–300.

Eisenberg, D. & Kauzmann, W. (1969). *The structure and properties of water.* Oxford: Oxford University Press.

Eller, B. M. (1977). Leaf pubescence: the significance of lower surface hairs for the spectral properties of the upper surface. *Journal of Experimental Botany*, **28,** 1054–9.

Ellmore, G. S. & Ewers, F. W. (1986). Fluid flow in the outermost xylem increment of a ring-porous tree, *Ulmus americana. American Journal of Botany*, **73,** 1771–4.

El-Sharkawy, M. & Hesketh, J. (1965). Photosynthesis among species in relation to characteristics of leaf and CO_2 diffusion resistances. *Crop Science*, **19,** 517–21.

Engledow, F. L. & Wadham, S. M. (1923). Investigations on the yield of cereals. Part I. *Journal of Agricultural Science, Cambridge*, **21,** 391–409.

Evans, G. C. (1972). *The quantitative analysis of plant growth.* Oxford: Blackwell.

Evans, J. R., Sharkey, T. D., Berry, J. A. & Farquhar, G. D. (1982). Carbon isotope discrimination measured concurrently with gas exchange to investigate CO_2 diffusion in leaves of higher plants. *Australian Journal of Plant Physiology*, **13,** 281–92.

Evans, L. T. (1975). *Crop physiology*. Cambridge: Cambridge University Press.

Evans, L. T., Wardlaw, I. F. & Fischer, R. A. (1975). Wheat. In Evans (1975), pp. 101–49.

Fahn, A. (1986). Structural and functional properties of trichomes of xeromorphic leaves. *Annals of Botany*, **57,** 631–7.

Fanjul, L. & Jones, H. G. (1982). Rapid stomatal responses to humidity. *Planta*, **154,** 135–8.

Farquhar, G. D. (1978). Feedforward responses of stomata to humidity. *Australian Journal of Plant Physiology*, **5,** 787–800.

Farquhar, G. D., Ehleringer, J. R. & Hubick, K. T. (1989). Carbon isotope discrimination and photosynthesis. *Annual Review of Plant Physiology and Plant Molecular Biology*, **40,** 503–37.

Farquhar, G. D., O'Leary, M. H., & Berry, J. H. (1982). On the relationship between carbon isotope discrimination and intercellular carbon dioxide concentration in leaves. *Australian Journal of Plant Physiology*, **9,** 121–37.

Farquhar, G. D., Schulze, E.-D. & Küppers, M. (1980*a*). Responses to humidity by stomata of *Nicotiana glauca* L. and *Corylus avellana* L. are consistent with the optimisation of carbon dioxide uptake with respect to water loss. *Australian Journal of Plant Physiology*, **7,** 315–27.

Farquhar, G. D. & Sharkey, T. D. (1982). Stomatal conductance and photosynthesis. *Annual Review of Plant Physiology*, **33,** 317–45.

Farquhar, G. D., von Caemmerer, S. & Berry, J. A. (1980*b*). A biochemical model of photosynthetic CO_2 assimilation in leaves of C_3 species. *Planta*, **149,** 78–90.

Farquhar, G. D. & von Caemmerer, S. (1982). Modelling of photosynthetic response to environmental conditions. In *Encyclopedia of Plant Physiology*, vol. 12B, *Physiological Plant Ecology*, eds O. L. Lange, P. S. Nobel, C. B. Osmond & H. Ziegler, pp. 549–88. Heidelberg–Berlin–New York: Springer-Verlag.

Fick, G. W., Loomis, R. S. & Williams, W. A. (1975). Sugar beet. In Evans (1975), pp. 259–95.

Field, C., Berry, J. A. & Mooney, H. A. (1982). A portable system for measuring carbon dioxide and water vapour exchange of leaves. *Plant, Cell and Environment*, **5,** 179–86.

Finlay, K. W. & Wilkinson, G. N. (1963). The analysis of adaptation in a plant breeding programme. *Australian Journal of Agricultural Research*, **14,** 742–54.

Firn, R. D. & Digby, J. (1980). The establishment of tropic curvatures in plants. *Annual Review of Plant Physiology*, **31,** 131–48.

Fischer, R. A. & Turner, N. C. (1978). Plant productivity in the arid and semiarid zones. *Annual Review of Plant Physiology*, **29,** 277–317.

Fiscus, E. L. (1975). The interaction between osmotic- and pressure-induced water flow in plant roots. *Plant Physiology*, **55,** 917–22.

Fisher, M. J., Charles-Edwards, D. A. & Ludlow, M. M. (1981). An analysis of the effects of repeated short-term soil water deficits on stomatal conductance to carbon dioxide and leaf photosynthesis by the legume *Macroptilium atropurpureum* cv. Siratro. *Australian Journal of Plant Physiology*, **8,** 347–57.

Fleagle, R. G. & Businger, J. A. (1963). *An introduction to atmospheric physics.* New York: Academic Press.

Fowler, D., Cape, J. N. & Unsworth, M. H. (1989). Deposition of atmospheric pollutants on forests. *Philosophical Transactions of the Royal Society, London B*, **324,** 247–65.

Foyer, C. H. (1984). *Photosynthesis.* New York: Wiley.

France, J. & Thornley, J. H. M. (1984). *Mathematical models in agriculture.* London: Butterworth.

Freedman, B. (1989). *Environmental ecology.* San Diego: Academic Press.

Fritschen, L. J. & Gay, L. W. (1979). *Environmental instrumentation*. New York: Springer.

Gaastra, P. (1959). Photosynthesis of crop plants as influenced by light, carbon dioxide, temperature, and stomatal diffusion resistance. *Mededelingen van de Landbouwhoogeschool te Wageningen*, **59,** 1–68.

Gaff, D. F. (1980). Protoplasmic tolerance of extreme water stress. In Turner & Kramer (1980), pp. 207–30.

Gale, J. & Hagan, R. M. (1966). Plant antitranspirants. *Annual Review of Plant Physiology*, **17,** 269–82.

Garner, W. S. & Allard, H. A. (1920). Effect of the relative length of day and night and other factors of the environment on growth and reproduction in plants. *Journal of Agricultural Research*, **18,** 553–706.

Gates, D. M. (1980). *Biophysical ecology*. New York: Springer-Verlag.

Gates, D. M. & Papain, L. E. (1971). *Atlas of energy budgets of plant leaves*. New York: Academic Press.

Gay, A. P. & Hurd, R. G. (1975). The influence of light on stomatal density in the tomato. *New Phytologist*, **75,** 37–46.

Geiger, R. (1965). *The climate near the ground*, 4th edn. Cambridge, Mass.: Harvard University Press.

Genty, B., Briantais, J.-M. & Baker, N. R. (1989). The relationship between the quantum yield of photosynthetic electron transport and quenching of chlorophyll fluorescence. *Biochimica et Biophysica Acta*, **990,** 87–92.

Gibbs, M. & Latzko, E. (eds) (1979). *Photosynthesis*, vol. II. *Photosynthetic carbon metabolism and related processes*. Berlin: Springer-Verlag.

Giersch, C., Lämmel, D. & Farquhar, G. D. (1990). Control analysis of photosynthetic CO_2 fixation. *Photosynthesis Research*, **24,** 151–65.

Gifford, R. M. (1977). Growth pattern, carbon dioxide exchange and dry weight distribution in wheat growing under differing photosynthetic environments. *Australian Journal of Plant Physiology*, **4,** 99–110.

Gifford, R. M. & Musgrave, R. B. (1973). Stomatal role in the variability of net CO_2 exchange rates by two maize inbreds. *Australian Journal of Biological Sciences*, **26,** 35–44.

Gilchrist, W. (1984). *Statistical modelling*. Chichester: Wiley.

Givnish, T. J. (1986). Optimal stomatal conductance, allocation of energy between leaves and roots, and the marginal cost of transpiration. In *On the economy of plant form and function*, ed. T. J. Givnish, pp. 171–213. Cambridge: Cambridge University Press.

Gold, H. J. (1977). *Mathematical modelling of biological systems – an introductory guidebook*. New York: Wiley.

Gollan, T., Passioura, J. B. & Munns, R. (1986). Soil water status affects the stomatal conductance of fully turgid wheat and sunflower leaves. *Australian Journal of Plant Physiology*, **13,** 459–64.

Gollan, T., Turner, N. C. & Schulze, E.-D. (1985). The responses of stomata and leaf gas exchange to vapour pressure deficits and soil water content, III. In the schlerophyllous woody species *Nerium oleander*. *Oecologia*, **65,** 356–62.

Gorham, J., Hardy, C., Wyn Jones, R. G., Joppa, L. & Law, C. N. (1987). Chromosomal location of a K/Na discrimination character in the D genome of wheat. *Theoretical and Applied Genetics*, **74,** 484–8.

Gowing, D. J. C., Davies, W. J. & Jones, H. G. (1990). A positive root-sourced signal as an indicator of soil drying in apple *Malus* × *domestica* Borkh. *Journal of Experimental Botany*, **41,** 1535–40.

Grace, J. (1977). *Plant response to wind.* London: Academic Press.

Grace, J. (1981). Some effects of wind on plants. In Grace *et al.* (1981), pp. 31–56.

Grace, J. (1989). Tree lines. *Philosophical Transactions of the Royal Society, London B*, **324,** 233–45.

Grace, J., Ford, E. D. & Jarvis, P. G. (eds) (1981). *Plants and their atmospheric environment.* Oxford: Blackwell.

Grace, J., Pitcairn, C. E. R., Russell, G. & Dixon, M. (1982). The effects of shaking on the growth and water relations of *Festuca arundinacea* Schreb. *Annals of Botany*, **49,** 207–15.

Graham, D. (1980). Effects of light on dark respiration. In Davies (1980), pp. 525–79.

Granier, A. (1987). Une nouvelle méthode pour la mesure du flux sève brute dans le tronc des arbres. *Annales des Sciences Forestières*, **44,** 1–14.

Grant, D. R. (1970). Some measurements of evaporation in a field of barley. *Journal of Agricultural Science, Cambridge*, **75,** 433–43.

Grierson, D. & Covey, S. (1988). *Plant molecular biology.* Glasgow: Blackie.

Grime, J. P. (1979). *Plant strategies and vegetation processes.* Chichester: Wiley.

Grime, J. P. (1989). Whole-plant responses to stress in natural and agricultural systems. In Jones *et al.* (1989*a*), pp. 157–80.

Guinn, G. & Mauney, J. R. (1980). Analysis of CO_2 exchange assumptions: feedback control. In Hesketh & Jones (1980), vol. II, pp. 1–16.

Gupta, R. K. & Saxena, S. K. (1971). Ecological studies on the protected and overgrazed rangelands in the arid zone of West Rajasthan. *Journal of the Indian Botanical Society*, **50,** 289–300.

Gusta, L. V., Burke, M. J. & Kapoor, A. C. (1975). Determination of unfrozen water in winter cereals at subfreezing temperatures. *Plant Physiology*, **56,** 707–9.

Gusta, L. V. & Fowler, D. B. (1979). Cold resistance and injury in winter cereals. In Mussell & Staples (1979), pp. 160–78.

Gutknecht, J. (1968). Salt transport in *Valonia*: inhibition of potassium uptake by small hydrostatic pressures. *Science*, **160,** 68.

Hack, H. R. B. (1974). The selection of an infiltration technique for estimating the degree of stomatal opening. *Annals of Botany*, **38,** 93–114.

Hall, A. E. (1979). A model of leaf photosynthesis and respiration for predicting carbon dioxide assimilation in different environments. *Oecologia*, **143,** 299–316.

Hall, A. E. & Kaufmann, M. R. (1975). Stomatal response to environment with *Sesamum indicum* L. *Plant Physiology*, **55,** 455–9.

Hall, A. E., Schulze, E.-D. & Lange, O. L. (1976). Current perspectives of steady-state stomatal responses to environment. In *Water and plant life*, eds O. L. Lange, L. Kappen & E.-D. Schulze, pp. 169–87. Berlin: Springer-Verlag.

Hall, D. O. (1989). Carbon flows in the biosphere: present and future. *Journal of the Geological Society, London*, **146,** 175–81.

Hall, D. O., Mynick, H. E. & Williams, R. H. (1991). Carbon sequestration vs. fossil fuel substitution: alternative roles for biomass in coping with greenhouse warming. *Report* 255, *Centre for Energy and Environmental Studies, Princeton University*.

Hall, H. K. & McWha, J. A. (1981). Effects of abscisic acid on growth of wheat (*Triticum aestivum* L.). *Annals of Botany*, **47,** 427–33.

Hansen, J., Fung, L., Lacis, A., Rind, D., Lebdeff, S., Ruedy, R., Russell, G. & Stone, P. (1988). Global climate changes as forecast by the Goddard Institute for Space Studies three dimensional model. *Journal of Geophysical Research*, **93,** 9341–64.

Hansen, J., Johnson, D., Lacis, A., Lebedeff, S., Lee, P., Rind, D. & Russell, G. (1981). Climate impact of increasing atmospheric carbon dioxide. *Science*, **213,** 957–66.

Hansen, P. (1970). ^{14}C-studies on apple trees. VI. The influence of the fruit on the photosynthesis of the leaves and the relative photosynthetic yields of fruit and leaves. *Physiologia Plantarum*, **23,** 805–10.

Hartsock, T. L. & Nobel, P. S. (1976). Watering converts a CAM plant to daytime CO_2 uptake. *Nature*, **262,** 574–6.

Haseba, T. (1973). Water vapour transfer from leaf-like surfaces within canopy models. *Journal of Agricultural Meteorology, Tokyo*, **29,** 25–33.

Hatch, M. D. & Boardman, N. K. (eds) (1981). *The biochemistry of plants*, vol. VIII, *Photosynthesis*. New York: Academic Press.

Havaux, M. & Lannoye, R. (1985). Drought resistance of hard wheat cultivars measured by a rapid chlorophyll fluorescence test. *Journal of Agricultural Science, Cambridge*, **104,** 501–4.

Heck, W. W., Taylor, O. C. & Tingey, D. T. (1988). *Assessment of crop loss from air pollutants*. London & New York: Elsevier Applied Science.

Heikkila, J. J., Papp, J. E. T., Schultz, G. A. & Bewley, J. D. (1984). Induction of heat shock protein messenger RNA in maize mesocotyls by water stress, abscisic acid and wounding. *Plant Physiology*, **76,** 270–4.

Hellebust, J. A. (1976). Osmoregulation. *Annual Review of Plant Physiology*, **27,** 485–505.

Henderson, A. (1987). Literature on air pollution and lichens XXV. *Lichenologist*, **19,** 205–10.

Henson, I. E. (1981). Changes in abscisic acid content during stomatal closure in pearl millet (*Pennisetum americanum* (L.) Leeke). *Plant Science Letters*, **21,** 121–7.

Henson, I. E. (1985). Solute accumulation and growth in plants of pearl millet (*Pennisetum americanum* [L.] Leeke) exposed to abscisic acid or water stress. *Journal of Experimental Botany*, **36,** 1889–99.

Henzell, R. G., McCree, K. J., van Bavel, C. H. M. & Schertz, K. F. (1976).

Sorghum genotype variation in stomatal sensitivity to leaf water deficit. *Crop Science*, **16,** 660–2.

Herrington, L. P. (1969). *On temperature and heat flow in tree stems.* Bulletin 73. New Haven: Yale University, School of Forestry.

Hesketh, J. D. & Jones, J. W. (1980). *Predicting photosynthesis for ecosystem models*, vols I & II. Boca Raton, Florida: CRC Press.

Hewitt, C. N., Kok, G. L. & Fall, R. (1990*a*). Hydroperoxides in plants exposed to ozone mediate air pollution damage to alkene emitters. *Nature*, **344,** 56–8.

Hewitt, C. N., Monson, R. K. & Fall, R. (1990*b*). Isoprene emissions from the grass *Arundo donax* L. are not linked to photorespiration. *Plant Science*, **66,** 139–44.

Ho, T.-H. D. & Sachs, M. M. (1989). Environmental control of gene expression and stress proteins in plants. In Jones *et al.* (1989*a*), pp. 157–80.

Hocking, P. J. (1980). The composition of phloem exudate and xylem sap from tree tobacco (*Nicotiana glauca* Grah.). *Annals of Botany*, **45,** 633–43.

Holmes, M. G. & Smith, H. (1975). The function of phytochrome in plants growing in the natural environment. *Nature*, **254,** 512–14.

Holmgren, P. (1968). Leaf factors affecting light-saturated photosynthesis in ecotypes of *Solidago virgaurea* from exposed and shaded habitats. *Physiologia Plantarum*, **21,** 676–98.

Hooker, R. H. (1905). Correlation of the weather and crops. *Journal of the Royal Statistical Society*, **70,** 1–51.

Houghton, J. T., Jenkins, G. J. & Ephraums, J. J. (1990). *Climate change: the IPCC assessment.* Cambridge: Cambridge University Press.

Hsiao, T. C. (1973). Plant responses to water stress. *Annual Review of Plant Physiology*, **24,** 519–70.

Huber, W. & Sankhla, N. (1980). Effect of abscisic acid on betaine accumulation in *Pennisetum typhoides* seedlings. *Zeitschrift für Pflanzenphysiologie*, **97,** 179–82.

Hubick, K. T., Shorter, R. & Farquhar, G. D. (1988). Heritability and genotype × environment interactions of carbon isotope discrimination and transpiration efficiency in peanut (*Arachis hypogaea* L.). *Australian Journal of Plant Physiology*, **15,** 799–813.

Hughes, J. E., Morgan, D. C., Lambton, P. A., Black, C. R. & Smith, H. (1984). Photoperiodic time signals during twilight. *Plant, Cell and Environment*, **7,** 269–77.

Hughes, S. G., Bryant, J. A. & Smirnoff, N. (1989). Molecular biology: application to studies of stress tolerance. Jones *et al.* (1989*a*), pp. 131–55.

Hunt, R. (1978). *Plant growth analysis.* Studies in Biology No. 96. London: Edward Arnold.

Hunt, R. (1990). *Basic growth analysis.* London: Unwin-Hyman.

Hunt, R., Hand, D. W., Hannah, M. A. & Neal, A. M. (1991). Response to CO_2 enrichment in twenty-seven herbaceous species. *Physiological Ecology*, **5** (in press).

Hüsken, D., Steudle, E. & Zimmermann, U. (1978). Pressure probe technique for measuring water relations of cells in higher plants. *Plant Physiology*, **61,** 158–63.

Hutton, J. T. & Norrish, K. (1974). Silicon content of wheat husks in relation to water transpired. *Australian Journal of Agricultural Research*, **25,** 203–12.

Hylton, C. M., Rawsthorne, S., Smith, A. M. & Jones, A. D. (1988). Glycine decarboxylase is confined to the bundle-sheath cells of leaves of C_3–C_4 intermediate species. *Plant*, **175,** 452–9.

Idso, S. B., Jackson, R. D., Ehrler, W. L. & Mitchell, S. T. (1969). A method for determination of infrared emittance of leaves. *Ecology*, **50,** 899–902.

Idso, S. B., Jackson, R. D., Pinter, P. J., Jr., Reginato, R. J. & Hatfield, J. L. (1981). Normalizing the stress-degree-day parameter for environmental variability. *Agricultural Meteorology*, **24,** 45–55.

Innes, P. & Blackwell, R. D. (1981). The effect of drought on the water use and yield of two spring wheat genotypes. *Journal of Agricultural Science, Cambridge*, **96,** 603–10.

Innes, P., Blackwell, R. D. & Quarrie, S. A. (1984). Some effects of genetic variation in drought-induced abscisic acid accumulation on the yield and water use of spring wheat. *Journal of Agricultural Science, Cambridge*, **102,** 341–51.

Irving, P. M. (1988). Overview of the US National acid precipitation assessment programme. In Mathy (1988), pp. 39–48.

Itai, C., Richmond, A. & Vaadia, Y. (1968). The role of root cytokinins during water and salinity stress. *Israel Journal of Botany*, **17,** 187–95.

Iwanoff, L. (1928). Zur Methodik der Transpirations-bestimmung am Standort. *Berichte der Deutschen Botanischen Geselleshaft*, **46,** 306–10.

Jackson, J. E. & Hamer, P. J. C. (1980). The causes of year-to-year variation in the average yield of Cox's Orange Pippin apple in England. *Journal of Horticultural Science*, **55,** 149–56.

Jackson, J. E. & Palmer, J. W. (1979). A simple model of light transmission and interception by discontinuous canopies. *Annals of Botany*, **44,** 381–3.

Jackson, R. D. (1982). Canopy temperature and crop water stress. *Advances in Irrigation*, **1,** 43–85.

Jackson, R. D., Idso, S. B., Reginato, R. J. & Pinter, P. J., Jr (1981). Canopy temperature as a crop water stress indicator. *Water Resources Research*, **17,** 1133–8.

Jaffe, M. J., Huberman, M., Johnson, J. & Telewski, F. W. (1985). Thigmomorphogenesis: the induction of callose formation and ethylene evolution by mechanical perturbation in bean stems. *Physiologia Plantarum*, **64,** 271–9.

Jane, F. W. (1970). *The structure of wood*, 2nd edn. London: Adam & Charles Black.

Jarman, P. D. (1974). The diffusion of carbon dioxide and water vapour through stomata. *Journal of Experimental Botany*, **25,** 927–36.

Jarvis, P. G. (1976). The interpretation of the variations in leaf water potential and stomatal conductance found in canopies in the field. *Philosophical Transactions of the Royal Society, London B*, **273,** 593–610.

Jarvis, P. G. (1980). Stomatal response to water stress in conifers. In Turner & Kramer (1980), pp. 105–22.

Jarvis, P. G. (1981). Stomatal conductance, gaseous exchange and transpiration. In Grace *et al.* (1981), pp. 175–204.

Jarvis, P. G. (1985). Coupling of transpiration to the atmosphere in horticultural crops: the omega factor. *Acta Horticulturae* **171,** 187–205.

Jarvis, P. G., James, G. B. & Landsberg, J. J. (1976). Coniferous forest. In Monteith (1976), pp. 171–240.

Jarvis, P. G. & Mansfield, T. A. (eds) (1981). *Stomatal physiology*. Cambridge: Cambridge University Press.

Jarvis, P. G. & Morison, J. I. L. (1981). The control of transpiration and photosynthesis by the stomata. In Jarvis & Mansfield (1981), pp. 248–79.

Jarvis, P. G. & McNaughton, K. G. (1986). Stomatal control of transpiration: scaling up from leaf to region. *Advances in Ecological Research*, **15,** 1–49.

Jarvis, P. G. & Slatyer, R. O. (1970). The role of the mesophyll cell wall in leaf transpiration. *Planta*, **90,** 303–22.

Johnson, C. B. (ed.). (1981). *Physiological processes limiting plant productivity*. London: Butterworth.

Johnson, H. B. (1975). Plant pubescence: an ecological perspective. *Botanical Review*, **41,** 233–58.

Jones, H. G. (1973*a*). Limiting factors in photosynthesis. *New Phytologist*, **72,** 1089–94.

Jones, H. G. (1973*b*). Moderate-term water stresses and associated changes in some photosynthetic parameters in cotton. *New Phytologist*, **72,** 1095–105.

Jones, H. G. (1976). Crop characteristics and the ratio between assimilation and transpiration. *Journal of Applied Ecology*, **13,** 605–22.

Jones, H. G. (1977*a*). Aspects of the water relations of spring wheat (*Triticum aestivum* L.) in response to induced drought. *Journal of Agricultural Science, Cambridge*, **88,** 267–82.

Jones, H. G. (1977*b*). Transpiration in barley lines with differing stomatal frequencies. *Journal of Experimental Botany*, **28,** 162–8.

Jones, H. G. (1978). Modelling diurnal trends of leaf water potential in transpiring wheat. *Journal of Applied Ecology*, **15,** 613–26.

Jones, H. G. (1979*a*). Effects of weather on spring barley yields in Britain. *Journal of the National Institute of Agricultural Botany*, **15,** 24–33.

Jones, H. G. (1979*b*). Stomatal behaviour and breeding for drought resistance. In Mussell & Staples (1979), pp. 408–28.

Jones, H. G. (1980). Interaction and integration of adaptive responses to water stress: the implications of an unpredictable environment. In Turner & Kramer (1980), pp. 353–65.

Jones, H. G. (1981*a*). PGRs and plant water relations. In *Aspects and prospects of plant growth regulators*, ed. B. Jeffcoat, pp. 91–100. Letcombe: British Plant Growth Regulator Group.

Jones, H. G. (1981*b*). Carbon dioxide exchange of developing apple fruits. *Journal of Experimental Botany*, **32,** 1203–10.

Jones, H. G. (1981*c*). The use of stochastic modelling to study the influence of stomatal behaviour on yield–climate relationships. In Rose & Charles-Edwards (1981), pp. 231–44.

Jones, H. G. (1983). Estimation of an effective soil water potential at the root surface of transpiring plants. *Plant, Cell and Environment*, **6,** 671–4.

Jones, H. G. (1985*a*). Partitioning stomatal and non-stomatal limitations to photosynthesis. *Plant, Cell and Environment*, **8,** 95–104.

Jones, H. G. (1985*b*). Physiological mechanisms involved in the control of leaf water status: implications for the estimation of tree water status. *Acta Horticulturae*, **171,** 291–6.

Jones, H. G. (1987). Repeat flowering in apple caused by water stress or defoliation. *Trees: Structure and Function,* **1,** 135–8.

Jones, H. G. (1990). Physiological aspects of the control of water status in horticultural crops. *HortScience*, **25,** 19–26.

Jones, H. G., Flowers, T. J. & Jones, M. B. (eds) (1989*a*). *Plants under stress.* Cambridge: Cambridge University Press.

Jones, H. G., Hamer, P. J. C. & Higgs, K. H. (1988). Evaluation of various heat-pulse methods for estimation of sap flow in orchard trees: comparison with micrometeorological estimates of evaporation. *Trees: Structure and Function*, **2,** 250–60.

Jones, H. G. & Higgs, K. H. (1980). Resistance to water loss from the mesophyll cell surface in plant leaves. *Journal of Experimental Botany*, **31,** 545–53.

Jones, H. G. & Higgs, K. H. (1989). Empirical models of the conductance of leaves in apple orchards. *Plant, Cell and Environment*, **12,** 301–8.

Jones, H. G., Higgs, K. H. & Bergamini, A. (1989*b*). The use of ultrasonic detectors for water stress determination in fruit trees. *Annales des Sciences Forestières*, **46,** 338s–41s.

Jones, H. G. & Osmond, C. B. (1973). Photosynthesis by thin leaf slices in comparison with whole leaves. *Australian Journal of Biological Sciences*, **26,** 15–24.

Jones, H. G. & Peña, J. (1987). Relationships between water stress and ultrasound emission in apple (*Malus domestica* Borkh.). *Journal of Experimental Botany*, **37,** 1245–54.

Jones, H. G. & Slatyer, R. O. (1972). Estimation of the transport and carboxylation components of the intracellular limitation to leaf photosynthesis. *Plant Physiology*, **50,** 283–8.

Jones, M. M. & Rawson, H. M. (1979). Influence of rate of development of leaf water deficits upon photosynthesis, leaf conductance, water use efficiency, and osmotic potential in sorghum. *Physiologia Plantarum*, **45,** 103–11.

Jordan, B. R., Chow, W. S., Strid, A. & Anderson, J. M. (1991). Reduction in cab and psb A transcripts in response to supplementary ultraviolet-B radiation. *FEBS Letters* (in press).

Jordan, B. R., Partis, M. D. & Thomas, B. (1986). The biology and molecular biology of phytochrome. *Oxford Surveys of Plant Molecular and Cell Biology*, **3,** 315–62.

Jordan, W. R. & Ritchie, J. T. (1971). Influence of soil water stress on evaporation, root absorption, and internal water status of cotton. *Plant Physiology*, **48,** 783–8.

Joshi, M. C., Boyer, J. S. & Kramer, P. J. (1965). CO_2 exchange, transpiration and transpiration ratio of pineapple. *Botanical Gazette*, **126,** 174–9.

Kacser, H. & Burns, J. A. (1973). The control of flux. *Symposia of the Society for Experimental Biology*, **27,** 65–107.

Kaiser, W. M. (1982). Correlations between changes in photosynthetic activity and changes in total protoplast volume in leaf tissue from hygro-, meso-, and xerophytes under osmotic stress. *Planta*, **154**, 538–45.

Kaiser, W. M. (1987). Effects of water deficit on photosynthetic capacity. *Physiologia Plantarum*, **71,** 142–9.

Kanemasu, E. T., Stone, L. R. & Powers, W. L. (1976). Evapotranspiration model tested for soybean and sorghum. *Agronomy Journal*, **68,** 569–611.

Kearney, T. H. & Shantz, H. L. (1911). The water economy of dryland crops. *Year Book of the US Department of Agriculture*, **10,** 351–62.

Keeley, J. E., Osmond, C. B. & Raven, J. A. (1984). *Stylites*, a vascular land plant without stomata absorbs CO_2 by its roots. *Nature*, **310,** 694–5.

Kendrick, R. E. & Kronenberg, G. H. M. (1986). *Photomorphogenesis in plants.* Dordrecht: Martinus Nijhoff.

Kerr, J. P. & Beardsell, M. F. (1975). Effect of dew on leaf water potentials and crop resistances in a paspalum pasture. *Agronomy Journal*, **67,** 596–9.

King, R. W., Wardlaw, I. F. & Evans, L. T. (1967). Effect of assimilate utilization on photosynthetic rate in wheat. *Planta*, **77,** 261–76.

Kluge, M. & Ting, I. P. (1978). *Crassulacean acid metabolism.* Berlin: Springer-Verlag.

Knutsen, R. M. (1974). Heat production and temperature regulation in eastern skunk cabbage. *Science*, **186,** 746–7.

Körner, Ch., Allison, A. & Hilscher, H. (1986*a*). Altitudinal variation of leaf diffusive conductance and leaf anatomy in heliophytes of montane New Guinea and their interrelation with microclimate. *Flora*, **174,** 91–135.

Körner, Ch., Bannister, P. & Mark, A. F. (1986*b*). Altitudinal variation in stomatal conductance, nitrogen content and leaf anatomy in different plant life forms in New Zealand. *Oecologia*, **69,** 677–88.

Körner, Ch. & Mayr, R. (1981). Stomatal behaviour in alpine plant communities between 600 and 2600 metres above sea level. In Grace *et al.* (1981), pp. 205–18.

Körner, Ch., Scheel, J. A. & Bauer, H. (1979). Maximum leaf diffusive conductance in vascular plants. *Photosynthetica*, **13,** 45–82.

Kowal, J. M. & Kassam, A. H. (1973). Water use, energy balance and growth of maize at Samaru, Northern Nigeria. *Agricultural Meteorology*, **12,** 391–406.

Koziol, M. J. & Whatley, F. R. (1984). *Gaseous air pollutants and plant metabolism.* London: Butterworth.

Kozlowski, T. T. (ed.) (1968). *Water deficits and plant growth*, vol. I. New York: Academic Press.

Kozlowski, T. T. (ed.) (1972). *Water deficits and plant growth*, vol. III. New York: Academic Press.

Kozlowski, T. T. (ed.) (1974). *Water deficits and plant growth*, vol. IV. New York: Academic Press.

Kozlowski, T. T. (ed.) (1981). *Water deficits and plant growth*, vol. VI. New York: Academic Press.

Kramer, P. J. (1983). *Water relations of plants*. Orlando, Florida: Academic Press.

Kreith, F. (1973). *Principles of heat transfer*, 3rd edn. Scranton, Pennsylvania: International Text Book Company.

Kyle, D. J., Osmond, C. B. & Arntzen, C. J. (eds) (1988). *Photoinhibition*. Amsterdam: Elsevier.

Laisk, A., Oja, V. & Kull, K. (1980). Statistical distribution of stomatal apertures of *Vicia faba* and *Hordeum vulgare* and Spannungsphase of stomatal opening. *Journal of Experimental Botany*, **31,** 49–58.

Lakso, A. N. (1979). Seasonal changes in stomatal response to leaf water potential in apple. *Journal of the American Society for Horticultural Science*, **104,** 58–60.

Landsberg, H. E. (1961). Solar radiation at the earth's surface. *Solar Energy*, **5,** 95–8.

Landsberg, J. J. (1981). The use of models in interpreting plant response to weather. In Grace *et al.* (1981), pp. 369–89.

Landsberg, J. J., Beadle, C. L., Biscoe, P. V., Butler, D. R., Davidson, B., Incoll, L. D., James, G. B., Jarvis, P. G., Martin, P. J., Neilson, R. E., Powell, D. B. B., Slack, E. M., Thorpe, M. R., Turner, N. C., Warrit, B. & Watts, W. R. (1975). Diurnal energy, water and CO_2 exchanges in an apple (*Malus pumila*) orchard. *Journal of Applied Ecology*, **12,** 659–84.

Landsberg, J. J., Blanchard, T. W. & Warrit, B. (1976). Studies on the movement of water through apple trees. *Journal of Experimental Botany*, **27,** 579–96.

Lang, A. R. G., Evans, G. N. & Ho, P. Y. (1974). The influence of local advection on evapotranspiration from irrigated rice in a semi-arid region. *Agricultural Meteorology*, **13,** 5–13.

Lange, O. L., Nobel, P. S., Osmond, C. B. & Ziegler, H. (eds) (1982). *Encyclopedia of plant physiology: physiological plant ecology*, vols. I to IV. New York: Springer-Verlag.

Larcher, W. (1983). *Physiological plant ecology*. Berlin: Springer-Verlag.

Larqué-Saavedra, A. & Wain, R. L. (1974). Abscisic acid levels in relation to drought tolerance in varieties of *Zea mays* L. *Nature*, **251,** 716–17.

Lauscher, F. (1976). Weltweite Typen der Höhen abhängigkeit des Niederschlags. *Wetter und Leben* **28,** 80–90.

Lawlor, D. W. (1987). *Photosynthesis: metabolism, control and physiology*. London: Longman.

Lee, R. E., Chen, C.-P. & Denlinger, D. L. (1987). A rapid cold-hardening process in insects. *Science*, **238,** 1415–17.

Legg, B. J. (1981). Aerial environment and crop growth. In Rose & Charles-Edwards. (1981), pp. 129–49.

Legg, B. J., Day, W., Lawlor, D. W. & Parkinson, K. J. (1979). The effects of drought on barley growth: models and measurements showing the relative importance of leaf area and photosynthetic rate. *Journal of Agricultural Science, Cambridge*, **92,** 703–16.

Lemeur, R. & Blad, B. L. (1974). A critical review of light models for estimating the shortwave radiation regime of plant canopies. *Agricultural Meteorology*, **14,** 255–86.

Lemon, E. (1967). Aerodynamic studies of CO_2 exchange between the atmosphere and the plant. In *Harvesting the sun*, eds A. San Pietro, F. A. Greer & T. J. Army. New York: Academic Press.

Leng, E. R. (1982). Status of sorghum production in relation to other cereals. In *Sorghum in the eighties*. Hyderabad: ICRISAT.

Lenz, F. & Daunicht, H. J. (1971). Einfluß von Wurzel und Frucht auf die Photosynthese bei *Citrus*. *Angewandte Botanik*, **45,** 11–20.

Levitt, J. (1980). *Responses of plants to environmental stresses*, 2nd edn, vols I & II. New York: Academic Press.

Lewis, M. C. & Callaghan, T. V. (1976). Tundra. In Monteith (1976), pp. 399–433.

Leyton, L. (1975). *Fluid behaviour in biological systems*. Oxford: Clarendon Press.

Liang, G. H., Dayton, A. D., Chu, C. C. & Casady, A. J. (1975). Heritability of stomatal density and distribution on leaves of grain sorghum. *Crop Science*, **15,** 567–70.

Linacre, E. T. (1969). Net radiation to various surfaces. *Journal of Applied Ecology*, **6,** 61–75.

Linacre, E. T. (1972). Leaf temperatures, diffusion resistances, and transpiration. *Agricultural Meteorology*, **10,** 365–82.

Livingston, B. E. & Brown, W. H. (1912). Relation of the daily march of transpiration to variations in the water content of foliage leaves. *Botanical Gazette*, **53,** 309–30.

Lockhart, J. A. (1965). An analysis of irreversible plant cell elongation. *Journal of Theoretical Biology*, **8,** 264–75.

Long, S. P. & Drake, B. G. (1990). The effect of rising atmospheric CO_2 concentration on the quantum yield of photosynthetic C-assimilation in a C_3 species. Proceedings of the AFRC Photosynthesis Meeting, Imperial College, London, 4–6 April, 1990.

Long, S. P., Incoll, L. D. & Woolhouse, H. W. (1975). C_4 photosynthesis in plants from cool temperature regions with particular reference to *Spartina townsendii*. *Nature*, **257,** 622–4.

Long, S. P. & Woodward, F. I. (eds) (1988). *Plants and temperature. Symposia of the Society for Experimental Biology, XLII*. Cambridge: Company of Biologists.

Lorimer, G. H. & Andrews, T. J. (1981). The C_2 chemo- and photorespiratory carbon oxidation cycle. In Hatch & Boardman (1981), pp. 329–74.

Luckwill, L. C. (1981). *Growth regulators in crop production*. London: Edward Arnold.

Ludlow, M. M. (1980). Adaptive significance of stomatal responses to water stress. In Turner & Kramer (1980), pp. 123–38.

Ludlow, M. M. & Jarvis, P. G. (1971*a*). Methods of measuring photorespiration in leaves. In Šesták *et al.* (1971), pp. 294–315.

Ludlow, M. M. & Jarvis, P. G. (1971*b*). Photosynthesis in Sitka spruce (*Picea sitchensis* (Bong.) Carr.). I. General characteristics. *Journal of Applied Ecology*, **8,** 925–53.

Ludlow, M. M. & Wilson, G. L. (1971*a*). Photosynthesis of tropical pasture plants. I. Illuminance, carbon dioxide concentration, leaf temperature and leaf–air vapour pressure difference. *Australian Journal of Biological Sciences*, **24,** 449–70.

Ludlow, M. M. & Wilson, G. L. (1971*b*). Photosynthesis of tropical pasture plants. III. Leaf age. *Australian Journal of Biological Sciences*, **24,** 1077–87.

Ludlow, M. M. & Wilson, G. L. (1972). Photosynthesis of tropical pasture plants. IV. Basis and consequences of differences between grasses and legumes. *Australian Journal of Biological Sciences*, **25,** 1133–45.

Lüttge, U. & Pitman, M. G. (eds) (1976). *Transport in plants*, vol. II. Berlin: Springer-Verlag.

Lyons, J. M., Graham, D. & Raison, J. K. (eds) (1979). *Low temperature stress in crop plants. The role of the membrane*. New York: Academic Press.

MacRobbie, E. A. C. (1987). Ionic relations of guard cells. In Zeiger *et al.* (1987), pp. 125–62.

McCree, K. J. (1970). An equation for the rate of respiration of white clover plants grown under controlled conditions. In *Prediction and measurement of photosynthetic productivity*, Proceedings of the IBP/PP Technical Meeting, Trebon, pp. 221–9. Wageningen: PUDOC.

McCree, K. J. (1972*a*). The action spectrum, absorptance and quantum yield of photosynthesis in crop plants. *Agricultural Meteorology*, **9,** 191–216.

McCree, K. J. (1972*b*). Test of current definitions of photosynthetically active radiation against leaf photosynthesis data. *Agricultural Meteorology*, **10,** 443–53.

McFarlan, D. (ed.) (1990). *The Guinness book of records 1991*. Enfield, Middlesex: Guinness Publishing Limited.

McLeod, A. R., Fackrell, J. E. & Alexander, K. (1985). Open-air fumigation of field crops: Criteria and design for a new experimental system. *Atmospheric Environment*, **19,** 1639–49.

McNaughton, K. G. (1989). Micrometeorology of shelter belts and forest edges. *Philosophical Transactions of the Royal Society, London B*, **324,** 351–68.

McNaughton, K. G. & Jarvis, P. G. (1983). Predicting effects of vegetation

changes on transpiration and evaporation. In *Water deficits and plant growth*, vol. VII, ed. T. T. Kozlowski, pp. 1–47. New York: Academic Press.
McPherson, H. G. (1969). Photocell-filter combinations for measuring photosynthetically active radiation. *Agricultural Meteorology*, **6,** 347–56.
Malkin, S. & Fork, D. C. (1981). Photosynthetic units of sun and shade plants. *Plant Physiology*, **67,** 580–3.
Malone, M., Leigh, R. A. & Tomos, A. D. (1989). Extraction and analysis of sap from individual wheat leaf cells: the effect of sampling speed on the osmotic pressure of extracted sap. *Plant, Cell and Environment*, **12,** 919–26.
Mansfield, T. A. (ed.) (1976). *Effects of air pollutants on plants.* Cambridge: Cambridge University Press.
Markhart III, A. H., Sionit, N. & Siedow, J. N. (1981). Cell wall water dilution: an explanation of apparent negative turgor potentials. *Canadian Journal of Botany*, **59,** 1722–5.
Marshall, B. & Woodward, F. I. (eds) (1985). *Instrumentation for environmental physiology.* Cambridge: Cambridge University Press.
Martin, J. T. & Juniper, B. E. (1970). *The cuticles of plants.* London: Edward Arnold.
Martínez-Lozano, J. A., Tena, F., Onrubia, J. E. & De La Rubia, J. (1984). The historical evaluation of the Ångstrom formula and its modifications: review and bibliography. *Agricultural and Forest Meteorology*, **33,** 109–28.
Maskell, E. J. (1928). Experimental researches in vegetable assimilation XVIII. *Proceedings of the Royal Society, London B*, **102,** 488–533.
Mathy, P. (ed.) (1988). *Air pollution and ecosystems.* Dordrecht: D. Reidel.
Maximov, N. A. (1929). *The plant in relation to water.* London: Allen and Unwin.
Mayer, H. (1987). Wind-induced tree sway. *Trees: Structure and Function*, **1,** 195–206.
Mayhead, G. J. (1973). Some drag coefficients for British forest species derived from wind tunnel studies. *Agricultural Meteorology*, **12,** 123–30.
Mayo, O. (1980). *The theory of plant breeding.* Oxford: Oxford University Press.
Meidner, H. & Mansfield, T. A. (1968). *Physiology of stomata.* London: McGraw-Hill.
Meidner, H. & Sheriff, D. W. (1976). *Water and plants.* London: Blackie.
Menz, K. M., Moss, D. N., Cannell, R. Q. & Brun, W. A. (1969). Screening for photosynthetic efficiency. *Crop Science*, **9,** 692–5.
Menzel, C. M. (1983). The control of floral initiation in lychee: a review. *Scientia Horticulturae*, **21,** 201–15.
Meyer, W. S., Reicosky, D. C. & Schaeffer, N. L. (1985). Errors in measurement of leaf diffusive conductance associated with leaf temperature. *Agricultural and Forest Meteorology*, **36,** 55–64.
Milburn, J. A. (1979). *Water flow in plants.* London: Longmans.
Milthorpe, F. L. & Moorby, J. (1979). *An introduction to crop physiology*, 2nd edn. Cambridge: Cambridge University Press.

Miranda, A. C., Jarvis, P. G. & Grace, J. (1984). Transpiration and evaporation from heather moorland. *Boundary-Layer Meteorology*, **28,** 227–43.

Miskin, K. E. & Rasmusson, D. C. (1970). Frequency and distribution of stomata in barley. *Crop Science*, **10,** 575–8.

Monsi, M. & Saeki, T. (1953). Über den Lichtfaktor in den Pflanzengesellschaften und seine Bedeutung für die Stoffproduktion. *Japanese Journal of Botany*, **14,** 22–52.

Monson, R. K. & Moore, B. D. (1989). On the significance of C_3–C_4 intermediate photosynthesis to the evolution of C_4 photosynthesis. *Plant, Cell and Environment*, **12,** 689–99.

Monteith, J. L. (1957). Dew. *Quarterly Journal of the Royal Meteorological Society*, **83,** 322–41.

Monteith, J. L. (1963). Dew: facts and fallacies. In *The water relations of plants*, eds A. J. Rutter & F. H. Whitehead, pp. 37–56. Oxford: Blackwell.

Monteith, J. L. (1965*a*). Light distribution and photosynthesis in field crops. *Annals of Botany*, **29,** 17–37.

Monteith, J. L. (1965*b*). Evaporation and environment. *Symposia of the Society for Experimental Biology*, **19,** 205–34.

Monteith, J. L. (1972). Solar radiation and productivity in tropical ecosystems. *Journal of Applied Ecology*, **9,** 747–66.

Montieth, J. L. (ed.) (1975). *Vegetation and the atmosphere*, vol. I, *Principles*. London: Academic Press.

Monteith J. L. (ed.) (1976). *Vegetation and the atmosphere*, vol. II, *Case studies*. London: Academic Press.

Monteith, J. L. (1977). Climate and the efficiency of crop production in Britain. *Philosophical Transactions of the Royal Society, London B*, **281,** 277–94.

Monteith, J. L. (1978). Reassessment of maximum growth rates for C_3 and C_4 crops. *Experimental Agriculture*, **14,** 1–5.

Monteith, J. L. (1981*a*). Evaporation and surface temperature. *Quarterly Journal of the Royal Meteorological Society*, **107,** 1–27.

Monteith, J. L. (1981*b*). Coupling of plants to the atmosphere. In Grace *et al.* (1981), pp. 1–29.

Monteith, J. L. & Unsworth, M. H. (1990). *Principles of environmental physics*, 2nd edn. London: Edward Arnold.

Mooney, H. A., Björkman, O. & Collatz, G. J. (1978). Photosynthetic acclimation to temperature in the desert shrub, *Larrea divaricata*. I. Carbon dioxide exchange characteristics of intact leaves. *Plant Physiology*, **61,** 406–10.

Mooney, H. A., Björkman, O., Ehleringer, J. & Berry, J. A. (1976). Photosynthetic capacity of *in situ* Death Valley plants. *Carnegie Institution Year Book*, **75,** 410–13.

Mooney, H. A., Ehleringer, J. & Björkman. I. (1977). The energy balance of leaves of the evergreen desert shrub *Atriplex hymenelytra*. *Oecologia*, **29,** 301–10.

Mooney, H. A., Gulmon, S. L., Ehleringer, J. & Rundel, P. W. (1980). Atmospheric water uptake by an Atacama Desert shrub. *Science*, **209,** 693–4.

Moorby, J. (1981). *Transport systems in plants.* London: Longman.
Moore, A. L. & Beechey, R. B. (eds) (1987). *Plant mitochondria.* New York: Plenum Press.
Moore, T. C. (1980). *Biochemistry and physiology of plant hormones.* New York: Springer-Verlag.
Morgan, J. M. (1980). Possible role of abscisic acid in reducing seed set in water stressed wheat plants. *Nature*, **285,** 655–7.
Morgan, J. M. (1984). Osmoregulation. *Annual Review of Plant Physiology*, **35,** 299–319.
Morison, J. I. L. (1987). Intercellular CO_2 concentration and stomatal response to CO_2. In Zeiger *et al.* (1987), pp. 229–51.
Mortensen, L. M. (1987). Review: CO_2 enrichment in greenhouses. Crop responses. *Scientia Horticulturae*, **33,** 1–25.
Mundy, J. & Chua, N.-H. (1988). Abscisic acid and water-stress induce the expression of a novel rice gene. *EMBO Journal*, **7,** 2279–86.
Murray, F. W. (1967). On the computation of saturation vapour pressure. *Journal of Applied Meteorology*, **6,** 203–4.
Mussell, H. & Staples, R. C. (eds) (1979). *Stress physiology of crop plants.* New York: Wiley.
Neales, T. F., Hartney, V. J. & Patterson, A. A. (1968). Physiological adaptation to drought in the carbon assimilation and water loss of xerophytes. *Nature*, **219,** 469–72.
Neales, T. F. & Incoll, L. D. (1968). The control of leaf photosynthesis rate by the level of assimilate concentration in the leaf: a review of the hypothesis. *Botanical Review*, **34,** 107–25.
Neales, T. F., Masia, A., Zhang, J. & Davies, W. J. (1989). The effects of partially drying part of the root system of *Helianthus annuus* on the abscisic acid content of the roots, xylem sap and leaves. *Journal of Experimental Botany*, **40,** 1113–20.
Neel, P. L. & Harris, R. W. (1971). Motion-induced inhibition of elongation and induction of dormancy in liquidambar. *Science*, **173,** 58–9.
Ng, P. A. P. & Jarvis, P. G. (1980). Hysteresis in the response of stomatal conductance in *Pinus sylvestris* L. needles to light: observations and a hypothesis. *Plant, Cell and Environment*, **3,** 207–16.
Nobel, P. S. (1978). Surface temperatures of Cacti – Influences of environmental and morphological factors. *Ecology*, **59**, 986–96.
Nobel, P. S. (1980). Morphology, surface temperatures, and northern limits of Columnar Cacti in the Sonoran Desert. *Ecology*, **61,** 1–7.
Nobel, P. S. (1982). Wind as an ecological factor. In Lange *et al.* (1982), vol. I, pp. 475–500.
Nobel, P. S. (1988). *Environmental biology of Agaves and Cacti.* Cambridge: Cambridge University Press.
Nobel, P. S. (1991). *Physicochemical and environmental plant physiology.* New York: Academic Press.
Nobel, P. S. & Jordan, P. W. (1983). Transpiration stream of desert species:

resistances and capacitances for a C_3, a C_4, and a CAM plant. *Journal of Experimental Botany*, **34,** 1379–91.

Nobel, P. S., Zaragoza, L. J. & Smith, W. K. (1975). Relation between mesophyll surface area, photosynthetic rate and illumination during development of leaves of *Plectranthus parviflorus* Henckel. *Plant Physiology*, **55,** 1067–70.

Nobs, M. A. (1976). Hybridizations in *Atriplex. Carnegie Institution Year Book*, **75,** 421–3.

Norman, J. M. & Campbell, G. S. (1989). Canopy structure. In Pearcy *et al.* (1989), pp. 301–25.

Norman, J. M. & Jarvis, P. G. (1974). Photosynthesis in sitka spruce (*Picea sitchensis* (Bong.) Carr.). III. Measurements of canopy structure and interception of radiation. *Journal of Applied Ecology*, **11,** 375–98.

Oertli, J. J., Lips, S. H. & Agami, M. (1990). The strength of schlerophyllous cells to resist collapse due to negative turgor pressure. *Acta Oecologia*, **11,** 281–9.

Ogren, W. L. (1984). Photorespiration: pathways, regulation and modification. *Annual Review of Plant Physiology*, **35,** 415–42.

Oke, T. R. (1987). *Boundary layer climates*, 2nd edn. London: Methuen.

Osmond, C. B. (1978). Crassulacean acid metabolism: a curiosity in context. *Annual Reviews of Plant Physiology*, **29,** 379–414.

Paltridge, G. W. & Denholm, J. V. (1974). Plant yield and the switch from vegetative to reproductive growth. *Journal of Theoretical Biology*, **44,** 23–34.

Parkhurst, D. F. (1977). A three dimensional model for CO_2 uptake by continuously distributed mesophyll in leaves. *Journal of Theoretical Biology*, **67,** 471–88.

Parkinson, K. J. & Day, W. (1980). Temperature corrections to measurements made with continuous flow porometers. *Journal of Applied Ecology*, **17,** 457–60.

Parkinson, K. J. & Legg, B. J. (1972). A continuous flow porometer. *Journal of Applied Ecology*, **9,** 669–75.

Passioura, J. B. (1984). Hydraulic resistance of plants. I: Constant or variable? *Australian Journal of Plant Physiology*, **11**, 333–9.

Pate, J. S. (1975). Pea. In Evans (1975), pp. 191–224.

Patten, B. C. (1971). *Systems analysis and simulation in ecology*. London & New York: Academic Press.

Patterson, D. T. (1980). Light and temperature adaptation. In Hesketh & Jones (1980), vol. 1, pp. 205–35.

Payne, R. W. *et al.* (1987). *GENSTAT 5 Reference Manual*. Oxford: Clarendon Press.

Pearcy, R. W., Berry, J. A. & Bartholomew, B. (1972). Field measurements of the gas exchange capacities of *Phragmites communis* under summer conditions in Death Valley. *Carnegie Institution Year Book*, **71,** 161–4.

Pearcy, R. W., Björkman, O., Harrison, A. T. & Mooney, H. A. (1971). Photosynthetic performance of two desert species with C_4 photosynthesis in Death Valley, California. *Carnegie Institution Year Book*, **70,** 540–50.

Pearcy, R. W. Ehleringer, J., Mooney, H. A. & Rundel, P. W. (eds) (1989). *Plant physiological ecology: field methods and instrumentation.* London & New York: Chapman & Hall.

Pearcy, R. W. & Troughton, J. H. (1975). C_4 photosynthesis in tree form *Euphorbia* species from Hawaiian rainforest sites. *Plant Physiology*, **55,** 1054–6.

Pearman, G. I. (ed.) (1980). *Carbon dioxide and climate: Australian research.* Canberra: Australian Academy of Science.

Penman, H. L. (1948). Natural evaporation from open water, bare soil and grass. *Proceedings of the Royal Society, London A*, **193,** 120–45.

Penman, H. L. (1953). The physical basis of irrigation control. *Report of 13th International Horticultural Congress*, **2,** 913–14.

Penman, H. L. & Schofield, R. K. (1951). Some physical aspects of assimilation and transpiration. *Symposia of the Society of Experimental Biology*, **5,** 115–29.

Penning de Vries, F. W. T., van Laar, H. H. & Chardon, M. C. (1983). Bioenergetics of growth seeds, fruits and storage organs. In *Potential productivity of field crops under different environments*, pp. 37–59. Los Baños, Philippines: International Rice Research Institute.

Pickard, B. G. (1971). Action potentials resulting from mechanical stimulation of pea epicotyls. *Planta*, **97**, 106–115.

Pielou, E. C. (1969). *An introduction to mathematical ecology*. New York: Wiley Interscience.

Pinthus, M. J. (1973). Lodging in wheat, barley and oats: the phenomenon, its causes, and its preventative measures. *Advances in Agronomy*, **25,** 209–63.

Polunin, N. (1960). *Introduction to plant geography*. London: Longman.

Pope, D. J. & Lloyd, P. S. (1975). Hemispherical photography, topography and plant distribution. In *Light as an ecological factor*, vol. II, eds G. C. Evans, R. Bainbridge & O. Rackham, pp. 385–408. Oxford: Blackwell.

Powell, D. B. B. & Thorpe, M. R. (1977). Dynamic aspects of plant-water relations. In *Environmental effects on crop physiology*, eds J. J. Landsberg & C. V. Cutting, pp. 259–79. London: Academic Press.

Powles, S. B. & Osmond, C. B. (1978). Inhibition of the capacity and efficiency of photosynthesis in bean leaflets illuminated in a CO_2-free atmosphere at low oxygen: a possible role for photorespiration. *Australian Journal of Plant Physiology*, **5,** 619–29.

Priestley, C. H. B. & Taylor, R. J. (1972). On the assessment of surface heat flux and evaporation using large-scale parameters. *Monthly Weather Review*, **100,** 81–92.

Quarrie, S. A. (1981). Genetic variability and heritability of drought-induced abscisic acid accumulation in spring wheat. *Plant, Cell and Environment*, **4,** 147–51.

Quarrie, S. A. (1991). Implications of genetic differences in ABA accumulation for crop production. In Davies & Jones (1991).

Quarrie, S. A. & Jones, H. G. (1977). Effects of abscisic acid and water stress on

development and morphology of wheat. *Journal of Experimental Botany*, **28**, 192–203.
Quarrie, S. A. & Jones, H. G. (1979). Genotypic variation in leaf water potential, stomatal conductance and abscisic acid concentration in spring wheat subjected to artificial drought stress. *Annals of Botany*, **44,** 323–32.
Quisenbery, J. E., Cartwright, G. B. & McMichael, B. L. (1984). Genetic relationship between turgor maintenance and growth in cotton germplasm. *Crop Science*, **24**, 479–82.
Rabinowitch, E. I. (1951). *Photosynthesis*, vol. II. New York: Interscience.
Rains, D. W. (1989). Plant tissue and protoplast culture: applications to stress physiology and biochemistry. In Jones *et al.* (1989*a*), pp. 181–96.
Raschke, K. (1975). Stomatal action. *Annual Review of Plant Physiology*, **26,** 309–40.
Rasmussen, R. A. & Khalil, M. A. K. (1988). Isoprene over the Amazon basin. *Journal of Geophysical Research*, **93**, 1417–21.
Rauner, Ju. L. (1976). Deciduous forests. In Monteith (1976), pp. 241–64.
Raupach, M. R. (1989). Stand overstorey processes. *Philosophical Transactions of the Royal Society*, *London B*, **324,** 175–90.
Raven, J. A. (1970). Exogenous inorganic carbon sources in plant photosynthesis. *Biological Reviews*, **45,** 167–221.
Raven, J. A. (1972). Endogenous inorganic carbon sources in plant photosynthesis. I. Occurrence of the dark respiratory pathways in illuminated green cells. *New Phytologist*, **71,** 227–47.
Ray, T. B. & Black, C. C. (1979). The C_4 pathway and its regulation. In Gibbs & Latzko (1979), pp. 77–101.
Richter, H. (1978). A diagram for the description of water relations in plant cells and organs. *Journal of Experimental Botany*, **29,** 1197–203.
Rikin, A., Atsmon, D. & Gitler, C. (1983). Quantitation of chill-induced release of a tubulin-like factor and its prevention by abscisic acid in *Gossypium hirsutum* L. *Plant Physiology*, **71,** 747–8.
Ritchie, G. A. & Hinckley, T. M. (1975). The pressure chamber as an instrument for ecological research. *Advances in Ecological Research*, **9,** 165–254.
Ritchie, J. T. (1972). Model for predicting evaporation from a row crop with incomplete cover. *Water Resources Research*, **8,** 1204–13.
Ritchie, J. T. (1973). Influence of soil water status and meteorological conditions on evaporation from a corn canopy. *Agronomy Journal*, **65,** 893–7.
Ritchie, J. T. & Jordan, W. R. (1972). Dryland evaporative flux in a subhumid climate. IV. Relation to plant water status. *Agronomy Journal*, **64**, 173–6.
Roberts, J. A. & Hooley, R. (1988). *Plant growth regulators*. Glasgow: Blackie.
Roberts, T. M. (1984). Effects of air pollutants in agriculture and forestry. *Atmospheric Environment*, **18,** 629–52.
Robson, M. J. (1981). Respiratory efflux in relation to temperature of simulated swards of perennial ryegrass with contrasting soluble carbohydrate contents. *Annals of Botany*, **48,** 269–73.

Rorison, I. H. (1981). Plant growth in response to variation in temperature: field and laboratory studies. In Grace *et al.* (1981), pp. 313–32.

Rose, D. A. & Charles-Edwards, D. A. (eds) (1981). *Mathematics and plant physiology*. London: Academic Press.

Rosenberg, N. J., Blad, B. L. & Verma, S. B. (1983). *Microclimate: the biological environment*. New York: Wiley.

Rosenberg, N. J., McKenney, M. S. & Martin, P. (1989). Evapotranspiration in a greenhouse-warmed world: a review and a simulation. *Agricultural and Forest Meteorology*, **47,** 303–20.

Ross, J. (1975). Radiative transfer in plant communities. In Monteith (1975), pp. 13–55.

Sage, R. F., Sharkey, T. D. & Seemann, R. (1989). Acclimation of photosynthesis to elevated CO_2 in five C_3 species. *Plant Physiology*, **89,** 590–96.

Salisbury, F. B. & Ross, C. W. (1985). *Plant physiology*, 3rd edn. Belmont, California: Wadsworth.

Salter, P. J. & Goode, J. E. (1967). *Crop responses to water at different stages of growth*. Farnham Royal: Commonwealth Agricultural Bureaux.

Sandford, A. P. & Grace, J. (1985). The measurement and interpretation of ultrasound from woody stems. *Journal of Experimental Botany*, **36**, 298–311.

Scaife, A., Cox, E. F. & Morris, G. E. L. (1987). The relationship between shoot weight, plant density and time during the propagation of four vegetable species. *Annals of Botany*, **59**, 325–34.

Schroeder, J. I. & Hedrich, R. (1989). Involvement of ion channels and active transport in osmoregulation and signalling of higher plant cells. *Trends in Biochemical Sciences*, **14**, 187–92.

Schulte-Hostede, S., Darrall, N. M., Blank, L. W. & Wellburn, A. R. (eds) (1988). *Air pollution and plant metabolism*. Barking, Essex: Elsevier Applied Science.

Schulze, E.-D. (1986). Carbon dioxide and water vapour exchange in response to drought in the atmosphere and soil. *Annual Review of Plant Physiology*, **37**, 247–74.

Schulze, E.-D., Čermák, J., Matyssek, R., Penka, M., Zimmermann, R., Vasicek, F., Gries, W. & Kucera, J. (1985). Canopy transpiration and water fluxes in the xylem of the trunk of *Larix* and *Picea* trees – a comparison of xylem flow, porometer and cuvette. *Oecologia*, **66,** 475–83.

Schulze, E.-D. & Hall, A. E. (1981). Short-term and long-term effects of drought on steady state and time integrated plant processes. In Johnson (1981), pp. 217–35.

Schulze, E.-D., Lange, O. L., Buchbom, U., Kappen, L. & Evenari, M. (1972). Stomatal responses to changes in humidity in plants growing in the desert. *Planta*, **108,** 259–70.

Seetharama, N., Subba Reddy, B. V., Peacock, J. M. & Bidinger, F. R. (1982). Sorghum improvement for drought resistance. In *Principles and methods of crop improvement for drought resistance with emphasis on rice*, Los Baños, Philippines: International Rice Research Institute.

Sellers, W. D. (1965). *Physical climatology*. Chicago: University of Chicago Press.

Šesták, Z., Catský, J. & Jarvis, P. G. (eds) (1971). *Plant photosynthetic production. Manual of methods*. The Hague; Dr W. Junk N. V.

Shackel, K. A. & Hall, A. E. (1979). Reversible leaflet movements in relation to drought adaptation of cowpeas. *Vigna unguiculata* (L.) Walp. *Australian Journal of Plant Physiology*, **6,** 265–76.

Shantz, H. L. & Piemeisel, L. N. (1927). The water requirement of plants at Akron, Colorado. *Journal of Agricultural Research*, **34,** 1093–190.

Sharkey, T. D. & Raschke, K. (1981). Effect of light quality on stomatal opening in leaves of *Xanthium strumarium* L. *Plant Physiology*, **68**, 1170–4.

Sharp, R. R. & Davies, W. J. (1985). Root growth and water uptake by maize plants in drying soil. *Journal of Experimental Botany*, **36**, 1441–56.

Sharpe, P. J. H., Wu, H-i. & Spence, R. D. (1987). Stomatal mechanics. In Zeiger *et al.* (1987), pp. 91–123.

Shomer-Ilan, A., Beer, S. & Waisel, Y. (1979). Biochemical basis of ecological adaptation. In Gibbs & Latzko (1979), pp. 190–201.

Shropshire, W. Jr. & Mohr, H. (eds) (1983). *Photomorphogenesis. Encyclopedia of plant physiology*, vol. 16. Heidelberg: Springer-Verlag.

Silvey, V. (1981). The contribution of new wheat, barley and oat varieties to increasing yield in England and Wales 1947–78. *Journal of the National Institute of Agricultural Botany*, **15**, 399–412.

Simmelsgaard, S. E. (1976). Adaptation to water stress in wheat. *Physiologia Plantarum*, **37,** 167–74.

Sionit, N., Strain, B. R. & Hellmers, H. (1981). Effects of different concentrations of atmospheric CO_2 on growth and yield components of wheat. *Journal of Agricultural Science, Cambridge*, **79,** 335–9.

Slack, E. M. (1974). Studies of stomatal distribution on the leaves of four apple varieties. *Journal of Horticultural Science*, **49,** 95–103.

Slatyer, R. O. (1960). Aspects of the tissue water relationships of an important arid zone species (*Acacia aneura* F. Muell) in comparison with two mesophytes. *Bulletin of Research Council of Israel*, **8D**, 159–68.

Slatyer, R. O. (1967). *Plant water relationships*. London: Academic Press.

Slatyer, R. O. (1970). Comparative photosynthesis, growth and transpiration of two species of *Atriplex*. *Planta*, **93,** 175–89.

Slatyer, R.O. (1977). Altitudinal variation in the photosynthetic characteristics of Snow Gum, *Eucalyptus pauciflora* Sieb. ex Spreng. III. Temperature response of material grown in contrasting thermal environments. *Australian Journal of Plant Physiology*, **4,** 301–12.

Slavik, B. (1974). *Methods of studying plant water relations*. New York: Springer-Verlag.

Smith, H. (1973). Light quality and germination. Ecological implications. In *Seed ecology*, ed. W. Heydecker, pp. 219–31. London: Butterworth.

Smith, H. (1975). *Phytochrome and photomorphogenesis.* London: McGraw-Hill.

Smith, H. (1981*a*) Adaptation to shade. In Johnson (1981), pp. 159–73.

Smith, H. (1981*b*). Light quality as an ecological factor. In Grace *et al.* (1981), pp. 93–110.

Smith, H., Casal, J. J. & Jackson, G. M. (1990). Reflection signals and the perception by phytochrome of the proximity of neighbouring vegetation. *Plant, Cell and Environment*, **13**, 73–8.

Smith, H. & Holmes, M. G. (1977). The function of phytochrome in the natural environment. III. *Photochemistry and Photobiology*, **25,** 547–50.

Smith, J. A. C., Schulte, P. J. & Nobel, P. S. (1987). Water flow and water storage in *Agave deserti*: osmotic implications of crassulacean acid metabolism. *Plant, Cell and Environment*, **10**, 639–48.

Smith, T. A. (1985). Polyamines. *Annual Review of Plant Physiology*, **36**, 117–43.

Smith, W. K. & Geller, G. N. (1979). Plant transpiration at high elevation: theory, field measurements, and comparisons with desert plants. *Oecologia*, **41,** 109–22.

Solárová, J. (1980). Diffusive conductances of adaxial (upper) and abaxial (lower) epidermes: response to quantum irradiance during development of primary *Phaseolus vulgaris* L. leaves. *Photosynthetica*, **14,** 523–31.

Solárová, J., Pospišilová, J. & Slavik, B. (1981). Gas exchange regulation by changing of epidermal conductance with antitranspirants. *Photosynthetica*, **15,** 365–400.

Sperry, J. S., Donnelly, J. R. & Tyree, M. T. (1988*a*). A method for measuring hydraulic conductivity and embolism in xylem. *Plant, Cell and Environment*, **11**, 35–40.

Sperry, J. S., Tyree, M. T. & Donnelly, J. R. (1988*b*). Vulnerability of xylem to embolism in a mangrove vs an inland species of Rhizophoraceae. *Physiologia Plantarum*, **74,** 276–83.

Spitters, C. J. T., Toussaint, H. A. J. M. & Goudriaan, J. (1986). Separating the diffuse and direct component of global radiation and its implications for modeling canopy photosynthesis. I. Components of incoming radiation. *Agricultural and Forest Meteorology*, **38**, 217–29.

Stålfelt, M. G. (1955). The stomata as a hydrophotic regulator of the water deficit of the plant. *Physiologia Plantarum*, **8,** 572–93.

Stanhill, G. (1981). The size and significance of differences in the radiation balance of plants and plant communities. In Grace *et al.* (1981), pp. 57–73.

Stewart, C. R. & Hanson, A. D. (1980). Proline accumulation as a metabolic response to water stress. In Turner & Kramer (1980), pp. 173–89.

Stewart, G. R. (1989). Desiccation injury, anhydrobiosis and survival. In Jones *et al.* (1989*a*), pp. 115–30.

Stone, E. C. (1957). Dew as an ecological factor. II. The effect of artificial dew on the survival of *Pinus ponderosa*. *Ecology*, **38,** 414–22.

Stroike, J. E. & Johnson, V. A. (1972). *Winter wheat cultivar performance in an international array of environments.* Bulletin 251. Lincoln: University of Nebraska Agricultural Experiment Station.

Swinbank, W. C. (1951). The measurement of the vertical transfer of heat. *Journal of Meteorology*, **8,** 135–45.

Szeicz, G. (1974). Solar radiation in plant canopies. *Journal of Applied Ecology*, **11,** 1117–56.

Szeicz, G., Monteith, J. L. & dos Santos, J. M. (1964). Tube solarimeter to measure radiation among plants. *Journal of Applied Ecology* **1,** 169–74.

Tanner, C. B. (1981). Transpiration efficiency of potato. *Agronomy Journal*, **73,** 59–64.

Teeri, J. A. (1980). Adaptation of kinetic properties of enzymes to temperature variability. In Turner & Kramer (1980), pp. 251–60.

Teeri, J. A. & Stowe, L. G. (1976). Climatic patterns and the distribution of C_4 grasses in North America. *Oecologia*, **23,** 1–12.

Teeri, J. A., Stowe, L. G. & Murawski, D. A. (1978). The climatology of two succulent plant families: Cactaceae and Crassulaceae. *Canadian Journal of Botany*, **56**, 1750–8.

Terashima, I., Wong, S. D., Osmond, C. B. & Farquhar, G. D. (1988). Characterisation of non-uniform photosynthesis induced by abscisic acid in leaves having different mesophyll anatomies. *Plant Cell Physiology*, **29**, 385–94.

Tevini, M. & Teramura, A. H. (1989). UV-B effects on terrestrial plants. *Photochemistry and Photobiology*, **50,** 479–87.

Thames, J. L. (1961). Effects of wax coatings on leaf temperatures and field survival of *Pinus taeda* seedlings. *Plant Physiology*, **36,** 180–2.

Thom, A. S. (1975). Momentum, mass and heat exchange of plant communities. In Monteith (1975), pp. 57–109.

Thomas, J. F. & Harvey, C. N. (1983). Leaf anatomy of four species grown under continuous CO_2 enrichment. *Botanical Gazette*, **144**, 303–9.

Thompson, D. A. & Matthews, R. W. (1989). The storage of carbon in trees and timber. *Research Information Note 160*. Farnham, UK: Forestry Commission.

Thompson, N., Barrie, J. A. & Ayles, M. (1982). The meteorological office rainfall and evaporation calculation system: MORECS (July 1981). Hydrological Memorandum 45, Meteorological Office. Bracknell: Meteorological Office.

Thornley, J. H. M. (1976). *Mathematical models in plant physiology*. London: Academic Press.

Thornley, J. H. M. & Johnson, I. R. (1990). *Plant and crop modelling: a mathematical approach to plant and crop physiology*. Oxford: Oxford University Press.

Thorpe, M. R. (1974). Radiant heating of apples. *Journal of Applied Ecology*, **11,** 755–60.

Ting, I. P. & Dugger, W. M. (1968). Factors affecting ozone sensitivity and susceptibility of cotton plants. *Journal of the Air Pollution Control Association*, **18**, 810–13.

Tomos, A. D. (1987). Cellular water relations of plants. In *Water Science*

Reviews, vol. 3, ed. F. Franks, pp. 186–267. Cambridge: Cambridge University Press.

Trebst, A. & Avron, M. (eds) (1977). *Photosynthesis*, vol. I, *Photosynthetic electron tranport and photophosphorylation.* Berlin: Springer-Verlag.

Trewartha, G. T. (1968). *An introduction to climate*, 4th edn. New York: H. H. McGraw.

Tucker, G. B. (1980). On assessing the climatic response to a continuing increase in the carbon dioxide content. In Pearman (1980), pp. 21–32.

Tuller, S. E. & Chilton, R. (1973). The role of dew in the seasonal moisture balance of a summer-dry climate. *Agricultural Meteorology*, **11,** 135–42.

Turk, K. J. & Hall, A. E. (1980). Drought adaptation of cowpea. IV. Influence of drought on water use, and relations with growth and seed yield. *Agronomy Journal*, **72,** 434–9.

Turner, N. C. (1988). Measurement of plant water status by the pressure-chamber technique. *Irrigation Science*, **9,** 289–308.

Turner, N. C. (1974). Stomatal response to light and water under field conditions. In *Mechanisms of regulation of plant growth*, eds R. L. Bieleski, A. R. Ferguson & M. M. Cresswell, pp. 423–32. Bulletin 12. Wellington: The Royal Society of New Zealand.

Turner, N. C., Begg, J. E. & Tonnet, M. L. (1978). Osmotic adjustment of sorghum and sunflower crops in response to water deficits and its influence on the water potential at which stomata close. *Australian Journal of Plant Physiology*, **5,** 597–608.

Turner, N. C. & Jones, M. M. (1980). Turgor maintenance by osmotic adjustment: a review and evaluation. In Turner & Kramer (1980), pp. 87–103.

Turner, N. C. & Kramer, P. J. (eds) (1980). *Adaptation of plants to water and high temperature stress.* New York: Wiley.

Tyree, M. T. (1988). A dynamic model for water flow in a single tree: evidence that models must account for hydraulic architecture. *Tree Physiology*, **4,** 195–217.

Tyree, M. T. & Dixon, M. A. (1983). Cavitation events in *Thuja occidentalis* L.? Ultrasonic acoustic emissions from the sapwood can be measured. *Plant Physiology*, **72,** 1094–9.

Tyree, M. T. & Dixon, M. A. (1986). Water stress induced cavitation and embolism in some woody plants. *Physiologia Plantarum*, **66,** 397–405.

Tyree, M. T. & Karamanos, A. J. (1981). Water stress as an ecological factor. In Grace *et al.* (1981), pp. 237–61.

Tyree, M. T. & Sperry, J. S. (1989). Vulnerability of xylem to cavitation and embolism. *Annual Review of Plant Physiology and Molecular Biology*, **40,** 19–38.

Tyree, M. T. & Yianoulis, P. (1980). The site of water evaporation from sub-stomatal cavities, liquid path resistances and hydroactive stomatal closure. *Annals of Botany*, **46,** 175–93.

United Nations Environment Programme (1989). *Environmental effects panel report* (ISBN 92 807 1245 4). Nairobi, Kenya: UNEP.

Unsworth, M. H., Biscoe, P. V. & Pinckney, H. R. (1972). Stomatal responses to sulphur dioxide. *Nature*, **239,** 458–9.
Unsworth, M. H., Black, V. J. (1981). Stomatal responses to pollutants. In Jarvis & Mansfield (1981), pp. 187–203.
Unwin, D. M. (1980). *Microclimate measurement for ecologists*. London: Academic Press.
Valko, P. (1982). Format for presentation of data. *Report on Subtask D/V of the International Energy Agency Solar R & D Programme.*
van Bavel, C. H. M. & Hillel, D. I. (1976). Calculating potential and actual evaporation from a bare soil surface by simulation of concurrent flow of water and heat. *Agricultural Meteorology*, **17,** 453–76.
van den Honert, T. H. (1948). Water transport in plants as a catenary process. *Discussions of the Faraday Society*, **3,** 146–53.
van Gardingen, P. R., Jeffree, C. E. & Grace, J. (1989). Variation in stomatal aperture in leaves of *Avena fatua* L. observed by low-temperature scanning electron microscopy. *Plant, Cell and Environment*, **12,** 887–98.
Vince-Prue, D. (1975). *Photoperiodism in plants*. London: McGraw-Hill.
Vince-Prue, D. & Cockshull, K. E. (1981). Photoperiodism and crop production. In Johnson (1981), pp. 175–97.
Vince-Prue, D., Cockshull, K. E. & Thomas, B. (eds) (1984). *Light and the flowering process*. London: Academic Press.
Vogel, S. (1984). The lateral thermal conductivity of leaves. *Canadian Journal of Botany*, **62**, 741–4.
von Caemmerer, S. & Farquhar, G. D. (1981). Some relationships between the biochemistry of photosynthesis and the gas exchange of leaves. *Planta*, **153,** 376–87.
von Mohl, H. (1856). Welche Ursachen bewirken die Erweiterung und Verengung der Spaltöffnungen. *Botanische Zeitung*, **14,** 697–704.
von Willert, D. J., Brinckmann, E., Scheitler, B. & Eller, B. M. (1985). Availability of water controls Crassulacean acid metabolism in succulents of the Richtersveld (Namib desert, South Africa). *Planta*, **164,** 44–55.
Wallace, J. S., Batchelor, C. H. & Hodnett, M. G. (1981). Crop evaporation and surface conductance calculated using soil moisture data from central India. *Agricultural Meteorology*, **25,** 83–96.
Wang, J. Y. (1960). A critique of the heat unit approach to plant response studies. *Ecology*, **41,** 785–9.
Wareing, P. F. & Phillips, I. D. J. (1981). *The control of growth and differentiation in plants*, 3rd edn. Oxford: Pergamon Press.
Warren Wilson, J. (1957). Observations of the temperatures of arctic plants and their environment. *Journal of Ecology*, **45,** 499–531.
Warrit, B., Landsberg, J. J. & Thorpe, M. R. (1980). Responses of apple leaf stomata to environmental factors. *Plant, Cell and Environment*, **3**, 13–22.
Watson, D. J. (1958). The dependence of net assimilation rate on leaf area index. *Annals of Botany*, **22,** 37–54.

Watts, S., Rodriguez, J. L., Evans, S. E. & Davies, W. J. (1981). Root and shoot growth of plants treated with abscisic acid. *Annals of Botany*, **47,** 595–602.

Weast, R. C. (1969). *Handbook of chemistry and physics*, 50th edn. Cleveland, Ohio: Chemical Rubber Publishing Company.

Weyers, J. D. B. & Meidner, H. (1990). *Methods in stomatal research.* London: Longman.

Williams, B. A. & Austin, R. B. (1977). An instrument for measuring the transmission of shortwave radiation by plant canopies. *Journal of Applied Ecology*, **14,** 987–92.

Willmer, C. M. (1983). *Stomata.* London: Longman.

Wilson, D. (1982). Response to selection for dark respiration rate of mature leaves in *Lolium perenne* and its effects on growth of young plants and simulated swards. *Annals of Botany*, **49,** 303–12.

Winston, G. W. (1990). Physicochemical basis for free radical formation in cells: Production and defenses. In Alscher & Cumming (1990), pp. 57–86.

Wolfenden, J. & Mansfield, T. A. (1990). Physiological disturbances in plants caused by air pollutants. *Proceedings of the Royal Society of Edinburgh B*, **97,** (in press).

Wong, S. C. (1980). Effects of elevated partial pressure of carbon dioxide on assimilation and water use efficiency in plants. In Pearman (1980), pp. 159–66.

Wong, W. C., Cowan, I. R. & Farquhar, G. D. (1979). Stomatal conductance correlates with photosynthetic capacity. *Nature*, **282,** 424–6.

Woodward, F. I. & Bazzaz, F. A. (1988). The response of stomatal density to CO_2 partial pressure. *Journal of Experimental Botany*, **39,** 1771–81.

Woodward, F. I. & Sheehy, J. E. (1983). *Principles and measurements in environmental biology*. London: Butterworth.

Wright, E. A., Lucas, P. W., Cottam, D. A. & Mansfield, T. A. (1986). Physiological responses of plants to SO_2, NO_x and O_3: implications for drought resistance. In *Direct effects of dry and wet deposition on forest ecosystems – in particular canopy interactions*, ed. P. Mathy, pp. 187–200. Brussels: Commission of the European Communities.

Wyn Jones, R. G. & Storey, R. (1981). Betaines. In *Biochemistry and physiology of drought resistance*, eds L. G. Paleg & D. Aspinall, pp. 172–204. London: Academic Press.

Yates, F. & Cochran, W. G. (1938). The analysis of groups of experiments. *Journal of Agricultural Science, Cambridge*, **28,** 556–80.

Yoshino, M. (1975). *Climate in a small area.* Tokyo: University of Tokyo Press.

Young, J. E. (1975). Effects of the spectral composition of light sources on the growth of a higher plant. In *Light as an ecological factor*, vol. II, eds G. C. Evans, R. Bainbridge & O. Rackham, pp. 135–60. Oxford: Blackwell.

Zeiger, E., Farquhar, G. D. & Cowan, I. R. (eds) (1987). *Stomatal function.* Stanford: Stanford University Press.

Zhang, J., Gowing, D. J. & Davies, W. J. (1990). ABA as a root signal in root to shoot communication of soil drying. In Davies & Jeffcoat (1990), pp. 163–74.

Zimmermann, M. H. (1983). *Xylem structure and the ascent of sap.* Berlin–Heidelberg–New York–Tokyo; Springer-Verlag.
Zimmermann, M. H. & Brown, C. R. (1971). *Trees, structure and function.* Berlin: Springer-Verlag.
Zimmermann, M. H. & Milburn, J. A. (eds) (1975). *Transport in plants*, vol.I, *Phloem transport*. Berlin: Springer-Verlag.
Zimmermann, U., Becker, F. & Coster, H. G. L. (1977). The effect of pressure on the electrical breakdown in the membranes of *Valonia utricularis*. *Biochimica et Biophysica Acta*, **464,** 399–416.

INDEX